Artificial Intelligence, Machine Learning, and Data Science Technologies

Demystifying Technologies for Computational Excellence: Moving towards Society 5.0

Series Editors: Vikram Bali and Vishal Bhatnagar

This series encompasses research work in the field of Data Science, Edge Computing, Deep Learning, Distributed Ledger Technology, Extended Reality, Quantum Computing, Artificial Intelligence, and various other related areas, such as natural-language processing and technologies, high-level computer vision, cognitive robotics, automated reasoning, multivalent systems, symbolic learning theories and practice, knowledge representation and the semantic web, intelligent tutoring systems, AI and education.

The prime reason for developing and growing out this new book series is to focus on the latest technological advancements - their impact on the society, the challenges faced in implementation, and the drawbacks or reverse impact on the society due to technological innovations. With the technological advancements, every individual has personalized access to all the services, all devices connected with each other communicating amongst themselves, thanks to the technology for making our life simpler and easier. These aspects will help us to overcome the drawbacks of the existing systems and help in building new systems with latest technologies that will help the society in various ways proving Society 5.0 as one of the biggest revolutions in this era.

Artificial Intelligence, Machine Learning, and Data Science Technologies
Future Impact and Well-Being for Society 5.0
Edited by Neeraj Mohan, Ruchi Singla, Priyanka Kaushal, and Seifedine Kadry

Transforming Higher Education through Digitalization
Insights, Tools, and Techniques
Edited by S. L. Gupta, Nawal Kishor, Niraj Mishra, Sonali Mathur, and Utkarsh Gupta

A Step towards Society 5.0
Research, Innovations, and Developments in Cloud-Based Computing Technologies
Edited by Shahnawaz Khan, Thirunavukkarasu K., Ayman AlDmour, and Salam Salameh Shreem

For more information on this series, please visit: www.routledge.com/Demystifying-Technologies-for-Computational-Excellence-Moving-Towards-Society-5.0/book-series/CRCDTCEMTS

Artificial Intelligence, Machine Learning, and Data Science Technologies

Future Impact and Well-Being for Society 5.0

Edited by
Neeraj Mohan, Ruchi Singla, Priyanka Kaushal,
and Seifedine Kadry

CRC Press
Taylor & Francis Group
Boca Raton London New York

CRC Press is an imprint of the
Taylor & Francis Group, an **informa** business

First edition published 2022
by CRC Press
6000 Broken Sound Parkway NW, Suite 300, Boca Raton, FL 33487-2742

and by CRC Press
2 Park Square, Milton Park, Abingdon, Oxon OX14 4RN

Library of Congress Cataloging-in-Publication Data
Names: Mohan, Neeraj, editor.
Title: Artificial intelligence, machine learning, and data science technologies:
future impact and well-being for society 5.0 / edited by
Neeraj Mohan, Ruchi Singla, Priyanka Kaushal, and Seifedine Kadry.
Description: First edition. I Boca Raton: CRC Press, 2022. I
Series: Demystifying technologies for computational excellence I
Includes bibliographical references and index.
Identifiers: LCCN 2021018852 (print) I LCCN 2021018853 (ebook) I
ISBN 9780367720919 (hbk) I ISBN 9780367720926 (pbk) I
ISBN 9781003153405 (ebk)
Subjects: LCSH: Artificial intelligence–Industrial applications. I
Artificial intelligence–Social aspects.
Classification: LCC TA347.A78 A86 2022 (print) I
LCC TA347.A78 (ebook) I DDC 006.3–dc23
LC record available at https://lccn.loc.gov/2021018852
LC ebook record available at https://lccn.loc.gov/2021018853

ISBN: 978-0-367-72091-9 (hbk)
ISBN: 978-0-367-72092-6 (pbk)
ISBN: 978-1-003-15340-5 (ebk)

DOI: 10.1201/9781003153405

Typeset in Times
by Newgen Publishing UK

Contents

Preface

This book provides a comprehensive, conceptual, and detailed overview of the wide range of applications of Artificial Intelligence, Machine Learning, and Data Science. These technologies are very successful in bringing a revolution across the world. These technologies have become ubiquitous and greatly contribute to change.

These technologies have an impact on various domains such as healthcare, business, industry, security etc. Moreover, they are recognized as a tool to deal with complexities.

The book aims at low-cost solutions that could be implemented even in developing countries. It highlights the significant impact these technologies have on various industries and on us as humans. It provides a virtual picture of forthcoming better human life shadowed by the new technologies and their applications. The impact of Data Science on various business applications has been discussed. The book will also include an overview of the different AI applications and their correlation between each other.

The book is useful for graduate/ post graduate students, researchers, academicians, institutions, and professionals who are interested in exploring latest technologies and advancements in the area of Artificial Intelligence, Machine Learning, and Data Science.

Editor Biographies

Neeraj Mohan is a passionate researcher and academician. His research interest areas are Network Traffic Management and Digital Image Processing. He works as Assistant Professor in Computer Science & Engineering Department in I.K. Gujral Punjab Technical University, Kapurthala (Punjab) India. He has a rich and quantitative academic experience of 19 years at various positions. He did his doctoral degree at I.K. Gujral Punjab Technical University, Kapurthala (Punjab) India. He is an active researcher with more than 50 research papers in reputable journals and conferences. He has one published patent. He has guided one PhD and 17 M.Tech theses. Two more Ph.D. thesis are in pipeline.

Ruchi Singla offers a unique skillset and expertise of a PhD in Wireless Communication from Thapar University, Patiala, with a diploma in Entrepreneurship from Ashton College, Vancouver, and a certification course in "Growth Strategies and Management" from IIM, Kolkata. She is the Professor in Department of Electronics and Communication and is also the Dean Research for the research division. She is currently coordinating the start-ups, paper publications, research projects, consultancy, and events related to research and innovation. She has 18 years of experience in teaching and research domains. Her areas of interest include antennas and biosensors. She has approximately 35 research publications in reputable international and national journals having good Impact factor. Additionally, she has three published patents. She has also worked on various DST funded projects like a greenhouse monitoring system and electrotiles.

Priyanka Kaushal is Associate Professor in Applied Sciences department at Chandigarh Engineering College, Landran (Mohali), Punjab, India. She did her master's degree at Himachal Pradesh University, Shimla, India and Doctoral degree from IK Gujral Punjab Technical University, Punjab, India. She has published 20+ research papers in reputable journals and conferences. She has delivered many talks in seminars and workshops. She has acted as reviewer for many reputable journals such as *Advanced Materials Science* (SCI Indexed), *Mechanics of Advanced Materials and Structures* (Taylor & Francis, SCI indexed), *Materials Today: Proceedings* (Elsevier, SCI Indexed) etc. She has two published patents.

Seifedine Kadry gained a Bachelor degree in Applied Mathematics in 1999 from Lebanese University, MS degree in Computation in 2002 from Reims University (France) and EPFL (Lausanne), PhD in 2007 from Blaise Pascal University (France), HDR degree in Engineering Science in 2017 from Rouen University. At present his research focuses on education using technology, system simulation, operation research, system prognostics, stochastic systems, and probability and reliability analysis. He is an ABET program evaluator. He is the editor-in-chief of *Research Journal of Mathematics and Statistics* and *ARPN Journal of Systems and Software*. He is the associate editor of *International Journal Of Applied Sciences* (IJAS) and editorial board member of *International Journal of Mathematics and Statistics*, *Journal of Theoretical and Applied Information Technology* and *International Journal of Engineering and Applied Sciences*. He has published more than 270 research items, which have been cited 1700 times, his h-index is 18 and i10-index is 40.

1 Breast Cancer Diagnosis Using Machine Learning and Fractal Analysis of Malignancy-Associated Changes in Buccal Epithelium

Dmitriy Klyushin, Kateryna Golubeva,
Natalia Boroday, and Dmytro Shervarly

CONTENTS

1.1 INTRODUCTION

Currently, there is a constant increase in the incidence and death of breast cancer in women around the world. Consequently, the problem of early diagnosis and screening of breast cancer is very urgent. Early diagnosis of breast cancer involves the examination of large populations, so the screening method must be highly sensitive, specific and safe.

The currently accepted "gold standard" for breast cancer diagnostics includes clinical examination, mammography and aspiration biopsy. It allows for high diagnostic accuracy, but mammography implies radiation exposure, and aspiration biopsy is associated with tumor injury. This is contrary to the requirement for screening to be

DOI: 10.1201/9781003153405-1

safe, so it is highly desirable to develop an effective screening method that is non-invasive and harmless. As such, we propose to use a non-invasive study of malignancy-associated changes (MAC) in the interphase nuclei of the buccal epithelium.

The purpose of the chapter is to describe a novel effective method for the screening of breast cancer based on the investigation of fractal properties of chromatin in Feulgen-stained nuclei of buccal epithelium using machine learning. The chapter consists of six sections. Section 1.1 is the introduction where the aim of the chapter and importance of the proposed method are described. Section 1.2 contains a short survey of papers on the malignancy-associated changes. In Section 1.3 we describe the morphometric research and image analysis of MAC in buccal epithelium. Section 1.4 describes the fractal analysis of the chromatin. Section 1.5 contains the results of diagnosis. Section 1.6 concludes our study and states some open problems.

1.2 MALIGNANT-ASSOCIATED CHANGES IN BUCCAL EPITHELIUM

The first reports on malignancy-associated changes (MAC) occurred in the 1960s, when the content of X-chromatin in somatic cells was widely studied and its lability was revealed during various functional changes in the body and general somatic pathology. In the presence of a tumor in the body, there are significant changes in the content of X-chromatin in the buccal epithelium and neutrophils of peripheral blood. It was shown that changes in the number of cells with X-chromatin are caused by disorders of the functional state of the heterocyclic X-chromosome.

Of particular interest are works showing changes in the epitheliocytes of the buccal epithelium in patients with tumors. Thus, in the 1960s H. Nieburgs and his co-authors (Nieburgs et al. 1962, Nieburgs 1968) reported a characteristic redistribution of chromatin masses in somatic cells in 77% of cancer patients and called these changes tumor-associated changes. The latter were characterized by an increase in the size of the nuclei of epitheliocytes, an increase in the size of the zones of "bounded" chromatin, which were surrounded by light zones. The same changes were observed in the cells of the liver, kidneys and other organs. In the paper (Obrapalska et al. 1973) it was reported that MAC was observed in the buccal epithelium of 74% of patients with malignant tumors. An increase in the content of DNA in the nuclei of epitheliocytes in patients with malignant melanoma in comparison with almost healthy women has been shown. At the same time, a decrease in the number of chromatin positive cells (X-chromatin) was found in patients with malignant melanoma compared with that in patients with benign nevi and in controls. An increase in the content of DNA and the size of the interphase nuclei of the buccal epithelium was found in patients with breast cancer. But some authors in cytospectrophotometric detection of the amount of DNA in the epitheliocytes of buccal epithelium in men with bronchial epithelioma did not find a significant difference between this indicator in sick and almost healthy men (Ogden et al. 1990).

Later, in the 1990s, in screening examinations of the population, in experimental conditions, and in a medical clinic the buccal epithelium of the oral cavity was used as a convenient object of study to detect early forms of disease (Rathbone et al. 1994, Rosin et al. 1994, Prasad et al. 1995, etc.). This object reflects the general health status of a person. Some researchers on the percentage of electronegative nuclei of this epithelium and the velocity of the nuclei during microelectrophoresis revealed the functional state of man and his biological age, the development of fatigue, the action of harmful environmental factors, the state of the periodontium. Other researchers have evaluated the genetic effects of environmental pollution and the genotoxicity of xenobiotics by accounting for micronuclei in the epitheliocytes of the oral mucosa. Elevated levels of cells with micronuclei were found in the epithelial exfoliative cells of the oral cavity of patients with various types of allergies. Under the action of various genotoxic carcinogens (chewing tobacco in different mixtures, nasal and betel, tobacco smoking), formaldehyde levels of exfoliative cells with micronuclei can increase tenfold compared to the control (Nair et al. 1991, Rosin et al. 1992; Tolbert et al. 1992 etc.). These data suggest that the increase in the level of micronuclei may be a kind of "dosimeter" of various pathological conditions of the body. The level of micronuclei in the oral cavity of cancer patients probably increases after antitumor chemotherapy or ionizing radiation (Tolbert et al. 1992). In exfoliative cells in persons, who consume different chewing gums, which are characterized by carcinogenic activity (mava, tamol, masher, nuts nat), the level of micronuclei increases (Adhraryn et al. 1991 etc.). It is very important that a high level of correlation was observed between the level of exfoliative cells with micronuclei and the number of other cytogenetic disorders in peripheral blood lymphocytes (sister chromatid exchanges and chromosomal aberrations) of subjects exposed to mutagens (Sarto et al. 1990).

The state of chromatin was used to assess the degree of differentiation of buccal epithelium in gastric and duodenal ulcers. Indicators that characterize the quantitative and functional state of buccal epitheliocytes are also used. This is the index of maturation, intoxication, differentiation and karyopyknotic index. The change in the nature of differentiation, which is inherent in the norm of a particular area of buccal epithelium, indicates local or systemic disorders. The presence of signs of cellular atypia with a high probability indicates the development of precancerous and tumor changes in buccal epithelium and in 96% of cases allows you to reliably diagnose these diseases by cytological methods. Changes in the differentiation of the epithelium of buccal epithelium can also be the result of metabolic and hormonal disorders, the action of mechanical factors and chemicals (Schonwetter 1995).

In the papers (Palcic et al. 1994 etc.) the authors reported that quantitative cytological studies of the DNA content and texture of chromatin in the nucleus can detect changes associated with malignancy, which were called H. Nieburgs' malignancy-associated changes (MAC). These changes were detected in normal cells of macroscopically unchanged areas located at some distance from the malignant tumor. They probably arose as a reaction of normal cells to growth caused by malignant transformation in a particular organ (lungs, cervix, breast). Based on the data obtained, it was hypothesized that the changes associated with tumors are clearly expressed near malignant tumors and weakly expressed or absent near tumors that do not have progressive growth. With the removal of tumors, the changes associated with the degree of malignancy disappear, while incomplete removal of the tumor does not significantly affect these changes. The hypothesis was expressed that quantitative cytospectrophotometry can detect changes associated with malignant growth to diagnose these tumors, including early cancers, and to assess the prognosis of this process.

In the 1990s, an attempt was made to use changes in the buccal epithelium to characterize the effect of the tumor on its condition. For example, Ogden et al. tried to characterize and substantiate the possibility of tumor influence on the functional state of the buccal epithelium in order to use the obtained data to characterize the course of processes that occurred in organs far from the tumor and identify patterns that characterize the course of these processes. Disturbances expressed by changes in nuclear material, heterogeneity of chromatin substances and changes in nuclear membranes were observed in 77% of patients with tumors of different localizations (carcinomas, lymphomas, seminomas). The criteria for assessing tumor-associated changes (MAC) were cytophotometric studies of DNA content, nucleus size and cytoplasm of tumor cells, and the nature of the distribution of chromatin in the nucleus. But the authors failed to identify clear patterns inherent in the tumor process, in addition to increasing the size of the nuclei of tumor cells and changes in the nuclear-cytoplasmic relationship. However, they did not deny that the detected disorders were related to the effect of tumors on the functional state of the oral mucosa.

Blood has been considered the traditional method of studying genomic DNA for genetic analysis, but recently the method using DNA isolated from cells of the oral mucosa has become more widespread. In benign hyperplastic processes, the number of Langerhans cells increases significantly compared to their number in the normal mucous membrane. In malignant tumors, it decreases the more the lower the level of their differentiation.

Thus, an analysis of the literature showed that the relationship between tumor and organism is very complex and due to the many connections that form between the tumor and the body's control systems (nervous, endocrine and immune) under the influence of exogenous and endogenous factors, which cause appropriate reactions from the body. From the general set of indicators of homeostasis of a cancer patient, important information about the effect of the tumor can be obtained by studying the functional state of the buccal epithelium, which has been shown to be in close anatomical and physiological relationship with various organs and systems. This is confirmed by the fact that a number of diseases of the internal organs are accompanied by changes in the mucous membrane, which may appear even earlier than other clinical symptoms in this pathology.

In 2009 it was discovered that the DNA packing in the cell nuclei has fractal properties, i.e. the DNA is twisted like a three-dimensional Peano curve (Lieberman-Aiden et al., 2009). The fractal analysis of cells became the field of intensive investigations. The fractal dimension is considered an effective measure of heterogeneity of cells of complex endometrial hyperplasia and well-differentiated endometrioid carcinoma (Bikou et al, 2016), and also as a prognostic factor for survival in melanoma (Bedin et al. 2010), leucemia (Adam et al, 2006 etc.) and other diseases (Losa 2012, Metze 2010, 2013). Note also that numerous investigations (Nikolaou and Papamarkos 2002, Ohri et al. 2004, Losa and Castelli 2005 etc.) have shown the significant potential of fractal dimension for estimation of the morphological data. However, these investigations were focused on tumor cells but not cells of buccal epithelium. Thus, we can assume that the fractal properties are reflected in the distribution of the chromatin in nuclei of buccal epithelium and a tumor can affect this distribution causing the malignancy-associated changes. As we shall see further, our hypothesis holds.

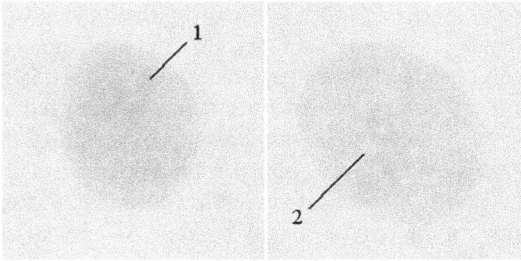

FIGURE 1.1 Distribution of chromatin in nuclei of buccal epithelium (1—heterochromatin, 2—euchromatin).

Registration of the characteristics of the interphase nucleus makes it possible to assess functional changes in the genetic apparatus of cells, and indicators of the structural organization of chromatin substance can be used as markers of these disorders in various pathological conditions, including tumors. There are also studies that indicate that the study of the interphase nuclei of the buccal epithelium may answer the question of the presence of a malignant tumor in the human body (Andrushkiw et al. 2007 etc.).

DNA cytophotometry is based on the established fact that the average amount of DNA in the nucleus in normal cells corresponds to a certain number of chromosomes, which is characteristic of this species. A number of researchers on various tumors have shown that an increase in the average DNA content in the nucleus is a characteristic feature of tumor cells (Boroday et al. 2016). Therefore, the increase in heterogeneity in DNA content can be considered as a diagnostic sign of the tumor and a criterion for its malignancy.

It should be noted that in the nucleus chromatin can be in a condensed (compact, tightly packed) and decondensed state. The degree of chromosome decondensation can be different. Areas of complete decondensation are called euchromatin. Areas of chromatin that are in a condensed state during the cell cycle are called heterochromatin. Due to the fact that heterochromatin is intensely stained with special dyes, it can be easily observed under a light microscope. At the same time, decondensed chromatin corresponds to poorly colored areas (Figure 1.1).

Changes in the nuclei are noticeable, but it is very difficult to make a visual assessment, especially when it comes to dozens of cells. Therefore, digital analysis should be used for objective quantification.

1.3 MATERIALS AND METHODS OF MORPHOMETRIC RESEARCH AND IMAGE ANALYSIS

We have studied the control group (29 people), the group of patients with breast cancer of the second stage (68 patients) and the group of patients with fibroadenomatosis (33 patients). All diagnoses were verified histologically. The morphological dataset consists of 20,256 images of interphase nuclei of buccal epithelium (6,752 nuclei scanned in three variants: without filter, through a yellow filter and through a violet filter).

The morphological materials are smears of epitheliocytes of the oral mucosa from the average depth of the spinous layer, dried at room temperature, fixated in the Nikiforov mixture and Feulgen-stained with cold hydrolysis in 5 n HCl for 15 min at t = 21–22°C. The Feulgen-stained chromatin was analyzed using the Olympus analyzer, consisting of the Olympus BX microscope, Camedia C-5050 digital zoom camera and a computer. On average, every preparation consists of 52 cells in every preparation. The DNA- fuchsine content in the nuclei of the epitheliocytes was computed as a product of the optical density by area. As a result of the first stage of analysis, we obtained images of the chromatin distribution in the form of a matrix containing 128×128 pixels.

Trying to reflect the fractal character of chromatin distribution and to provide the invariance with respect to rotation of the image, we constructed a space-filling curve (Sagan, 1994) passing through every pixel of an image and sequentially read the RGB values of the colors of the pixels of the scan not line by line. As a result, we can map a matrix of pixels to three vectors corresponding to the three channels of the RGB color model. As a space-filling curve we used the Serpinski curve (Sagan 1994).

Before applying the methods of fractal image analysis, it must be pre-processed. To do this, we used the Otsu method. This method is used to perform threshold binarization of halftone images. The algorithm assumes the presence of two classes of pixels (main and background) in the image and looks for the optimal threshold that divides them into two classes so that their intraclass variance was minimal.

This principle of the method was used to select kernels. Two clusters were considered: kernel pixels and background pixels. We will explain the method in more detail. Suppose there is an image, each point of which is represented by the value of the intensity of gray. So we have N image points and L grayscales. Assume that all points of the image can be divided into two classes, C_1 and C_2, which will distinguish points with intensities. Then the optimal value of the threshold T will be the value of k, at which the interclass variance reaches a maximum value:

$$T = k \Big|_{\sigma^2_{C_1 C_2} \max} \cdot$$

The interclass variance $\sigma^2_{C_1 C_2}$ is computed as

$$\sigma^2_{C_1 C_2} = \omega_1 \omega_2 (\mu_2 - \mu_1)^2,$$

where ω_1 is the probability to get the class C_1, ω_2 is the probability to get the class C_2, μ_1 is the mean brightness in the class C_1, μ_2 is the mean brightness in the class C_2.

To calculate the interclass variance, you must first construct an image histogram, i.e. a distribution that determines the number of n_i points with intensity I, from the total number of image points N. To calculate the probabilities and fall into classes C_1 and C_2, accordingly, use the formulas:

$$\omega_1 = \sum_{i=1}^{k} p_i, \omega_2 = 1 - \omega_1,$$

where p_i is the distribution $p_i = n_i / N$.

The mean brightness μ_1 and μ_2 in C_1 and C_2 respectively are computed as

$$\mu_1 = \frac{1}{\omega_1} \sum_{i=1}^{k} ip_i, \ \mu_2 = \frac{1}{\omega_2} \sum_{i=k+1}^{L} ip_i.$$

Next, sequentially comparing the values of the intensities of each pixel of the image with the selected threshold T determines the affiliation of the point to the corresponding class. All points that belong to the class C_2 will be painted white. The selection of the kernel by means of adaptive threshold image processing proved to be the best. The contour of the nucleus stands out quite clearly (Figure 1.2).

In the presence of any artifacts on the background of the image, the selection of the kernel may not be quite correct. When we talk about the image of the nuclei of the buccal epithelium, the artifacts' mean, for example, contamination—small spots of black or just dark color, painted background—occurs in cases where the drugs were poorly washed under running water; is an area of different regions (from a few pixels to several tens). The color brightness of such areas is similar to the brightness of the nucleus itself, so in the process of selecting the nucleus, these areas remain in the background. On the other hand, when selecting the contour of the nucleus, it is necessary to take into account such a feature as the heterogeneity of the distribution of chromatin in the nucleus. In practice, this means that in the core itself among the dark areas can be very light, which in terms of its brightness may even coincide with the average brightness of the general background.

Therefore, in the case when in the process of contour selection some areas in the kernel image were initialized as a background and accordingly painted white, it is necessary to restore the original color but only for the specified areas. This process is meant by smoothing the selected nuclei.

The idea of algorithms in both described cases is very similar. For example, to remove artifacts, the pixels of the image are sequentially considered first horizontally

FIGURE 1.2 Nuclei after adaptive threshold pre-processing.

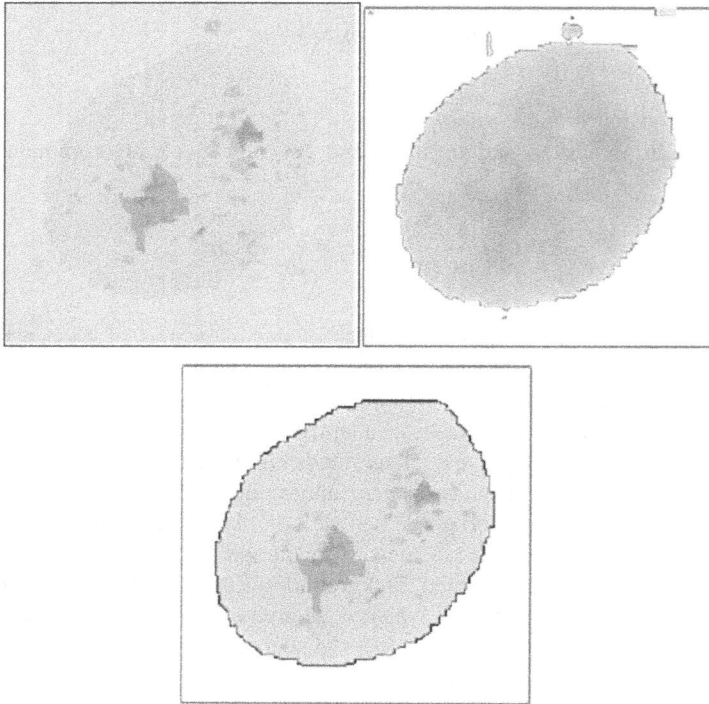

FIGURE 1.3   Artifact removing.

and then vertically, and there are areas whose area is less than a certain value of ε. If there is a small area of non-white color in the middle of the background, it means that you have come across an artifact, so we paint it white. After viewing all the rows, and then the columns of the image, we get rid of all the artifacts (Figure 1.3). It should be noted that this algorithm is not designed to distinguish large area artifacts, in which case you can use the selection of the kernel in manual mode.

Just like the artifact removal algorithm, the white spot search algorithm works on the kernel. If there is a small area of white among the nucleus, and it is known in advance that there can be no artifacts on the nucleus (this is one of the requirements when choosing nuclei), then we restore the original color in this area. Thus, areas are searched first in the rows of the image, and then in the columns (Figure 1.4).

After selecting the kernel contour, as well as after the operation of smoothing and removing artifacts, the kernels are ready for further use.

1.4   FRACTAL ANALYSIS OF CHROMATIN

There are several methods of computation of the fractal dimension of an image. We selected the Hurst exponent because it is very suitable for analysis of sequential. The

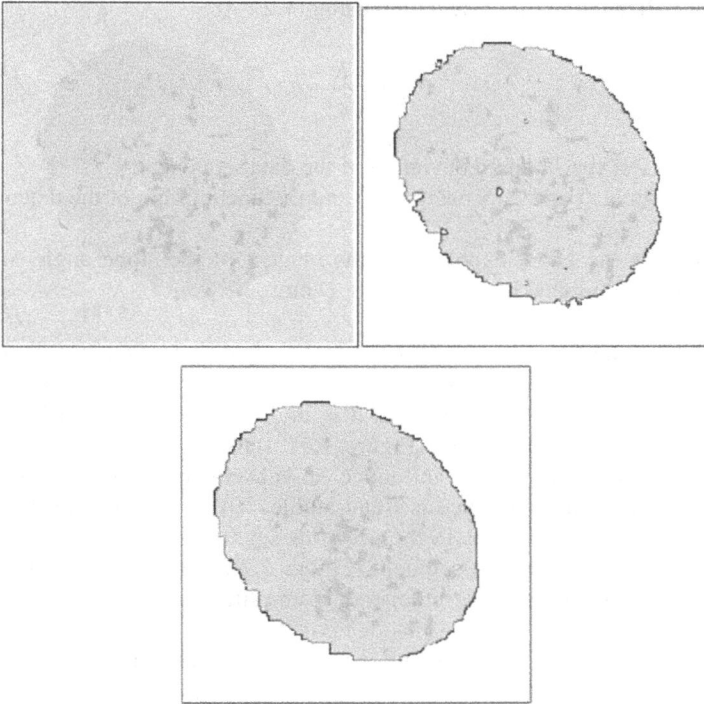

FIGURE 1.4 Nucleus smoothing.

Hurst exponent is connected with the fractal dimension D by the formula $H = 2 - D$. The algorithm of computation of the Hurst exponent is the following (Butakov and Grakovskiy 2005).

1. Compute the deviation of the values of the data sequence from the mean in current segment of a sequence:

$$\delta_{m,N} = \sum_{i=1}^{m}\left(x_i - \overline{x}_N\right),$$
(1.1)

   where N is the length of the segment (from 2 up to the end the sequence), m is the upper limit of summing (from 1 to $N-1$), x_i is a value of the data sequence, $\overline{x}_N$ is the mean of the segment. As a result, we have $N-1$ values $\delta_{2,N}, ..., \delta_{N-1,N}$.

2. Compute the range of deviations using (1.1):

$$R = \max_{m=2,...,N}\delta_{m,N} - \min_{m=2,...,N}\delta_{m,N}$$
(1.2)

3. Normalize the range of deviation using (1.2):

$$Q = \frac{R}{s},\qquad\qquad(1.3)$$

where s is the standard deviation of the data sequence.
4. Compute lg Q and lg N using (1.3) and the linear graph of the dependence of lg Q on lg N.
5. Compute the Hurst exponent as the tangent of the slope angle of the line approximating the dependence of lg Q on lg N.

According to the Hurst exponent, we can classify data sequence by their chaotic properties. If $0 < H < 0.5$, a sequence is considered as ergodic, i.e. if the sequence has increased in the previous segment, it is highly likely that it will decrease in the next segment, and vice versa (Figure 1.5). If $H = 0.5$, then the sequence is chaotic, i.e. their values do not influence on subsequent values (Figure 1.6). If $0.5 < H < 1.0$, then the sequence is trend-stable. If the sequence in the previous segment increases or decreases in the previous segment, it is highly likely that it will save this trend in the next segment (Figure 1.7). If $H > 1$, then a sequence is a fractal random process having independent amplitude jumps, using Lévy distribution (Figure 1.8).

We found the H values for each cell using the red, blue, and green channels. To describe statistical properties of the Hurst exponent we calculated for every patient the mean, the maximum and the minimum value of the Hurst exponents. Therefore, we have $\min_{BH}$, $\min_{GH}$, $\min_{RH}$ (minimums of Hurst exponents for three channels among all patient cells), $\max_{BH}$, $\max_{GH}$, $\max_{RH}$ (maximums of Hurst exponents for three channels among all patient cells), and $aver_{BH}$, $aver_{GH}$, $aver_{RH}$ (means of Hurst exponents for three channels among all patient cells).

As a result of analysis of the Hurst exponent, we have discovered that the most informative is the blue channel. That is why in addition to the first group of features

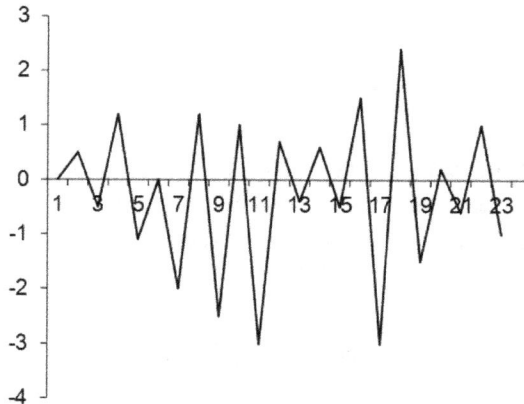

FIGURE 1.5   An example of ergodic time series.

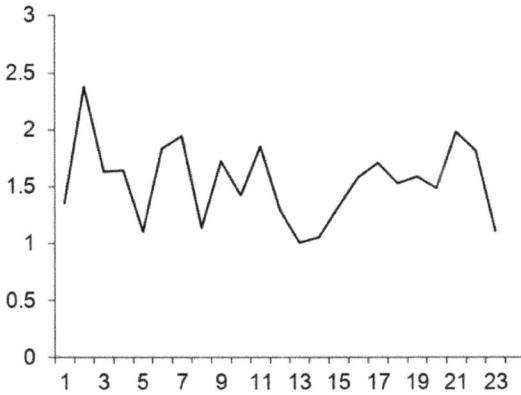

FIGURE 1.6 An example of chaotic time series.

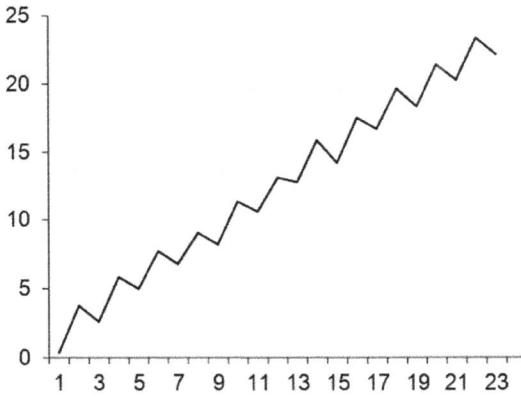

FIGURE 1.7 An example of trend-stable time series.

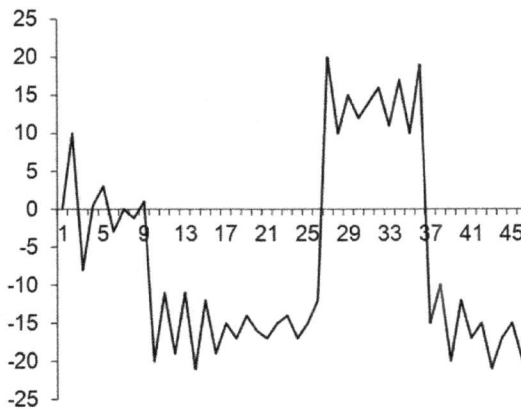

FIGURE 1.8 An example of fractal time series.

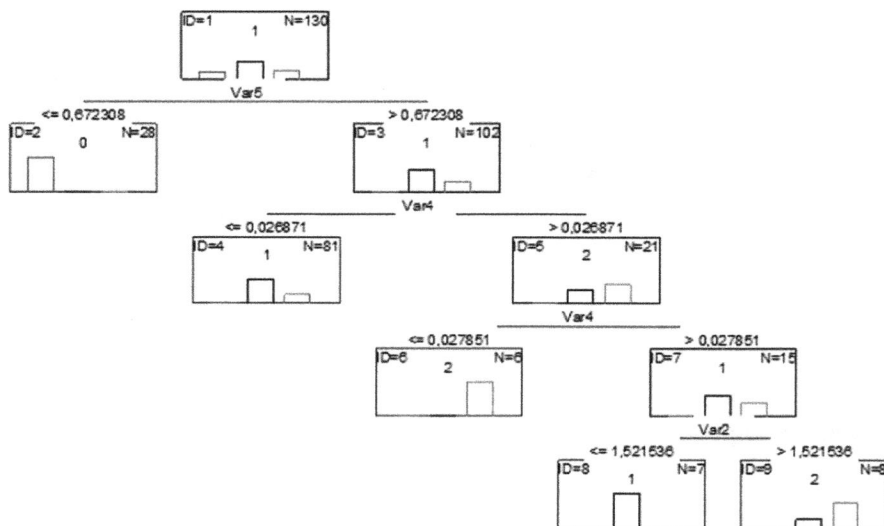

FIGURE 1.9   An example of the CART decision tree.

we found out for every patient the descriptive statistics of the Hurst exponent of all the cells in the blue channel (let us call it the second group of features):

1. Variance.
2. Mean square.
3. 75th empirical quartile.
4. Median.
5. Harmonic mean.
6. Geometrical mean.
7. Average cut-off value (including all but 5%).
8. Mean.
9. Excess values.

For classification of the data, we used the classification and regression tree (CART). The accuracy of the classification was justified by the cross-validation (one-leave-out). We constructed a decision tree using all but one observation, then tested this absent observation using the decision tree, and carried out this procedure for every patient (see example at Figure 1.9).

1.4.1 THE OVERALL ALGORITHM FOR THE SCREENING OF BREAST CANCER

1. If the harmonic mean of the Hurst coefficient is less than or equal to 0.672 then go to 3 else go to 2.
2. If the median of the Hurst coefficient is less than or equal to 0.027 than go to 4 else go to 5.
3. Normal. Stop.
4. Breast cancer. Stop.

5. If the median of the Hurst coefficient is less than or equal to 0.028 then go to 6 else go to 7.
6. Breast cancer. Stop.
7. If the mean square of the Hurst coefficient is less than or equal to 1.152 then go to 4 else go to 8.
8. Fibroadenomatosis. Stop.

The complexity of the building of CART decision tree is $O(mnlog_2n)$ where m is the number of the features and n is number the patients (Sani et al., 2018).

1.5 RESULTS AND DISCUSSION

As far as we considered three channels, it is natural to compare their effectiveness from the point of view of classification. One of the main properties of such data sequences are their type. Thus, at the first stage it is useful to classify the color channels by their Hurst exponents: ergodic, chaotic, trend stable and fractal (Tables 1.1–1.3).

TABLE 1.1
Classification of data sequences by red component

Type/group	Healthy	Breast cancer	Fibroadenomatosis
Ergodic	0	0	0
Chaotic	0	0	0
Trend stable	29	54	27
Fractal	0	14	6

TABLE 1.2
Classification of data sequences by green component

Type/group	Healthy	Breast cancer	Fibroadenomatosis
Ergodic	0	0	0
Chaotic	0	0	0
Trend stable	2	6	6
Fractal	27	62	27

TABLE 1.3
Classification of data sequences by blue component

Type/group	Healthy	Breast cancer	Fibroadenomatosis
Ergodic	0	0	0
Chaotic	0	0	0
Trend stable	29	43	24
Fractal	0	25	9

TABLE 1.4
Pairwise specificity, sensitivity and accuracy of differential diagnosis by the first group of features in the blue channel

	Control vs breast cancer	Control vs fibroadenomatosis	Breast cancer vs fibroadenomatosis
Specificity, %	82.76	96.55	63.63
Sensitivity, %	95.59	93.94	52.94
Accuracy, %	91.75	95.16	56.44

TABLE 1.5
Specificity, sensitivity and accuracy of screening by the first group of features in the blue channel

	Control vs breast cancer and fibroadenomatosis
Specificity, %	96.55
Sensitivity, %	94.00
Accuracy, %	94.94

TABLE 1.6
Pairwise specificity, sensitivity and accuracy of differential diagnosis by the second group of features in the blue channel

	Control vs breast cancer	Control vs fibroadenomatosis	Breast cancer vs fibroadenomatosis
Specificity, %	93.10	93.10	72.72
Sensitivity, %	86.76	90.91	69.12
Accuracy, %	88.66	91.94	70.30

Tables 1.1 to 1.3 indicate the blue component potentially can allow separation of the control group from the groups of patients suffering from breast cancer and fibroadenomatosis, because it provides the most clear distinction between healthy and breast cancer patients. Thus, further, we shall consider only the blue channel (Table 1.4).

Since it was not possible to distinguish cancer from fibroadenomatosis using the method described above, at the second stage the classes of patients with cancer and fibroadenomatosis were combined into a general class and the classification of healthy and sick patients was carried out (Table 1.5). This situation is typical for screening, when among a large population of people it is necessary to recognize a risk group without making a differential diagnosis of diseases. In this case, all (29) healthy patients and half of the combined class (51 patients) were taken. The selection

TABLE 1.7

Specificity, sensitivity and accuracy of screening by the second group of features in the blue channel

	Control vs breast cancer and fibroadenomatosis
Specificity, %	96.55
Sensitivity, %	94.00
Accuracy, %	94.95

of sick patients was made at random. This was done for reasons of a large number of sick patients and a small number of healthy ones.

In Tables 1.6 and 1.7 we present similar results obtained for descriptive statistics of the distribution of Hurst exponent (the second group of the features).

The pairwise classification of "healthy – cancer" and "healthy – fibroadenomatosis" allows us to build a very good model based on the Serpinsky curve. Classification of all three cell types based on scanning of the Serpinsky curves led to the following results.

1.6 CONCLUSION AND FUTURE SCOPE

Our investigation has shown that fractal dimension of the Feulgen-stained chromatin in interphase nuclei of buccal epithelium is an effective biomarker of breast cancer. The classification method developed on this basis has the excellent accuracy (94.95%), sensitivity (94%) and specificity (96.5%). For comparison, the sensitivity of the mammography is about 87% (Breast Cancer Surveillance Consortium, 2017) and the specificity after the mammography varies from 93% to 88% (Nelson et al., 2016) depending on the age (the younger the patient the higher the false positive rate). Our method allows effective detecting of the presence of malignant tumors in the human body during screening without invasive procedures. However, the differential diagnosis of cancer and fibroadenomatosis still remains an open problem.

REFERENCES

Adam R., Silva R., Pereira F. et al. 2006. The fractal dimension of nuclear chromatin as a prognostic factor in acute precursor B lymphoblastic leukemia. *Cellular Oncology*. 28, pp. 55–59. doi: 10.1155/2006/409593.

Adhraryn S.G., Dave B.J., Trivedi A.H. 1991. Cytogenetic surveillance of tobacco-areca nut (mava) chewers, including patient with oral cancers and premalignant conditions. *Mutation Research*. 261, 1, pp. 41–49. doi: 10.1016/0165-1218(91)90096-5.

Andrushkiw R.I., Boroday N.V., Klyushin D.A., Petunin Y.A. *Computer-aided cytogenetic method of cancer diagnosis*. New York: Nova Publishers, 2007. ISBN 10: 1-59454-882-X.

Bedin V. et al. 2010. Fractal dimension of chromatin is an independent prognostic factor for survival in melanoma. *BMC Cancer*. 10, 260. doi:10.1186/1471-2407-10-260.

Bikou O. et al. 2016. Fractal dimension as a diagnostic tool of complex endometrial hyperplasia and well-differentiated endometrioid carcinoma. *In Vivo*. 30, pp. 681–690.

Boroday N., Chekhun V., Golubeva E. and Klyushin D. 2016. In vitro and in vivo densitometric analysis of DNA content and chromatin texture in nuclei of tumor cells under the influence of a nano composite and magnetic field. *Advances in Cancer Research & Treatment*. 2016, 706183. pp. 1–11, doi: 10.5171/2016.706183.

Breast Cancer Surveillance Consortium (BCSC). 2017. Sensitivity, specificity, and false negative rate for 1,682,504 screening mammography examinations from 2007–2013. www.bcsc-research.org/statistics/screening-performance-benchmarks/screening-sens-spec-false-negative.

Butakov V., Grakovskiy A. 2005. Evaluation of arbitrary time series stochastic level by Hurst parameter. *Computer Modelling and New Technologies*. 9, 2, pp. 27–32.

Lieberman-Aiden E. et al. 2009. Comprehensive mapping of long-range interactions reveals folding principles of the human Genome. *Science*. 326, 5959, pp. 289–193. doi: 10.1126/science.1181369.

Losa G., Castelli C. 2005. Nuclear patterns of human breast cancer cells during apoptosis: characterization by fractal dimension and (GLCM) co-occurrence matrix statistics. *Cell and Tissue Research*. 322, pp. 257–267. doi: 10.1007/s00441-005-0030-2.

Losa G. 2012. Fractals and their contribution to biology and medicine. *Medicographia*. 34, pp. 365–374.

Metze K. 2010. Fractal dimension of chromatin and cancer prognosis. *Epigenomics*. 2, 5, pp. 601–604. doi: 10.2217/epi.10.50.

Metze K. 2013. Fractal dimension of chromatin: potential molecular diagnostic applications for cancer prognosis. *Expert Review of Molecular Diagnostics*. 13, 7, pp. 719–735. doi: 10.1586/14737159.2013.828889.

Nair U., Obe G., Nair J., Maru G.B. 1991. Evaluation of frequency of micronucleated oral mucosa cells as a marker for genotoxic damage in chewers of betel quid with or without tobacco. *Mutation Research*. 261, 2, pp. 163–168.

Nelson H.D., Fu R., Cantor A., Pappas M., Daeges M., Humphrey L. 2016. Effectiveness of breast cancer screening: systematic review and meta-analysis to update the 2009 U.S. Preventive Services Task Force Recommendation. Annals of Internal Medicine. 164, 4, pp. 244–255. doi: 10.7326/M15-0969.

Nieburgs H.E. 1968. Recent progress in the interpretation of malignancy associated changes (MAC). *Acta Cytologica*. 12, pp. 445–453.

Nieburgs H.F., Herman B.E., Reisman H. 1962. Buccal cell changes in patients with malignant tumors. *Laboratory Investigation*. 11, 1, pp. 80–88.

Nikolaou N., Papamarkos N. 2002. Color image retrieval using a fractal signature extraction technique. *Engineering Applications of Artificial Intelligence*. 15, 1, pp. 81–96. doi: 10.1016/S0952-1976(02)00028-3.

Obrapalska E., Cadel Z., Kostyrka J. 1973. Ocena cytologic zna nablonka jamy ustney u chorych na nowotwory zlosliwe. *Nowotwory*. 23, 1/2, pp. 25–29 (in Polish).

Ogden G.R., Cowpe J.G., Green M.W. 1990 The effect of distant malignancy upon quantitative cytologic assessment of normal oral mucosa. *Cancer*. 65, pp. 477–480. doi:/10.1002/1097-0142(19900201)65:3<477::AID-CNCR2820650317>3.0.CO;2-G.

Ohri S., Dey P., Nijhawan R. 2004. Fractal dimension in aspiration cytology smears of breast and cervical lesions. *Analytical and Quantitative Cytology and Histology*. 26, pp. 109–112.

Palcic B. 1994. Nuclear texture: can it be used as a surrogate endpoint biomarker? *Journal of Cellular Biochemistry*. 19, 1, pp. 40–46.

Prasad M.P., Mukundan M.A., Krishnaswamy K. 1995. Micronuclei and carcinogen DNA adducts as intermediate end points in nutrient intervention trial of precancerous lesions in the oral cavity. *European Journal of Cancer*. 31B(3), pp. 155–160. doi: 10.1016/0964-1955(95)00013-8.

Rathbone M.J., Drummond B.K., Tucker I.G. 1994. The oral cavity as a site for systemic drug delivery. *Advanced Drug Delivery Reviews*. 13, 1–2, pp. 1–23.

Rosin M. 1992. The use of the micronucleus test on exfoliated cells to identify anticlastogenic action in humans. *Mutation Research*. 287, 2, pp. 265–276. doi: 10.1016/0027-5107(92)90071-9.

Rosin M.P., Ragab N.F., Anwar W. et al. 1994. Localized induction of micronuclei in the oral mucosa of xeroderma pigmentosum patients. *Cancer Letters*. 81, 1, pp. 39–45. doi: 10.1016/0304-3835(94)90162-7.

Sagan H. 1994. *Space-filling curves*. — Springer-Verlag: New York–Berlin, ISBN 978-0-387-94265-0.

Sani H.M., Lei C. and Neagu D. 2018. Computational complexity analysis of decision tree algorithms. In: Bramer M., Petridis M. (eds.) Artificial intelligence XXXV. SGAI 2018. *Lecture Notes in Computer Science*. Springer: Cham. 11311: pp. 191–197. doi: 10.1007/978-3-030-04191-5_17.

Sarto F., Tomanin R., Giacomelli L. et al. 1990. The micronucleus assay in human exfoliated cell of the nose and mouth: application to occupational exposures to chromic acid and ethylene oxide. *Mutation Research*. 244, 2, pp. 345–351. doi: 10.1016/0165-7992(90)90083-V.

Schonwetter B.S., Stolzenberg E.D., Zasloff M.A. 1995. Epithelial antibiotics induced at sites of inflammation. *Science*. 257, 5204, pp. 1645–1648. doi: 10.1111/j.1600-0757.2010.00373.x.

Tolbert P.E,. Shy C.M., Allen J.W. 1992. Micronuclei and other nuclear anomalies in buccal smears development. *Mutation Research*, 271, 1, pp. 69–77. doi: 10.1016/0165-1161(92)90033-i.

2 Artificial Intelligence for Sustainable Health Care Advancements

Ayesha Banu

CONTENTS

DOI: 10.1201/9781003153405-2

2.1   INTRODUCTION

The possibility of machines to actually think and simulate human behavior was first introduced by Alan Turing who differentiated humans from machines by developing the Turing test. Later, John McCarthy coined the term artificial intelligence (AI). The main focus of AI is to mimic human cognitive functions. [Yoav Mintz, Ronit Brodie: 2019]. AI has no particular definition that is universally agreed. This term mostly refers to the multiple technologies of computing such as learning, reasoning, adaptation, interaction and sensory understanding that bear a resemblance to human intelligence [https://epsrc.ukri.org/research/ourportfolio/researchareas/ait/]. AI is sometimes used as the umbrella for machine learning (ML), deep learning (DL) and robotics, which gives a system the power to learn and harness the experience and performs some specific tasks faster, better and more efficiently than humans. ML is also called the superset of DL and also subset of AI as shown in Figure 2.1.

In ML, the system learns iteratively based on its past experiences whereas DL is the inspiration of human neural network where the system is given huge amounts of data until it learns by example. This is a more complex form of AI. Robotics is a machine performing any task beyond human capability with more accuracy and precision [Ross 2019]. AI is one of the most interesting and upcoming research areas of this decade, which has totally revolutionized the way humans live and work today. AI is capable of learning from its own mistakes and performing suitable improvements and becomes more useful with use. Currently, there are many applications of AI that are assisting specialists to increase their efficiency [Kumar 2020]. These applications are more general in education, banking, business, social media and day-to-day life whereas they are more specific in the area of health and medicine. Recent research in AI and the application of AI algorithms is gaining increased interest and showing a positive impact on education. The advances in AI systems like ICAI (intelligent computer-assisted instruction) open new possibilities for teaching and learning in the field of education [Chen, Chen and Lin 2020]. In today's world of e-commerce, business management is in an epoch of data, AI applications continuously help in improving efficiency in decision-making and overall business operations. AI aims at elevating the reality of extensive data, promoting business intelligence using complicated algorithms and creating insight on market trends and consumer behavior, which gives businesses a real competitive edge [Al-Zahrani and Marghalani 2018].

AI is now spreading its wings across health care to an extent where AI doctors may possibly assist human physicians for improvisation in the future. The real

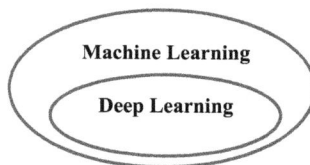

Machine Learning

Deep Learning

FIGURE 2.1   Artificial intelligence—machine learning—deep learning.

motivation of AI in the health domain is to learn features and acquire knowledge from huge volumes of medical data using complicated algorithms, and providing clinical assistance from the obtained knowledge. AI also has the ability to learn and perform self-corrections in order to improve its accuracy on the basis of previous learning feedback. An AI system provides assistance to physicians by giving updated medical information from various sources and also helps reduce errors in diagnosis that may be possible in clinical practice done by humans.

The primary focus of this chapter is on AI algorithms and its applications in health care systems. The rest of the chapter is organized as follows. Section 2.2 gives an insight of the different sources for health care data and Section 2.3 describes the types of AI technologies in relevance to health care system. Section 2.4 focuses on the application of AI for disease diagnosis and its treatment and section 2.5 emphasizes on how the diagnostic errors can be reduces with AI. Sections 2.6 and 2.7 explain the industrial revolution of AI in drug discovery and medical transcription. Section 2.8 gives an insight to how AI has helped to handle the COVID-19 pandemic. The benefits and challenges of AI in health care are discussed in Section 2.10 and the following Section 2.11 puts forward the future of AI in health care.

2.2 HEALTH CARE DATA

Before an AI system is deployed on any health care application, it must be trained on medical data such that the system is able to learn and expertise itself. The data pertaining to health care can be generated from but not limited to clinical activities, electronic recordings and many other sources like www. fda. gov/ downloads/drugs/ guidances/ ucm328691. pdf.

2.2.1 ELECTRONIC HEALTH RECORDS

The clinical data in an electronic form obtained from hospitals, health systems, medical libraries or repositories is usually referred to as an Electronic Medical Record (EMR). This data includes demographics, administrative information, medical prescriptions, laboratory tests and diagnosis, drugs, medical notes, treatment, and doctor and patient details. There also exist large collaborations like National Institutes of Health (NIH), De-identified Clinical Data Repository (DCDR) and the Stanford Center for Clinical Informatics, which provides access to clinical data repositories.

2.2.2 DISEASE REGISTRIES

These registries are clinical information systems that keep a track of a very narrow range and important data related to specific chronic disease conditions like Alzheimer's, asthma, diabetes, cancer, heart disease and many more. Disease registries also provide necessary information on patient conditions. The Global Alzheimer's Association Interactive Network (GAAIN) provides access to a vast data repository of Alzheimer's disease. The National Cardiovascular Data Registry (NCDR) is a suite of data registries holding the data of cardiovascular care they provide. The National Trauma Data Bank (NTDB) is the largest trauma registry data.

2.2.3 HEALTH SURVEYS

These surveys are one of the important ways of collecting health care data especially for the purpose of research. National health surveys are generally conducted to provide estimates of the most common chronic health conditions, i.e. National Health and Nutrition Examination Survey (NHANES).

2.3 MAJOR AI TECHNOLOGIES RELEVANT TO HEALTH CARE

AI is not a single technology, but is rather a combination of multiple technologies. Many of these technologies are found to be relevant to the field of health care, but the tasks they support may vary. The AI technologies that are considered highly important to health care systems are described below.

2.3.1 MACHINE LEARNING

ML is a statistical technique that analyses structured data mostly used for identifying patterns in the data or predicts its future. In medical terms these patterns can either be used in identifying the risk factors for infection or predicting infected patients in future. This is the most common technology of AI. Structured data is an organized collection of information with some specific defined purpose. In health care databases, the structured data is available in the form of patient details, lab investigation values, demographic data, imaging, genetic and financial information. In health care applications, the ML algorithms attempt to form clusters from patients' traits or estimate the probability of a particular disease outcome. There exist different ML algorithms like logistic regression and decision trees, etc. to accomplish this task [Jiang et al. 2017].

ML models can be classified into three types for understanding the inputs of data.

(i) The first and most straightforward model is the supervised learning model (SLM). The inputs for this model are labeled, and can be trained to correctly map between the inputs and the labels using features and weights of hidden layers. This model can be exposed to newly recorded data to make predictions after being trained. The accuracy of this model can be measured and refined. This learning can be applied for predictive modeling where relationships can be built taking patient traits as input and the outcomes of interest.

(ii) The second model is unsupervised learning, where there are no user-defined labels. This model has to discover the features on its own from the given inputs in order to perform the mapping with the outputs. This model has much less human intervention and is mostly used for extracting interesting features. The two important learning methods are clustering and principal component analysis (PCA). Clustering, groups the data with similar traits into one cluster and gives cluster labels for the patients. PCA is primarily used for dimensionality reduction, when the traits of the patient are saved with large number of dimensions.

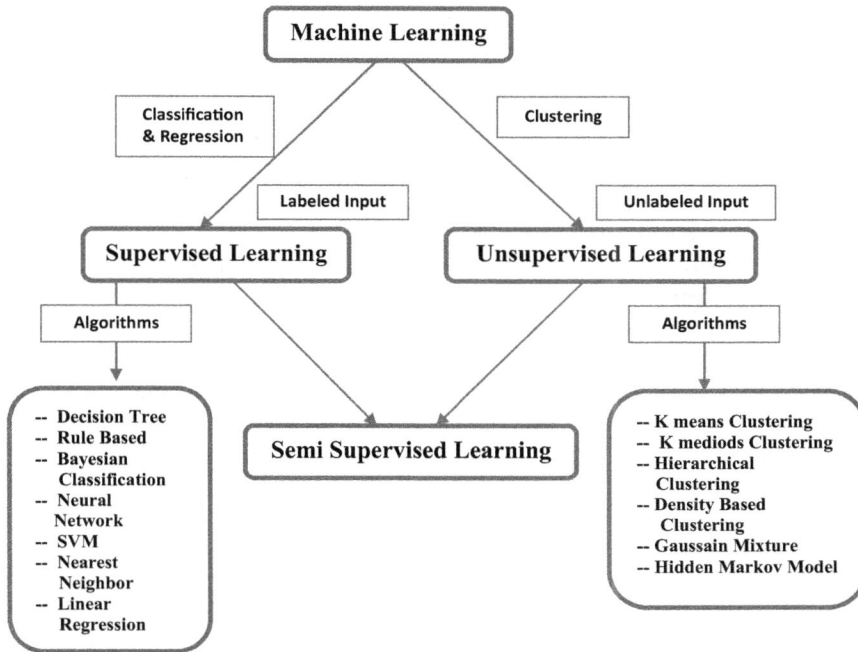

FIGURE 2.2 Machine learning algorithms in health care.

(iii) The third model is Hybrid, which is also called semi-supervised learning. The inputs of this model consist of both labeled and unlabeled data. This model is very close to the method in which human pathologists perform diagnosis using labeled data, like tumor width or density, and unlabeled data i.e. previous records [Carpenter 2020].

The major ML algorithms especially used in health care [Jenni and Chris 2019, Jabbar et.al. 2018, Smith 2020] can be shown in Figure 2.2.

2.3.2 DEEP LEARNING

DL is an extension of neural network model and considered to be the most complex form of ML. DL can also be called a neural network having multiple layers and thousands of features hidden within it. This helps in exploring complex non-linear patterns of data. In recent days, data is rapidly increasing in its volume and complexity. DL is gaining its popularity in handling such data. Recognizing potentially cancerous lesions in radiology images is one of the most common applications of DL in health care. Some other increasing applications of DL include radiomics, oncology-oriented image analysis and detection of relevant features in imaging data. In the field of medical research, the application of DL started in the year 2013 and more than doubled by 2016 and most of these algorithms are applied in image analysis because images are complex with a high volume. The four major DL algorithms are Deep Belief Network, Deep Neural Network, Convolutional Neural Network and Recurrent Neural Network [Davenport and Kalakota 2019].

2.3.3 NATURAL LANGUAGE PROCESSING

It comprises several applications like text analysis, speech recognition, translation and many others for making sense of human language. A very large proportion of health care data is generated from narrative text like laboratory reports, medical notes, discharge summaries etc. This data is mostly unstructured and also beyond understandability for any computer program. In this context, the primary goal of NLP is to extract meaningful information from these narrative texts and give assistance in clinical decision-making. In the domain of health care, the important applications of NLP include creating, understanding and classifying clinical documents, analyzing the unstructured medical notes, generating reports for examinations and transcribing patient interactions. Most of the NLP systems learn repeatedly by reabsorbing the previous interaction results as feedback to determine the accurate results and the results that did not meet the required expectations. The two major components of the NLP system are classification and text processing. Text processing is used to identify all the disease-related keywords from the medical notes. A subset of keywords are further selected and their effect on classification is examined for both normal and abnormal cases. After applying classification, a set of validated keywords are generated to improve the structured data and support the process of decision-making [Jiang et al. 2017].

2.3.4 ROBOTICS

Robotics has become an important and emerging research field from the last two decades. Robots are also called machines that are capable of performing various tasks autonomously and with a greater degree of freedom when compared to humans. The rapid development of ML and robotics has opened a new means to integrate these technologies and transform the way of living. Robotics handles the machines specially designed and programmed to perform labor intensive works. ML makes functioning possible even if machines are not being programmed to do so. The integration of these two gives robots the capability to perform tasks on their own and allow ML tasks to start in supervised fashion and move forward to semi-supervised and complete with unsupervised learning. In the field of health care and medical industries, the integrated duties done by robots may include patients' care, monitoring, e-health, rehabilitation, medical interventions and artificial prosthetics [Patel et al. 2017]. As per the report of "TNO Quality of Life on Robotics for Healthcare" medical robotics has an enormous value in health care with respect to health, economic and societal benefits. The solutions offered by robotics are very significant especially for patient groups. The important application areas of medical robotics include:

- Developing smart medical capsules.
- Surgical robotics for prosthetics.

- Treatment by robots.
- Mental and social therapy assisted by robots.
- Patient monitoring systems using robots.

2.4 AI FOR DIAGNOSIS AND TREATMENT

Diagnosis of disease and suggestion for treatment has been a focus of AI for the past five decades. In 1972 Stanford University in California started working on MYCIN, an early AI program to treat blood infections. MYCIN made an attempt to diagnose patients on the basis of the symptoms reported and results of medical tests. This program also suggests extra laboratory tests and gives a probable diagnosis. After diagnosing the disease, this program suggests a suitable treatment explaining the reasons leading to the diagnosis and suggestion of treatment. This program was found to operate at the same level of proficiency as medical specialists to identify and treat blood infections [Copeland: 2018]. Practically, MYCIN was never used in medicine due to legal and ethical issues not because of its performance weakness. In the mid-1980s, another medical expert system named CADUCEUS was launched by the University of Pittsburgh, which worked on inference engine similar to MYCIN and was intended to improve MYCIN and focus on blood-borne infectious bacteria. The CADUCEUS system could diagnose nearly 1000 different diseases [Banks 1986]. Several such AI medical systems evolved subsequently, out of which some gained importance and a few were not clinically practiced.

2.4.1 RECENT APPLICATIONS OF AI IN MEDICAL DIAGNOSTICS

The process of determining the condition of the disease is termed medical diagnosis. The information required for this medical diagnosis is taken from a patient's history, physical examinations and medical tests. AI plays an integral role in evolution of medical diagnosis. Currently, many of the applications of AI in medical diagnosis come under the following major categories [Faggella 2020].

2.4.2 CHATBOTS

Chatbots are computer programs specifically designed to interact with users through natural language. Chatbots basically work on the principles of AI. These chatbots have inbuilt conversational agent programs that ask the user a series of questions regarding their health issues and obtain the symptoms of disease. They also give recommendations to users regarding different symptoms for clarity. These obtained symptoms may be potentially ambiguous. Therefore, the chatbots map them with the documented symptoms and their respective codes from the medical databases. This helps the chatbots to diagnose the disease. Now the chatbots personalize the diagnosis for future reference and refer the patient to an appropriate doctor for consultation in case of any major disease. The chatbots already available for medical diagnosis are Florence, Your.MD and Babylon. Many surveys done in the area of chatbots for medical diagnosis have proved that they are very good health care providers that are considerably low in cost [Divya et al. 2018].

2.4.3 ONCOLOGY

Oncology is the field of medicine that diagnoses and treats cancer. AI contributes in resolving several biomedical problems. DL, a subset of AI, extracts features automatically and is more flexible to be applied in different research areas including cancer. It plays a very important role especially in the early detection of cancer. It was stated by Jeremy Howard, the CEO and founder of Enlitic that "If cancer can be detected early, then the probability of survival of the patients is 10 times higher." DL achieved highest accuracy in early diagnosis compared to many domain experts. The current DL application in oncology includes detecting cancer from gene expression data, which is considered as very complex data due to its high dimensionality. DL extracts meaningful features from this data and classifies the breast cancer cells. This technology also extracts genes helpful for cancer prediction and cancer biomarkers, to detect breast cancer.

DL can also be used for segmenting brain tumors in MR images, with much more stable results compared to segmenting brain tumors manually by physicians, since it is prone to errors due to motion and vision. This can also be used to measure tumor sizes during treatment and also detect new metastases if there are any that may have been overlooked. The algorithm is made to read more CT and MRI scans of patients to get more accurate results. Prognosis gives an approximation of how advanced the cancer is and what the chances of survival are. DL develops a prediction model for prognosis detection of patients suffering from cancer and receiving treatment. It was found that the survival predictive power of DL is superior to other prediction models [Ali 2019].

2.4.4 PATHOLOGY

Pathology is a study related to the different causes of disease and their respective effects. The disease diagnosis is based on laboratory analysis done on body fluids like urine, blood and also tissues. Pathology plays a vital role in advanced medicine and suggests suitable treatments to fight viruses. The initial way of pathologists to diagnose a disease involves images observation under a microscope done manually. The gold standard for diagnosis in pathology is microscopic morphology, but diagnostic variability is the main limitation bearing errors among pathologists. In this case AI can be introduced in the domain of pathology for consistent and more accurate diagnosis. To improve the speed and accuracy in diagnoses, DL is used to train the algorithm and make it capable of image recognition to diagnose tumors. The research was first started using hundreds of labeled images showing noncancerous as well as cancerous cells. The results extracted from the labeled regions were used as the model to train the algorithm. The popular AI algorithms used in pathology are CNN-Convolutional Neural Network, SVM- Support Vector Machine and KNN- K nearest neighbors [Faggella: 2020; Chang et al. 2018].

2.4.5 RADIOLOGY

AI techniques or algorithms, DL in particular, have shown significant advancement in image-recognition. Different methods starting from CNN to auto encoders have initiated numerous applications in the field of medical image analysis. Traditionally in radiology the trained physicians evaluate medical images visually in order to detect and characterize the diseases. Radiological imaging data is continuously growing at a faster rate compared to number of trained professionals and this has increased radiologists' workloads dramatically. Some corporate studies reported that a radiologist must interpret at least one image for every five seconds. As radiology involves visual perception and decision-making, such constrained conditions makes errors inevitable. Today, AI methods have excelled at recognizing complex patterns in image data and producing qualitative assessment of radiographic characteristics. Integration of an AI system with the imaging workflow definitely increases the efficiency, reduces diagnostics errors and achieves the expected objectives by providing radiologists with all pre-screened images and identified features. Recently, many clinical tasks have been automated and powered by AI, which shifted radiology from a perceptual domain to a computable domain evolving radiology in parallel to other application areas [Hosny et al. 2018].

2.5 AI FOR REDUCING DIAGNOSTIC ERRORS

In today's health care systems, the total volume of medical data associated with every patient is increasing tremendously. Whenever a patient visits any doctor or is admitted into any hospital or goes for any medical test, the size of data repository starts increasing. The ability to access this real-time data and produce accurate diagnoses without human errors has really become critical. A misdiagnosis of any illness or other health problems may lead to adverse effects and even patients' death in some cases. In this sensitive situation of health care, the application of AI is of great help to clinicians to lower the error rate. AI changes the way in which data is analyzed and results are obtained. Today AI is contributing to accurate diagnoses at a faster rate, and also at a reduced cost. Some popular ways in which AI reduce diagnostics errors are as follows [Daley 2019].

2.5.1 BUOY HEALTH

This is a symptom checker that uses sophisticated algorithms to diagnose the disease and suggest treatment to cure the illness. Buoy works in five processes: (1) There is a chatbot that talks to the patient about the symptoms and difficulties concerned to health. (2) Later, it gets feedback on the symptoms from cause to severity. (3) Buoy then chooses the best option for care and guides the patient based on its diagnosis. (4) After this, with the permission of the patient Buoy can follow up the process and progress of the treatment. (5) Finally, Buoy adds the patient to a community group to get in touch with people who have experienced similar symptoms and also help others from their own experience [www.buoyhealth.com/]. Harvard Medical School is the hospital using Buoy's AI for diagnosis and treatment of patients more quickly.

2.5.2 PathAI

PathAI develops ML technology that assists pathologists to make rapid and accurate diagnoses of cancer. The present goal of PathAI is error reduction in cancer diagnosis. It is also providing solutions and methods for individualized medical treatment to patients. It helps patients to get the benefits of novel therapies. PathAI is currently collaborating with Bristol-Myers Squibb the drug developers and the Bill & Melinda Gates Foundation to share the technology globally and expand AI into other health care industries [www.pathai.com/].

2.5.3 Enlitic

Enlitic is a company that is bridging the gap between human and AI to bring the latest advancements in medical diagnostics. It is developing DL medical tools especially for radiology. World class radiologists have collaborated with data scientists to analyze the most comprehensive clinical data. The DL platform makes it more convenient to analyze unstructured medical data like blood tests reports, radiology images, patient medical history and genomics. Enlitic is also pioneering new medical software that gives doctors a better insight to the data and enables them to perform diagnoses sooner and with more accuracy [www.enlitic.com/].

Freenome is a biotechnology company that uses AI in screenings and diagnostics. It is a pioneer in early cancer detection and also the multiomics platform [www.freenome.com/]. Zebra Medical Vision is working on transformation of patient care using the power of AI. It provides AI-enabled assistance to radiologists by receiving image scans and analyzing them automatically. It also produces several clinical findings. These findings are considered by the radiologists while doing diagnosis [www.zebra-med.com/]. BIDMC- Beth Israel Deaconess Medical Center is a teaching hospital of Harvard University, which is working on diagnosing toxic blood diseases at an early stage using AI. The doctors use sophisticated AI-enhanced microscopes to scan the blood samples and identify the harmful bacteria. The entire process completes faster than manual scanning. The machine is trained to search for bacteria using 25,000 sample blood images. The machines were found to learn identifying and predicting harmful bacteria in blood with a 95% accuracy rate [www.bidmc.org/].

2.6 AI FOR DRUG DISCOVERY

AI is one of the prime technologies in the era of the industrial revolution and it is expected to have a tremendous effect in reducing the cost and time required to discover new drugs. Many pharmaceutical industries perform the rational drug discovery through omics and structure-based drug development. Even though traditional drug discovery brings huge benefits if successful, it is still costly with high risk and also time consuming. At present, great changes are observed with the fusion of AI and previous technologies for drug discovery [Son 2018].

The primary goal in drug discovery is identifying the medicines that can help in preventing or treating any particular disease. Many of the drugs are small chemically synthesized molecules. In order to identify these molecules, the traditional approach is to screen the libraries of molecules to recognize the one which can potentially become a drug. Later, several rounds of tests are performed to develop the challenging compound. Currently rational structure-based drug design is becoming the most common approach which avoids the initial screening phase.

The chemists create new drugs by synthesizing and evaluating several compounds. In spite of this, it is still unknown which chemical structure has the properties to become an effective drug and show the desired biological effects. The process of refining a chemical compound into an effective drug is costly and time consuming as well. Now experts are showing inclination towards the data processing capabilities of AI to accelerate the process and reduce the cost of new drug discovery [www.merckgroup.com/en/]. Different companies depending on AI for developing next generation medicines are as follows:

2.6.1　BioXcel Therapeutics

This is a clinical-stage biopharmaceutical company using AI to identify enhanced therapies in the field of neuroscience and immuno-oncology and new drug development. Their primary focus is to utilize novel technologies and innovative research for developing valuable therapeutics that aim at transforming patients' lives. BioXcel Therapeutics works on unique AI platforms to reduce the therapeutic development costs. Several big data and ML algorithms are considered to identify new therapeutic indices and reduce the time taken for drug development. Additionally, the company employs AI for drug re-innovation and identifying the new applications for the existing drugs [www.bioxceltherapeutics.com/].

2.6.2　BERG Health

BERG is a clinical-stage biotechnology company powered by AI using a "Back to Biology" approach to health care. Their prime focus areas of research include oncology, neurology and rare diseases. By using the interrogative biology intelligence platform, the goal of BERG is to map diseases and develop treatments. By combining the AI technology with patient biology, this company aims at picking up the pace of clinical identification and searching for capable therapeutic targets to treat disease. Combining the approaches of interrogative biology with the traditional research methods, BERG can develop strong drugs that can fight rare diseases [www.berghealth.com/]. Having understood the power of AI in discovering new medicines, 43 pharma companies are currently using AI [Smith 2020].

2.7　AI FOR MEDICAL TRANSCRIPTION

Today most of the medical documentation across the globe is stored in Electronic Health Records (EHR) and online databases are helping clinicians in accessing patient information quickly. The doctors and surgeons mostly use traditional methods for documenting the treatment details, patient history or the procedure carried out for surgeries either writing on a paper or typing in any text editor. The doctors and medical staff usually spend a lot of time in the documenting process. Unfortunately, many of these documents are not integrated completely with the patient electronic record systems thus making the entire effort worthless. This time-consuming manual documentation has been digitalized using voice recording and transcription technologies. Medical transcription is all about converting dictated medical notes into typed health records. The transcribers spend hours of time in this conversion. Speech recognition becomes the biggest hurdle for transcribers in this process [Kulkarni et al. 2020].

Another development in medical transcription is the dictation–transcription software that uses voice recognition technology that listens to a doctor's summary and transcribes the dictation into words. Some of these tools automatically integrate with EHRs and populate the data into the health system. However, these tools fail to solve the documentation problem, as they simply change the task of typing with dictation and editing. Medical scribes and virtual scribes enhance complete support for documentation rather than compared to dictation–transcription tools. They manage note-taking, summarization and coding appropriate information into the suitable EHR fields. The medical scribing industry is among the best emergent sectors in health care. Today the challenges of clinical documentation are responded to by the new AI-powered scribes. These scribes use natural language processing to listen to the conversation between the patient and the doctor. The medically relevant terms are then parsed out and the information is summarized into medical notes and finally mapped to the suitable fields of EHR. Compared to other tools for dictation and transcription, the AI scribes work further on verbatim summarization at a more affordable price than medical scribes. This has resulted in AI-powered medical scribes capturing the entire medical transcription system [www.deepscribe.ai/].

Several companies claim to offer AI-based medical transcription software to hospitals and health care industries intended to help them in transcribing speech into text and update patient medical records in EHR. Nuance Communications recommends a software named Dragon Medical One (DMO), which helps doctors and health care providers to transcribe speech into EHR using NLP. It is also possible to integrate this software into any of the existing EHR systems. Nuance Communications claims that Nebraska Medicine has used DMO, which improved the competence of their health care providers when they integrated the software with their EHR. There was a 23% decrease in transcription costs and 71% of company physicians observed an improvement in their documentation quality and 50% stated that by using DMO at least 30 minutes of time was saved every day. M*Modal recommends the software Fluency Direct for Transcription (FDT), which claims that health care providers can combine the software into their existing EHR databases, such as Epic, Cerner and Athena health. Floyd Memorial Hospital incorporated this software into its EHR system, which increased the total number of medical documents written per day. Medical Transcription Billing, Corp. (MTBC) recommends software called TalkEHR, which includes a voice assistant feature called Allison. Many more companies have come up with the AI-based medical transcription software [Mejia 2018].

Amazon Transcribe Medical (ATM) is a newly launched ML service of Amazon Web Services (AWS), which can convert physician-dictated medical notes from speech to text automatically. This ATM uses both ML and NLP technologies for accurate speech recognition and health care providers can utilize this service for analysis of patient–doctor conversations or for HER entries. ATM has benefited pharmaceutical companies, transcribing patient or doctor calls to record medicine names accurately and identify key terms that describe the side effects. This can help the companies to analyze and detect safety issues related to drugs. ATM is trained specially to understand different styles of clinical languages and terminologies and it also provides auto-punctuation that health providers can speak naturally [Kent 2019].

2.8 AI FOR COVID-19 DIAGNOSIS

COVID-19, also called the coronavirus pandemic, was first identified in December 2019 in Wuhan, China. This was declared a public health emergency by the World Health Organization (WHO). This resulted in a worldwide crisis and put the entire health industry on its toes for diagnosis, prediction and treatment. Quick and accurate diagnosis of positive patients became the top priority to avoid further spread of the virus and provide on-time treatment. The health care industry now requires the support of AI, ML and DL technologies to fight against the new disease. AI is a prominent technology to track the spread of virus and identify the high-risk patients that can help in controlling this infection.

The National Institutes of Health (NIH) has made an ambitious effort in harnessing AI for COVID-19 diagnosis, treatment, and monitoring. This resulted in launching the Medical Imaging and Data Resource Center (MIDRC), which utilize the power of AI and medical imaging to fight against the pandemic. The National Institute of Biomedical Imaging and Bioengineering (NIBIB), a part of NIH, created novel tools that can help physicians for early detection of the virus and personalized therapies for the patients [www.nih.gov/news-events/].

Several researchers worked on all the possible means to identify major applications of AI for the COVID-19 pandemic and the diagnostic improvements that can benefit from AI. Early detection of COVID-19 using chest CT enables on-time treatment to patients and also helps in controlling the spread of disease. A new AI system is proposed for rapid COVID-19 diagnosis and an extensive statistical analysis of chest CT performance. This system is evaluated based on a large dataset with a collection of 10,000 CT volumes, community acquired pneumonia, influenza and non-pneumonia subjects. This is a deep convolutional neural network-based system applicable for multi-class diagnosis task. The diagnosis performance of chest CTs is compared to chest x-rays and the AI system was found to have outperformed many radiologists in more challenging tasks [Jin et al. 2020].

A research team from the Indian Institute of Technology (IIT) in Hyderabad developed a COVID-19 test kit powered by AI that would give the test result in 20 minutes and at cost of only ₹600 per test. This overcomes the biggest hurdle, i.e. the cost of the screening test. This test kit has undergone clinical trials at ESIC Medical College and Hospital in Hyderabad to determine its efficiency and sought approval from the Indian Council for Medical Research (ICMR) [https://iith.ac.in/news/2020/06/10/Covid19-detection-kit/].

2.9   THE ROLE OF CORPORATIONS IN AI IN HEALTH CARE

AI has started a new era of innovations in health care and influenced health care industries or corporations tremendously in this decade. AI can help in reducing the costs of on-going health operations and improve the quality of patient care. The success of AI in the medical field has disposed more and more companies towards health care and currently there are more than 100 companies using AI in this sector. Some of the popular, but not limited to, health care corporations are:

Watson Health is a dedicated health branch of International Business Machines-IBM Corporation aiming at bringing data, technology and expertise on to one platform to transform health. IBM helps organizations by providing tools and services required to solve clinical, operational and also financial problems across health care. Using cognitive computing, Watson Health has helped several well-known organizations including Mayo Clinic in clinical trials of breast cancer and Biorasi in bringing drugs to the market faster. IBM Watson Health combines human experts with augmented intelligence and helps researchers around the world in translating data and knowledge for intelligent decision-making [www.ibm.com/in-en/watson-health].

Google Health is now striving to improve the lives of people in all the important areas of health by merging with DeepMind's health team. The major research successes of Google Health include predicting acute kidney injuries accurately using AI nearly 48 hours early than currently diagnosed and detecting 5% more cancer cases, reducing false-positive exams more than 11% compared to unassisted radiologists. Automated Retinal Disease Assessment (ARDA) and Optical Coherence Tomography (OCT) scans of Google Health help doctors to quickly trace diabetic retinopathy and interpret eye scans [https://health.google/health-research/].

CloudMedX Health utilizes NLP and DL to acquire patient data from EHR and produces the clinical insights into health care professionals which help them to improve patient outcomes. The solutions of CloudMedX are already applied showing promising results in medical areas like liver cancer, congestive heart failure, orthopedic surgery and renal failure [https://cloudmedxhealth.com/].

2.10   BENEFITS AND CHALLENGES OF AI SERVICES IN HEALTH CARE

From the concepts of DL algorithms reading CT scans more rapidly than humans to NLP that can read and interpret the unstructured data in EHRs, the AI applications in health care are endless. AI starts from early detection of diseases and moves forward until suggesting the appropriate treatment for cure. The applications of AI span all the important areas in health care including oncology, radiology, pathology, surgery, patient care and many more in the past decade. But, any technology at the peak of its success also faces challenges that need to be addressed.

Researchers have done many surveys to understand the real pros and cons of AI services from the health care professionals' point of view as they are the end users of any tool or application. A survey done by MITTR in collaboration with GE Healthcare reported that that more than 82% of health-care industries in the US and UK have already deployed AI and created workflow improvements in operational and administrative activities [https://complexstories.com/work/mit-ge-health-ai/]. Another survey has been done on the practicality of AI services for health care and the willingness of health care professionals to adopt AI technologies. This study revealed that 67.8% of professionals were willing to use AI for independent diagnosis and decision-making support, and 28.9% of them showed interest in using AI in a restricted manner to help professionals in care processes. Only 3.3% of respondents were not willing to use AI services [Väänänen et al. 2020].

The major benefits of AI in three promising perspectives are:

1. **Patient self-service benefits:** More convenient choices are now available for patients to easily and rapidly complete tasks like enquiring about hospital and doctor details, scheduling doctor appointments, follow-up treatment, payments and other procedures. AI has made it easier to access health care facilities and increases the patient satisfaction also.
2. **Clinician's diagnosis benefits:** AI tools helps early and effective diagnosis of diseases comparatively at reduced amount of time that makes clinicians more confident in extending treatment. AI and robotics are coming up with assistants helping the professionals in complicated surgeries.
3. **AI benefits to pharma:** With the advent of AI, the time and cost for new drug discovery has reduced to a greater extent and this benefits the pharmacy to utilize the time in more research on different chemical combinations and drugs.

Despite its potential to bring new insights and help providers and patients in health care, AI may still face few challenges that need to be addressed to retain the rewards of this powerful technology.

- AI requires large amounts of training data gathered from millions of patients' records to make correct diagnosis but there are limited chances of mistakes if the data is incomplete or insufficient.
- AI with robots is eventually machines with a lack of emotions and the human touch. For patients with mental disabilities this emotional bonding is very much essential to keep the morale of patients high.
- The technological development of AI in health industry at various levels may result in serious risk of unemployment.
- EHR is the major data repository for all the AI applications. The adoption of AI in health care may rely on high-quality data in standard formats.
- Data protection and preserving privacy is the most promising challenges of AI in health care.
- In practice, both patients and professionals must be able to trust the AI systems to be implemented successfully in health care.

2.11 LOOKING AHEAD—FUTURE OF AI IN HEALTH CARE

AI today is being implemented in every aspect of daily life. The scope of AI applications is rapidly spreading its wings to cover every basic necessity of humans including complicated decision-making. There is no technological area that is untouched by AI applications. The quality, safety and implementation of health care are greatly improved by AI. It helps extract interesting patterns from large amounts of medical data considerably in less time compared to humans. AI can streamline diagnosis, treatment, drug discovery and patient care. The different areas of health care that can be benefit from using AI in future are:

- Currently the doctors or clinicians spend a significant amount of time in reading patient records, analyzing the test reports, finding the cause of disease and guidelines for treatment. AI will extract important information efficiently from her, which will save time and directly guide patient management. It can also convert a patient's verbal consultation into a summary letter. These applications can save time and be implemented quickly as they help in assisting clinicians.
- Using the patient's clinical record, AI can identify the risks and actions automatically. Since AI can monitor millions of inputs simultaneously, it plays a significant role in preventative medicine.
- These systems also bring specialist diagnostic expertise into patient care. The patients who are identified to have low risk would receive reassurance instantly whereas the high-risk patients can be immediately exposed to clinicians for treatment with less waiting time [Buch et al. 2018].
- With the capability of massive data analysis both DL and AI systems can achieve faster and more accurate diagnoses. Personalized health care and treatments can be made cheaper. More effective drugs can be customized for each disease.
- Future research in AI can be directed towards many areas in health care and these systems can be integrated into clinical practice, which helps in building a beneficial relationship between AI and clinicians. AI offers clinicians greater efficiency or cost benefits and in turn clinicians can offer AI all the essential exposure required to learn complex medical cases.
- Throughout this process it is very much essential to ensure that AI does not compromise the social and ethical constraints of humans. Even though AI is still in its early days, the achievements already made by AI in health care are impressive and moving towards a healthy future.

REFERENCES

Ali, A.-R. 2019. "Deep learning in oncology – applications in fighting cancer." *Business Intelligence and Analytics Healthcare Process Automation*, Feb 2019. https://emerj.com/ai-sector-overviews/deep-learning-in-oncology/.

Al-Zahrani, A. and Marghalani, A. 2018. "How artificial intelligent transform business" (April 25, 2018). Available at SSRN: https://ssrn.com/abstract=3226264 or http://dx.doi.org/10.2139/ssrn.3226264.

Banks, G. 1986. "Artificial intelligence in medical diagnosis: the INTERNIST/CADUCEUS approach." *Critical Reviews in Medical Informatics*. 1(1): 23–54.

Buch, V.H., Ahmed, I., and Maruthappu, M. 2018. "Artificial intelligence in medicine: current trends and future possibilities." *British Journal of General Practice* Mar; 68(668): 143–144. doi: 10.3399/bjgp18X695213.

Carpenter, A. 2020. "An introduction to machine learning in oncology, on the future of AI in cancer treatment and research." https://towardsdatascience.com/machine-learning-ai-applications-in-oncology-73a8963c4735.

Chen, L., Chen, P. and Lin, Z. 2020. "Artificial intelligence in education: a review." *IEEE Access* 8: 75264–75278.

Copeland, B.J. "MYCIN." *Encyclopedia Britannica*, November 21, 2018, www.britannica.com/technology/MYCIN.

Daley, S. "32 Examples of AI in healthcare that will make you feel better about the future," July 2019, https://builtin.com/ artificial-intelligence/artificial-intelligence-healthcare.

Davenport, T. and Kalakota, R. 2019. "The potential for artificial intelligence in healthcare." *Future Healthc J.* 6(2): 94–98. doi:10.7861/futurehosp.6-2-94.

Divya, S., Indumathi, V., Ishwarya, S., Priyasankari, M., and Kalpana Devi, S. 2018. "A self-diagnosis medical chatbot using artificial intelligence." *Journal of Web Development and Web Designing* 3(1): 1–7.

Faggella, D. 2020. "Machine learning for medical diagnostics – 4 current applications." March 2020, https://emerj.com/ai-sector-overviews/machine-learning-medical-diagnostics-4-current-applications/.

Hosny, A., Parmar, C., Quackenbush, J, Schwartz, L.H., and Aerts, H.J.W.L. 2018. "Artificial intelligence in radiology." *Nat Rev Cancer* 18(8): 500–510. doi:10.1038/s41568-018-0016-5.

Jiang, F., Jiang, Y., and Zhi, H., et al. 2017. "Artificial intelligence in healthcare: past, present and future." *Stroke and Vascular Neurology* 2: doi: 10.1136/svn-2017-000101.

Jin, C., Chen, W., and Cao, Y. et al. 2020. "Development and evaluation of an artificial intelligence system for COVID-19 diagnosis." *Nat Commun* 11: 5088. https://doi.org/10.1038/s41467-020-18685-1.

Jabbar, M.A., Samreen, S., and Aluvalu, R. 2018. "The future of health care: machine learning." *International Journal of Engineering & Technology* 7(4.6): 23–25.

Kumar, B. 2020. "AI applications assisting specialists to increase their efficiency." 360digitmg, April 2020, https://360digitmg.com/application-of-artificial-intelligence.

Kent, J. 2019. "Amazon introduces machine learning medical transcription service." December 2019, https://healthitanalytics.com/news/amazon-introduces-machine-learning-medical-transcription-service.

Kulkarni, S., Torse, D.A., and Kulkarni, D. 2020. "A cloud based medical transcription using speech recognition technologies." *International Research Journal of Engineering and Technology (IRJET)* 7(5): 6160–6163.

Mintz, Y. and Brodie, R. 2019. "Introduction to artificial intelligence in medicine." *Journal Minimally Invasive Therapy & Allied Technologies* 28(2): 73–81.

Mejia, N. 2018. "Machine learning for medical transcription – current applications." December 2018, https://emerj.com/ai-sector-overviews/machine-learning-for-medical-transcription-current-applications/.

Patel, A.R., Patel, R.S., Singh, N.M., and Kazi, F.S. 2017. "Vitality of robotics in healthcare industry: An Internet of Things (IoT) perspective." In: Bhatt C., Dey N., Ashour A. (eds) *Internet of Things and Big Data Technologies for Next Generation Healthcare. Studies in Big Data,* vol 23. Springer, Cham. https://doi.org/10.1007/978-3-319-49736-5_5.

Ross, D.B. 2019. "Artificial intelligence and healthcare." In: Claypoole T. F. (ed.) The Law of Artificial Intelligence and Smart Machines, ABA Publishing, Chicago.

Smith, S., "43 Pharma companies using artificial intelligence in drug discovery," September 2020, https://blog.benchsci.com/pharma-companies-using-artificial-intelligence-in-drug-discovery.

Son, Woo Sung. 2018. "Drug discovery enhanced by artificial intelligence." *Biomedical Journal of Scientific & Technical Research (BJSTR)* December 12, 2018, ISSN: 2574-1241 doi: 10.26717/BJSTR.2018.12.002189.

Väänänen, A., Haataja, K., and Toivanen, P. 2020. Survey to healthcare professionals on the practicality of AI services for healthcare [version 1; peer review: 1 approved with reservations] F1000Research 2020, 9:760 https://doi.org/10.12688/f1000research.23883.1.

Ye Yoon Chang, Chan Kwon Jung, Junwoo Isaac Woo et al. 2018. "Artificial intelligence in pathology." December 2018 *Journal of Pathology and Translational Medicine* 53(1), doi: 10.4132/jptm.2018.12.16.

3 Identification of Lung Cancer Malignancy Using Artificial Intelligence

Vinod Kumar and Brijesh Bakariya

CONTENTS

DOI: 10.1201/9781003153405-3

3.1   INTRODUCTION

Recent advances in IT suggests that artificial intelligence (AI) can forecast the like-lihood of cancer, diagnosis, and illness. Therapeutic decision-making soon is now a reasonable possibility. It has also been exciting to incorporate AI in many areas, such as cardiology, nephrology, critical care, and mental health. For a range of factors, cancer is an environment in which AI is projected to have a positive short-term effect, and in many cases boost the credibility of AI-based cancer health care methods [Asan et al., 2020]. First, cancer is a common choice, which places essential stress on phys-ical injury, mental distress, and recurrent financial costs to the patients and society. There is also an immediate need to boost cancer survival in all cancer sub-types, without exception.

Early detection and recovery are the secrets to improve the five-year survival rate. During the next five years, Indian cancer cases will increase by 12%; 1.5 million people are expected to stay non-communicable by 2025, from 1.39 million by 2020, based on current cancer trends by the Indian Council of Medical Research (ICMR) [Mathur et al., 2020]. Currently, there are 57,795 new cases of lung cancer added every year. It may reach 67,000 each year by 2020.

CT imaging has been shown to be a remarkably successful tool for early diagnosis of lung cancer, which can minimize mortality by up to 20%. Although technology has proven effective, a vast amount of research has investigated the use of other upcoming technology, including AI. Such technology will further enhance the identification, recognition, and size of nodules, thus reducing false positives [Asan et al., 2020].

Cancer is an environment where AI is projected to have a major impact in the short term and cancer is a sensible choice in several ways, for improving the "reputation" of AI-based public health care approaches. Second, cancer is widespread and brings significant pressures upon patients and the community as a whole in relation to specific disabilities, social discomfort, and economic effects. There is also a vital need to boost cancer-related findings virtually beyond question across all cancer forms. Second, at the human level, there is a strong need for a more effective diagnosis of diseases, prognostic and therapeutic methods; cancer is a major cause of morbidity and death and it is frightening that there is no other factor in terms of the perception of cancer as a lethal disorder [Pelc et al., 2020]. There are also big decision-making difficulties with the tumor-heterogeneity, specifically due to real data stress from a practitioner's perspective. It has been stated, quite alarmingly, that the modern cancer expert has spent over 20 hours reading every day to keep up with scientific advancement. The use of AI has been effective over the past years since AI uses several statistical, probability theory, and optimization methods and criteria to "improve" on previous cases and to identify "noisy" or multi-dimensional datasets in clinically useful latent patterns. This chapter summarizes existing and evolving developments in AI interventions at all levels of the cancer survivor's journey [Yang et al., 2017].

The image can be defined as f (x, y) two-dimensional, where x, y is the spatial coordinate, and (f) is the amplitude of any single pair of coordinates (x, y). The amplitude of this position on the gray surface or image is also known as amplitude (f). This digital image consists of a few x and y elements, all of which have the same position and value. These image components are also referred to as pixels. Three-dimensional function (x, y, z) will define the 3D images later and voxels are also referred to as single components. Image quality is a significant radiologic factor and the term local resolution means that two neighboring objects can be recognized in an image [Altaf et al., 2019]. The transitional resolution, which shows imaging quality and performance, is another significant resolution feature. DICOM is a common protocol for the handling and communication of information on associated medical images. DICOM is an Association of National Manufacturers of Electrical Equipment registered trademark. It is commonly used in radiology, cardio logical, oncological, obstetrical, and dental sciences. To retrieve health information and images, DICOM files can transfer systems and images in DICOM format to systems. A pixel size DICOM file includes location base, patient identity, self-image, and image attributes. Moreover, DICOM [Herrmann et al., 2018] files make it is difficult to distinguish between patient and image data. The DICOM is also still connected to the patient. There is no understanding of where basic image processing stops and where more image and computer vision research begins. Image processing is often defined as being a function that comprises one image, both the input and output of the operation.

3.2 BACKGROUND AND RELATED WORK

In each AI paradigm, datasets were an important component. The accuracy of the related data leads to analytical growth, preparation, and progress. For this development to be helpful, the evidence obtained should be justified and labelled by experts in computer vision. This segment includes information on the artificial intelligence of early work for the identification of lung cancer [Wang et al., 2019]. One of the most serious ailments in the globe is lung cancer. Researchers are taking action in different directions to boost lung cancer. CT scans through machine learning techniques and lung nodular identification from a variety of defects. A deeper neural network with progeny [Taher et al., 2016] extracted images of information. The technique by the author here shows that the larger the nodule, the higher the risk of cancer growth. Compared to the class labels, also known as the pattern classification, the test set includes features in the linear SVM that increase the quality of the tests. It gets CT images using compressed technology ROI segmentation [Deep Prakash, et al. 2017]. Each ROI image is broken into a different DWT (Discrete View Switch) strategy and also some SVM-detectable GLCM classification bands [Gupta et al., 2020]. A method based on nodule size is also suggested below [Kim et al., 2016], which shows excellent performance compared with ROI isolation for the detection of cancer cell types. The feature extraction [Kumar et al., 2015], including the Otsu division thresholds and gray-level repeat matrix (GLCM), is defined as the physical dimension level. It identifies cancer nodules from its view of the effect of such traits. It also grades initial stages to prevent cancer. The median filter is a method of pre-treatment for filtering salt and pepper image noise through [Murillo, 2018] and it is measured in a numerical format [MyaTun et al., 2014] via arranging the value of the entire pixel round. Some intersecting processes covering the Otsu threshold subsequently.

An analysis of the above concepts indicates that its GLCM or SVM helps in improved classifications of images of lung cancer [Kim et al., 2016]. Using these images in a CAD system, the critical characteristics are just different from several other definitions since that model is based on characteristics like NC ratio, circularity, and so forth and establishes an estimation threshold for cells coinciding only with rule-based classification.

Although the process of extracting features is much more equitable than other approaches, the concept classification does not have all the desired outcomes such as SVM and GLCM. In the initial stages of growing cancer, several methods can be used for staging approaches, trying to correct the image using Gabor [Kumar et al., 2015] filters, consisting of a harmonic function and Gaussian function to distinguish the watershed from characteristics like area, perimeter, and eccentricity and bringing out the radionics form that is used to manage different cancerous diseases [Kumar et al., 2015]. This concept provides an even more automated toboggan-based image-precision segmentation strategy, producing a starting point without any human involvement. The method for min–max normalization has been described as a tool to define the stages of disease.

It improves classification accuracy by using random forest algorithms. It contains some standard methods for classifying the stage as malign or benign as in the task [Murillo, 2018] that focuses on the co-occurrence matrix of structure (SCM), which helps to identify as lethal and benign in the classification technology. The image of nodules employing the GLCM technique is extracted. Laplace and Gaussian are grouped by SCM filters for better classification that graded as nodules malignancies at the stages of metastases [Kumar et al. 2021].

This has a strategy that provides improved ranking for all criteria, with ANV, MLP and decision-making grades, and SVM ratings. The combination of classification technologies achieves exact results compared with individual tests. The idea of machine tool technologies was the focus of several studies. It was discovered that only the SVM and even some machine-learning strategies help assess lung cancer with reliable results [Murillo, 2018]. The combination of classification technologies achieves exact results when compared with individual tests. Furthermore, the author examines important predictions of lung cancer and reveals some specific capabilities and weaknesses. Deep learning technologies are very useful to discover non-small cell lung cancer [Sujitha and Seenivasagam, 2021]. Fermenting characteristics of quantitative imaging with the aid of deep learner networks, physiological imaging techniques, and patient stratification are being examined. The web application also will present in-depth learning applications for lung cancer diagnosis using just a web application 2D-UNet model [Taher et al., 2015]. The 2D mask is extracted by specific technologies and advanced parameters from the specialized dataset and a deep neural network has been developed to recognize the subtypes of pulmonary cancer in histopathology images [Hussein et al., 2019]. Research shows that machine learning can be effective in the challenging task of image classification and has the potential to aid pathologists in classifying lung cancer. CNN model was connected to RNN using a single point seed position and validating pathology [Yiwen et al., 2019], involving medication and intervention care for NSCLC. Non-delta radiometric features derived from clinical tomography were evaluated as X-ray improvements of the findings obtained from its concerned images [Alahmari et al., 2018].

The lung cancer diagnostic strategy has also enhanced efficiency with a variety of characteristics by using the ROC curve. To improve efficiency, the study uses conventional radiological resources. It enhances tumor cancerous cell response from broad sets of data via Bayesian networks pre and post-radiotherapy [Luo et al., 2018; Patel et al., 2020]. Local control analysis is improved in combination with a complete Markov system and wrapper-based access. The author suggested a radiation model for the prediction of pulmonary cancer with an earlier mentioned radiology-based prognosis [Wu et al., 2018]. This scientific radiation model, in relatively small terms, is being used to resolve some features of diagnostic methods in combination with belief and quantitative learning. In addition, the nearest classification features are suggested for predicting treatment performance. While the other classification features are focused on foreseeing lung cancer with a few specialized high dimensionalities and choose those features for another approach to focusing on sub-predictive selected features [Rodrigues et al., 2018]. The neural networks are proposed to forecast the prognosis of a patient with localized cancer following radiation therapy [Cui et al., 2018]. Also, it takes the temporal correlation with the sequence dataset as a feature to enhance the predictive performance. In response to 1-D wrapping, a maximum distribution was introduced by two locally connected and fully connected.

However, their bulk model improves performance in contrast to the completely integrated multi-layer architecture [Liu et al., 2018; Nie et al., 2018]. The use of a MAP analysis technique illustrates the effectiveness of parallel SVM in the prediction and diagnosis of lung cancer (see Figure 3.1).

The main purpose of this chapter is to enhance performance, minimize diagnosis times, identify nodules, and classify features and functionality. Several studies have been conducted to determine the most dangerous and common type of lung cancer

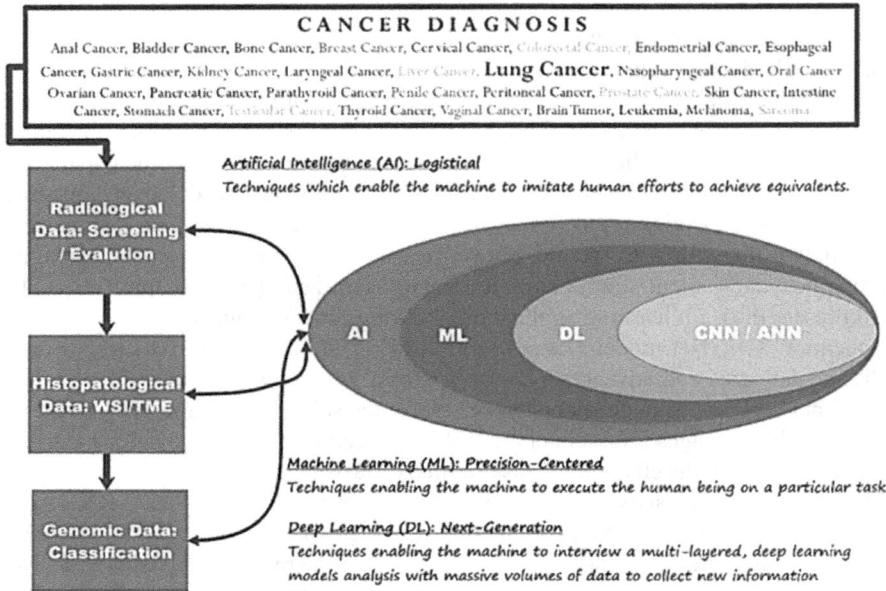

FIGURE 3.1 Cancer diagnosis using artificial intelligence, machine learning and deep learning.

have been completed. These lung nodules and their increased sensitivity, precision, and accuracy are then analyzed. The chapter's key aim is to evaluate diagnostic approaches to lung cancer and to improve this strategy.

Motivation: The key goals are to increase performance, minimize diagnosis times, and determine the location of nodules and features. Since then, numerous studies have been conducted to assess if lung cancer is the most dangerous and harmful. As a consequence, the sensitivity, precision, and accuracy of all lung nodules are analyzed together. The diagnosis of lung cancer should be enhanced and improved.

3.3 PROPOSED MODEL

The general outline of the model is shown in Figure 3.2 with five normal steps. The first step is to read and process the CT images with different filters according to their smoothness. The next step is to segment the image that extracts the right segment. A nodule is produced to extract several characteristics. After several features are collected, each role is chosen. A wide range of special classifiers for machines and deep learning were used as well. The second level then classifies half of the high-quality images, which appear to work in which the majority of pixels cost over 150 and the remaining slices are taken away. The third method is based on the proposed BLSTM and SVM for elimination and classification of features. In the following step, a majority vote indicates the shape of the titles and finally shows the title status.

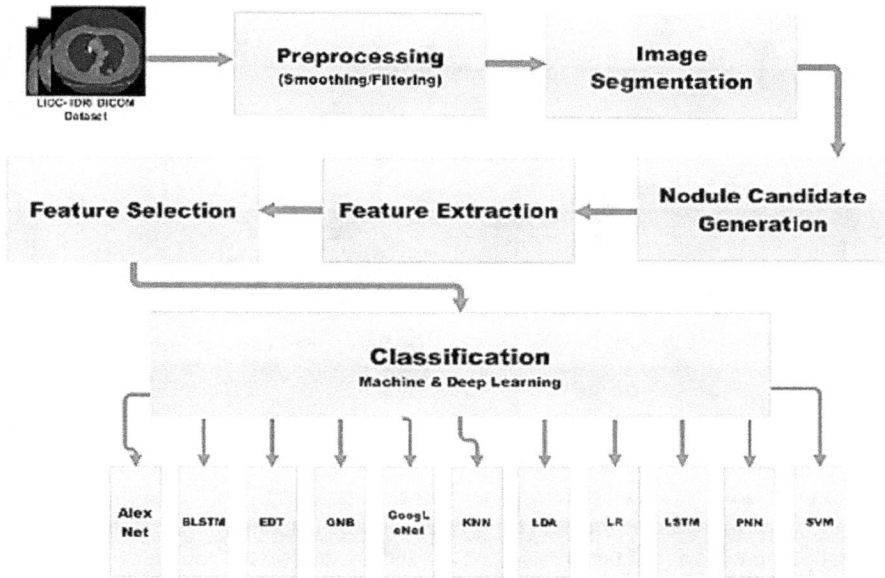

FIGURE 3.2 Diagram of the proposed architecture of the system.

3.3.1 DATASET (LIDC-IDRI)

In each artificial intelligence paradigm, datasets were an important component. The accuracy of the related data leads to analytical growth, preparation, and progress. For this development to be helpful, the evidence obtained should be justified and labelled by experts in computer vision. This segment includes information on the artificial intelligence of early work for the identification of lung cancer [Rabbani et al., 2018]. The dataset contains 1018 cases from seven research centers and eight medical diagnostic firms. An XML document of CT-scan annotations is given for each case [Chen et al., 2020]. Four skilled thoracic radiologists carry out certain annotations in a two-stage process. Every radiologist classifies the results separately into three groups ("nodule ≥ 3 mm," "nodule < 3 mm," and "nodule ≥ 3 mm") [Kim et al., 2018]. Then any radiologist investigates anonymously in the second step, their classification, and the classifications of other radiologists. The four radiologists separately analyze every nodule annotation. Average values above three are classed as malignant nodules and below three as benign nodules. One of four radiologists validated an overall grade of three with certain inconsistencies and their identities, and also as some nodules were omitted from its analysis. A DCM file can also be interpreted more easily through screening, and visualizing, as seen in Figure 3.3(a), by three or five radiologists, presenting a malignant or benign stage of cancer. Browsing using a Pylidc-tool [Loyman et al., 2020], as seen in Figure 3.3(b), shows nodules in the CT image of LIDC-IDRI-0082 as detailed: CT Slice thickness: 1.250 mm, pixel spacing: 0.703, no of nodule: 1, three annotations near slice 173 and annotation info (Subtlety-4, Internal structure-1, Calcification-6, Sphericity-3, Margin-4, Lobulation-2, Speculation-5, artificial intelligence Texture-5, Malignancy-5) [Wu et al., 2019].

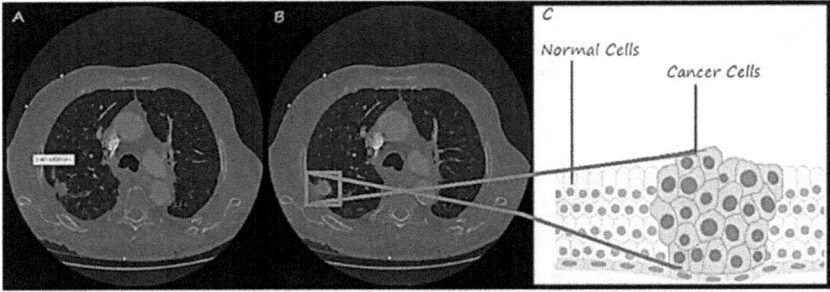

FIGURE 3.3(a), 3.3(b), 3.3(c)   [Patient Id: LIDC-IDRI-0082].

3.3.2   IMAGE ACQUISITION

As seen in Figure 3.2, the proposed approach is divided into three parts. In short, the first aim is to collect CT images from their image databases LIDC-IDRI. In the second step, the lung nodes are segmented by specialist markers [Yadav et al., 2020]. Finally, the analysis and performance of various ML and DL classifiers will be achieved.

3.3.3   IMAGE ENHANCEMENT

Image smoothing eliminates distortion or other small disturbances in the whole image that mean the edges are distorted and iteratively reduces noise. The preferred images are entered, and analyzed with some filters. As seen in Figure 3.2, a median, Gaussian, and Gaber filters were added in the pre-processing stage and the current best solution was updated and a new one was proposed. A median filter, a nonlinear 3×3 scale feature [Shah et al., 2020], is used to improve the image by minimizing impulses.

$$g(x,y) = \frac{1}{2\pi\sigma^2} \exp\left[\frac{-\left(x^2 + y^2\right)}{2\sigma^2}\right]$$

Equation 3.1: 2D Gaussian filter

The Gaussian filter is also useful for eliminating high-frequency elements to eliminate fluctuations in an image. This means that medfilt2 is used to reduce noise only. This low-pass filter eliminates turbulence and increases the smoothness and the exact strength of the surface. A standard deviation of the Gaussian distribution is the distance of the axes from the center and y of the axis from the center of the axis [Tabish et al., 2017]. After pre-processing, the processed image is segmented by a watershed segmentation. This image displays the identified nodes of cancer.

The Gaber function is an image evaluation linear filter that is useful now. The Gaussian and harmonic functionalities can be used. It also increases compression between surrounding areas of objects such as nodules. Equation 3.2 shows the expression of the 2D Gabor filter [Albu et al., 2019]. Here, in this equation, λ is the wavelength; for the orientation, θ is used from standard to parallel slices; φ is the phase offset; σ is standard deviation; γ is the ratio of spatial aspect. Other characteristics, including surface areas, perimeters, and eccentricity, as well as other features like centroid, diameter, and main intensity pixels, have been developed during the cancer detection development stage. Since then, the features have been extracted and the cancer node accurately evaluated. It's not apparent what's benign or malignant thereafter.

$$g(x,y) = \exp\left(\frac{x'^2 + \gamma^2 y'^2}{2\sigma^2}\right)\cos\left(2\pi\frac{x'}{\lambda} + \phi\right)$$

$$x' = x\cos\theta + y\sin\theta$$

Equation 3.2: 2D Gabor filter

$$y' = -x\sin\theta + y\cos\theta$$

The next move was to identify the advanced nodes that had previously been trained with deep learning, i.e. AlexNet, BLSTM, GoogLeNet, LSTM, and other classifiers such as machine learning, i.e. EDT, LR, LDA, SVM, PNN, GNB, and KNN. Training features are used to construct a trained model. The next move is a trained prognostic model to classify the unknown cancer nodes.

3.4   IMAGE SEGMENTATION

The best approach to isolate the grayscale information into target grey data is by splitting the threshold segmentation, i.e. local or global. The most widely used is the Otsu algorithm. An initial seed value that merges all identical pixels outside the badge is initially applied. The value is that it usually distinguishes from the related characteristics and reliably gives better data, although it also produces noise or distortions. The benefit is that usually the same characteristics are separated from the related areas and offer improved segmentation output information and the draw-back is that computing resources are costlier, which contributes to severe noise or regular spotting. Pre-filtered images are translated into another format, which partially removes the complexity of the front pixel mostly on edges. The process for lung fragmentation, goes through several phases while the strategies for lung segmentation go through various steps like pre-processing, implementation of median filter, Gaussian filter, then the seeding of the Gabor function and new processes are begun, as seen in Figures 3.4(a), 3.4(b), 3.4(c), and 3.4(d), to separate the lungs from the superior and inferior lobe.

3.4.1   WATERSHED SEGMENTATION

This approach is used to identify the object or context in a specific image location. Next, an image in color has been transformed into a grey image. The segmentation function with gradient magnitude has been developed. The object's generally low gradient was mostly within, while the high gradient was observed to be the boundary of the object. The first object was then labelled. To locate a foreground target, blobs of its object's pixels must all be related. Morphological techniques are used in the formation and removal of the dark patches and trim marks on each object [Manikandan

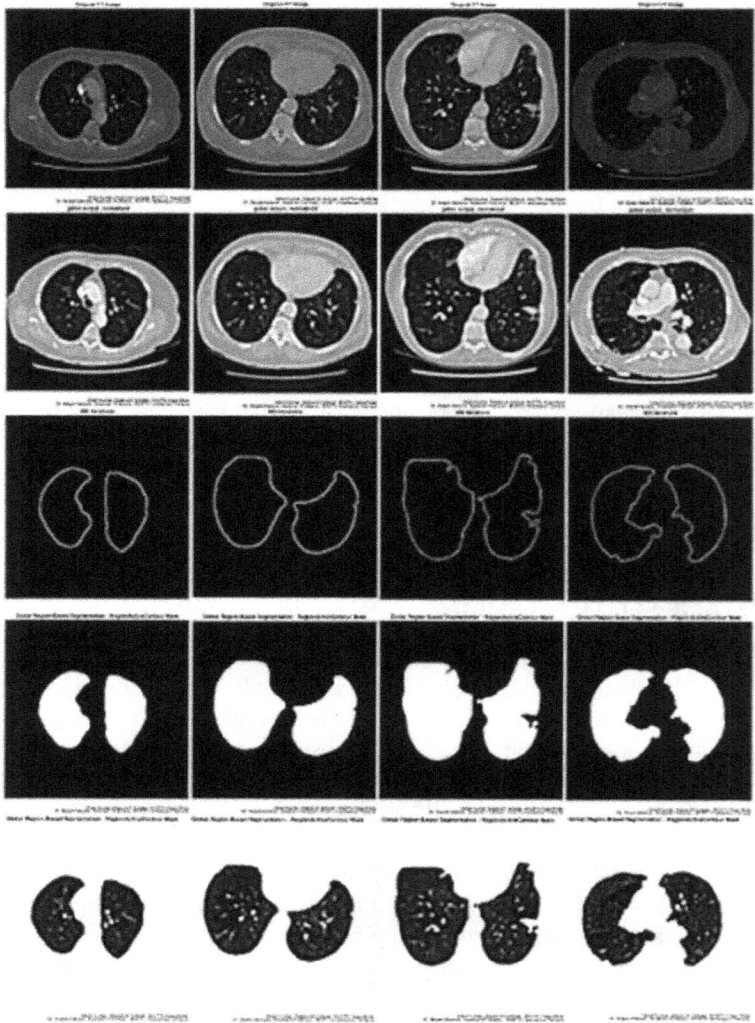

FIGURE 3.4(a)   P-Id: LIDC-IDRI-0001-00096.dcm. Figure 3.4(b) P-Id: LIDC-IDRI-0006-00057.dcm. Figure 3.4(c) P-Id: LIDC-IDRI-0013-00027.dcm. Figure 3.4(d) P-Id: LIDC-IDRI-0082-00007.dcm.

et al., 2016]. The backdrop pixel was in black and the effect of the foreground was measured by using the skeleton. The reason seems to have been that the context markers were too near the segmentation of the object. Figures 3.4(a), 3.4(b), 3.4(c), and 3.4(d) illustrate using watershed segmentation such an image to differentiate contacting points. Through viewing it all as a layer when lighter pixels become higher and dark pixels are lower, the watershed transformation identifies "catchment basins" as well as "watershed ridge edges" inside an image. However, it can define, or "label," foreground objects, including context points, segmentation using the watershed transition that performs better [Tripathi Priyanshu et al., 2019]. Such a simple technique is accompanied by marker-controlled watershed segmentation: Calculate the function of segmentation. It is an image of which can attempt to segment darkened parts. (a) Calculate labels for the foreground. There seem to be pixel blocks attached to every point. (b) Evaluate background markings. There can be pixels never found in a single object. (c) Adjust that segmentation feature to only provide limited detail at the front and rear markers. (d) Calculate the improved segmentation feature watershed transformation.

3.4.2 Otsu's Thresholding

Thresholding is the method of removing pixels from the foreground to the background. There are several methods to hit the limit and the Otsu process, introduced by Nobuyuki Otsu, is one of them. The Otsu method was a technique of comparing the values of a threshold where the difference in weight between the background and foreground pixels is the smallest. The fundamental concept should be to get to know all feasible threshold values and calculate the distribution of the foreground and background pixels. The minimum spread is then sought. The formula for determining the variance in underclass at a certain threshold t is defined by:

$$\sigma^2(t) = \omega_{bg}(t)\sigma_{bg}^2(t) + \omega_{fg}(t)\sigma_{fg}^2(t) \qquad \text{Equation 3.3: 2D Otsu thresholding}$$

This algorithm [Li et al., 2018] seeks the threshold to minimize the internal variance specified by a weighted total (background and foreground) of the two categories of variances. The gray colors typically vary from 0–255 (0–1 for floats). Thus, if this threshold is 100, all pixels below 100 will be the background, and all pixels above or equal to 100 will be the foreground of that same image.

Where $\omega_{bg}(t)$ and $\omega_{fg}(t)$ describes the probability of pixel number by threshold t and σ^2 is the variance of the pixel value. P_{all} = total count of pixels in an image, $P_{bg}(t)$ = the count of background pixels at threshold t, $P_{fg}(t)$ = the count of foreground pixels at threshold t. Hence, the weights are then indicated as

$$\omega_{bg}(t) = \frac{P_{bg}(t)}{P_{all}}, \quad \omega_{fg}(t) = \frac{P_{fg}(t)}{P_{all}}$$

The variance could be determined with the formula $\sigma^2(t) = \dfrac{\sum(x_i - \bar{x})^2}{N-1}$ where x_i is the value of a pixel at i in the class (bg or fg), $\bar{x}$ = the means of pixel values in the class (bg or fg), and N is the number of pixels.

3.5 FEATURE EXTRACTION

The extraction of features is an important step for image processing. This process specifies the appropriate processing information. Conformity or irregularity may also be observed in the lung. For tumor diagnosis and staging, these isolated properties have been used. Area, Major Axis Length, Minor Axis Length, Eccentricity, Convex Area, Filled Area, Perimeter, Solidity, Extent, Mean Intensity, Actual Area, Actual Perimeter, Actual Major Axis Length, and Compactness have become the separate notable features in this article. The purpose would be to use the fewest steps available to accurately classify an object to be categorized unequivocally. The accuracy of its main image, and how the images are pre-processed relies on the efficiency of some shape calculation. Erosion of objects, including small holes, and noises can lead to poor measuring results and inevitably misclassified outcomes. Shape information is just what exists after an object has been removed with its position, inclination, and size characteristics [Echegaray et al., 2016]. Shape features could be classified into the boundary and region features, which are described below [Johora et al., 2018].

3.5.1 DISTANCES

The shortest calculation of all distances has been between two specific pixels (x_1, y_1) and (x_2, y_2), as shown in Table 3.1.

Eucliden

$$d = \sqrt{(x_1 - x_2)^2 + (y_1 - y_2)^2}$$

Equation 3.4: Euclidean distance measure

Chessboard

$$d = \max(|x_1 - x_2|, |y_1 - y_2|)$$

3.5.2 AREA

The region consists of a sequence of pixels in a shape. This region is the sum of all of the pixels throughout the lung tumor potion. A scalar valuation is an area achieved. Table 3.1 shows the area.

3.5.3 CONVEX AREA

The object convex area is the surrounding area of a convex hull. The convex hull reveals the smallest convex polygon containing the tumor segment in the lung. This convex polygon gives the number of pixels in the convex area. The value is a scalar one.

3.5.4 PERIMETER

The perimeter seems to be the number of pixels on the object boundary. It has been the summation of the connected tumor region within the lung. That is a scalar value for the perimeter. If the boundary list is $x_1, \ldots, x_N$, then the perimeter is given by

$$Perimeter = \sum_{i=1}^{N-1} d_i = \sum_{i=1}^{N-1} |x_i - x_{i+1}|$$ *Equation 3.5: Perimeter calculation*

Distance d_i is equal to one for four-connected limits as well as one or $\sqrt{2}$ eight-connected boundaries. There has been one pixel in the length of the number of diagonal ties between $N_4 - N_8$ and one $N_8 - (N_4 - N_8)$ connection within the eight-related border as described in Table 3.1. The complete perimeter is thus:

$$Perimeter = (\sqrt{2} - 1) N_4 + (2 - \sqrt{2}) N_8$$ *Equation 3.6: Perimeter calculation*

3.5.5   CONVEX PERIMETER

A convex perimeter of such an object is just the convex hull covering the object. This implies that the convex boundary of the hull covering the sphere is the object's convex perimeter.

3.5.6   AXIS LENGTH

The major axis of the object is the (x, y) endpoints of the longest line, which can primarily be identified by calculating the distance from each border-pixel combination inside the object border with the maximum length of a major axis end-pixels (x1, y1) and (x2, y2) [Wang et al., 2018] is shown in Table 3.1.

3.5.6.1   Major Axis Length

The major axis length of an object consists of the pixel distance between the main axis endpoints:

$$MajorAxisLength = \sqrt{(x_2 - x_2)^2 + (y_2 - y_1)^2}$$ *Equation 3.7: Major axis*

The outcome is the object length estimation.

3.5.6.2   Minor Axis Length

The minor axis is the (x, y) endpoints of the longest line that can be formed by the object and which are perpendicular to the major axis. By measuring a pixel gap from both endpoints of the boundary pixels, you can find the minor axes endpoints (x_1, y_1) and (x_2, y_2).

$$MajorAxisLength = \sqrt{(x_2 - x_2)^2 + (y_2 - y_1)^2}$$ *Equation 3.8: Minor axis*

Here the outcome is the object width estimation.

3.5.7 COMPACTNESS

Compactness is a ratio of an object's area to the circle area with a perimeter of the same [Riti et al., 2016]. A circle is chosen since that is the most compact object. The maximum value for a circle is one. The compactness of a square is $\frac{\pi}{4}$

$$Compactness = \frac{4\pi.Area}{\left(Perimeter\right)^2}.$$ *Equation 3.9: Compactness calculation*

The calculation will decrease objects that have an elliptically shaped or an irregular border that is not smooth. For a circle, the measurement takes a minimum value of one. The complicated, irregular boundaries of the objects are more compact as described in Table 3.1.

$$Compactness = \frac{\left(Perimeter\right)^2}{4\pi.Area}.$$ *Equation 3.10: Compactness calculation*

3.5.8 ECCENTRICITY

Eccentricity is the ratio of the length of the minor (short) axis to the length of the object's (major) long axis [Kavitha et al., 2019]. As a result, an object eccentricity measurement is given in the value 0 to 1. The eccentricity has sometimes known the first eccentricities to distinguish between the second and third eccentricities.

$$Eccentricity = \frac{AxisLength_{short}}{AxisLength_{long}}.$$ *Equation 3.11: Eccentricity calculation*

This is used to determine the tumor's roundness as described in Table 3.1. For a regular object, the value of eccentricities equals one and in an irregular form, the value exceeds one.

3.5.9 STANDARD DEVIATION

It calculates the mean square deviation from the mean of the gray pixel scale value.

3.5.10 SKEWNESS

It is the asymmetry rating of pixel distribution in the designated ROI around the average.

3.5.11 KURTOSIS

It tests a distribution in contrast to a normal distribution's peak as well as flatness.

3.5.12 ENTROPY

It is a measurement of the maximum possible amount of the segmented ROI content.

3.5.13 CIRCULARITY OR ROUNDNESS

Often steps that are only susceptible to departing from a certain form of circularity are useful like convexity, for example (measures irregularities) and roundness (excludes local irregularities) [Riti Yosefina Finsensia et al., 2016] [Kavitha et al., 2019]. A calculation of circularity or roundness (area-to-perimeter ratio) that removes localized irregularities could be achieved as its ratio of a circle area with the same convex perimeter from the area of an object as elaborated in Table 3.1.

$$Roundness = \frac{4\pi.Area}{\left(ConvexPerimeter\right)^2}.$$ Equation 3.12: Roundness calculation

This number is 1 for a circulating object and less than 1 for an object that goes away from circularity relatively indifferent to irregular boundaries, but it is reasonably indifferent to irregular borders for an object that moves away from circularity.

3.5.14 CONVEXITY

Convexity is the proportion that differs between an object and a convex object [Wang et al., 2018]. The ratio of the perimeter of the convex hull of the object to the perimeter of the object itself can be measured as described in Table 3.1.

$$Convexity = \frac{ConvexPerimeter}{Perimeter}.$$ Equation 3.13: Convexity calculation

This is one for a convex object. It is smaller than one if the object is not convex, for example, if the boundary is irregular.

3.5.15 SOLIDITY

The density of the object is measured by solidity. As a ratio of an object's area to the area of an object's convex hull, a measure of solidity may be obtained as viewed in Table 3.1.

$$Solidity = \frac{Area}{ConvexArea}.$$ Equation 3.14: Solidity calculation

A value of one means a solid object, and an object with an irregular boundary or a-holes value below one means an irregular boundary.

3.5.16 Bounding Box

A rectangle that restricts the object is the bounding box or rectangle of an object [Wang et al., 2018]. The bounding box proportions are the greater and the smaller axes as described in Table 3.1. The boundary box area is

$$BoundingBoxArea = (MajorAxisLength) \times (MinorAxisLength).$$

Equation 3.15 Bounding box calculation

The bounding minimal area that restricts the shape is the minimum bounding box.

3.5.17 Mean Intensity

The value is a scalar one. It determines the mean pixel intensity in a certain area of interest, the portion of the tumor [Riti Yosefina Finsensia et al., 2016].

3.6 MACHINE AND DEEP LEARNING ALGORITHMS

Machine learning algorithms assist people and animals to learn as natural. These algorithms are used to study the raw data to your computer processes, while the number of trails available to learning performance without adopting a pre-set model machine. These models can recognize natural phenomena by giving insight and help to make more informed major decisions like forecasts, medical diagnosis, inventory, trading, energy consumption assessment, and more. Media sites concentrate on the teaching of computers and millions of possible ways to propose movies or songs. This is being achieved by businesses to listen to their consumers' shopping patterns. It includes two approaches: supervised learning, which builds a framework with knowing input and output data to forecast future results, and unsupervised learning, including input information with hidden patterns or intrinsic structures.

Up to this point, the author has indicated various types of machine learning and many of these algorithms will have their advantages and limitations, which make them perfect for use in such fields. Enhanced versions of such algorithms are defined as deep learning algorithms, which have become extremely popular lately. These approaches, though, come from biological neural networks that date back many years ago, before their contribution to data science. The author has used simply machine and deep learning algorithms in this research, and some of its most widely used LIDC-IDRI datasets for the research [Pradhan et al., 2020]. The output analyses, as well as the findings of this deep study, will assist researchers and specialists in similar fields to gain improved information and experience to create the best algorithms in line with their challenges in this notable number of algorithms based on data of different lung cancer datasets. Logistical Regression (LR), K-Nearest Neighbors (KNN), Gaussian Naive (GNB), Decision Trees (DT), Random Forestry (RF) are the machine learning algorithms discussed in this chapter. As the leading category of artificial neural networks in deep learning, the Bidirectional Long Term Memory (BLSTM) and Long Short Term Memory (LSTM) are analyzed.

TABLE 3.1
Features extraction from processed lung cancer DICOM images

3. Area	Major Axis Length	Minor Axis Length	Eccentricity	Convex Area	Filled Area	Perimeter	Solidity	Extent	Mean Intensity	Actual Area	Actual Perimeter	Actual Major Axis Length	Compact- ness	Label
076.00	11.01	08.91	0.59	077.00	076.00	28.09	0.99	0.77	0566.66	37.57	19.75	07.74	180.09	1
028.00	07.65	04.81	0.78	029.00	028.00	16.47	0.97	0.80	0539.82	13.84	11.58	05.38	086.65	0
124.00	13.42	11.88	0.47	129.00	124.00	36.92	0.96	0.73	0566.57	61.30	25.96	09.43	256.33	1
066.00	09.64	08.86	0.39	068.00	066.00	26.03	0.97	0.81	0566.88	32.63	18.30	06.77	162.49	1
001.00	01.15	01.15	0.00	001.00	001.00	00.00	1.00	1.00	0729.00	00.49	0.00	00.81	000.00	0
103.00	14.87	09.61	0.76	117.00	103.00	41.36	0.88	0.61	1098.53	50.92	29.08	10.45	201.16	1
069.00	11.58	07.68	0.75	070.00	069.00	27.33	0.99	0.77	0468.49	34.11	19.22	08.14	165.77	1
022.00	07.46	03.88	0.85	022.00	022.00	14.88	1.00	0.79	0627.91	10.88	10.46	05.24	071.64	0
014.00	08.08	02.31	0.96	014.00	014.00	13.36	1.00	1.00	0588.21	06.92	09.39	05.68	048.11	0
075.00	11.02	09.21	0.55	086.00	075.00	32.00	0.87	0.68	0696.04	37.08	22.50	07.75	166.53	1
026.00	10.15	03.54	0.94	029.00	026.00	19.26	0.90	0.72	0782.73	12.85	13.54	07.13	074.41	0
119.00	15.05	10.43	0.72	126.00	119.00	38.55	0.94	0.79	0743.34	58.83	27.10	10.58	240.74	1
170.00	21.42	10.72	0.87	181.00	170.00	51.79	0.94	0.58	0637.68	84.05	36.41	15.06	296.70	1
010.00	04.26	03.21	0.66	010.00	010.00	08.15	1.00	0.83	0622.30	04.94	05.73	02.99	044.01	0
084.00	12.32	08.91	0.69	089.00	084.00	30.82	0.94	0.70	0588.44	41.53	21.67	08.66	190.03	1
035.00	08.36	05.67	0.74	039.00	035.00	20.05	0.90	0.63	0488.66	17.30	14.10	05.88	098.18	0
024.00	08.02	04.01	0.87	026.00	024.00	16.47	0.92	0.75	0561.17	11.87	11.58	05.64	074.27	0
005.00	05.89	01.60	0.96	007.00	005.00	08.51	0.71	0.50	0502.20	02.47	05.98	04.14	021.57	0
108.00	15.79	09.10	0.82	113.00	108.00	36.91	0.96	0.64	0507.11	53.39	25.95	11.10	223.27	1
055.00	09.15	07.75	0.53	055.00	055.00	23.21	1.00	0.86	0594.16	27.19	16.32	06.43	143.40	1
115.00	13.14	11.23	0.52	117.00	115.00	35.29	0.98	0.80	0611.18	56.85	24.81	09.24	243.15	1

(continued)

TABLE 3.1 Continued
Features extraction from processed lung cancer DICOM images

3. Area	Major Axis Length	Minor Axis Length	Eccentricity	Convex Area	Filled Area	Perimeter	Solidity	Extent	Mean Intensity	Actual Area	Actual Perimeter	Actual Major Axis Length	Compactness	Label
071.00	10.85	08.53	0.62	072.00	071.00	27.61	0.99	0.79	0460.93	35.10	19.42	07.63	169.70	1
022.00	06.27	04.61	0.68	022.00	022.00	13.59	1.00	0.73	0594.95	10.88	09.55	04.41	074.96	0
010.00	04.69	03.00	0.77	010.00	010.00	08.70	1.00	0.83	0866.00	04.94	06.12	03.29	042.58	0
046.00	09.92	06.05	0.79	047.00	046.00	22.10	0.98	0.77	0565.11	22.74	15.54	06.97	122.91	0
072.00	11.42	08.12	0.70	073.00	072.00	27.61	0.99	0.82	0514.14	35.60	19.42	08.03	172.09	1
041.00	08.55	06.23	0.69	041.00	041.00	19.84	1.00	0.85	0550.76	20.27	13.95	06.01	115.61	0
024.00	07.63	04.18	0.84	024.00	024.00	15.46	1.00	0.69	0886.21	11.87	10.87	05.37	076.67	0
001.00	1.15	01.15	0.00	001.00	001.00	00.00	1.00	1.00	0867.00	00.49	00.00	00.81	000.00	0
055.00	10.46	07.29	0.72	61.00	055.00	27.33	0.90	0.61	0697.07	27.19	19.22	07.35	132.13	1
104.00	15.09	09.75	0.76	119.00	104.00	40.09	0.87	0.54	0690.89	51.42	28.19	10.61	206.29	1
003.00	03.46	01.15	0.94	003.00	003.00	03.92	1.00	1.00	0805.67	01.48	02.76	02.44	019.03	0
077.00	12.89	07.77	0.80	078.00	077.00	29.78	0.99	0.70	0560.94	38.07	20.94	09.06	177.22	1
046.00	10.95	05.51	0.86	049.00	046.00	23.49	0.94	0.77	0577.65	22.74	16.51	07.70	119.22	1

3.7 METRICS OF PERFORMANCE EVALUATION

Such tools are required after the implementation of machine learning algorithms to know and understand what they have accomplished in their work. These metrics are known as performance evaluation assessments. In the study, the author finds various elements of algorithm efficiency, and many metrics were introduced. Then we need an acceptable collection of metrics for the performance assessment for each learning task. In this chapter the author employs many standard metrics to obtain valuable data about algorithm efficiency and perform a similar analysis for classification problems. These include precision, recall, confusion matrix, accuracy, and ROC-AUC ranking. These tests are the following.

3.7.1 PRECISION / POSITIVE PREDICTIVE VALUE (PPV)

"How many required data are acceptable." This means that many of them seem to be correct and accurate, depending on the findings that an algorithm has expected to be supportive [Shakeel et al., 2020]. As per the formula, the number of true positives is divided by the summation of the number of true positives and the number of false positives.

$$PPV = \frac{TP}{TP + FP}$$ *Equation 3.16: Precision calculation*

3.7.2 NEGATIVE PREDICTIVE VALUE (NPV)

$$NPV = \frac{TN}{TN + FN}$$ *Equation 3.17: NPV calculation*

3.7.3 RECALL / SENSITIVITY / PROBABILITY OF DETECTION / HIT RATE / TRUE POSITIVE RATE (TPR)

It assesses how well a test makes a good option for patients who are diagnosed accurately and then classified as a true positive rate for the disease. An incredibly difficult test can hit those with the most erroneous negative results and thus may not show the most false-negative results: 90% of lung cancer patients would be able to obtain positive results and 10% of the people with and who are tested positive would have a negative outcome and who is positive would have a negative result claiming the erroneous negative side. The result of this test is highly sensitive. It includes "How many types of data are chosen." Indeed, how many have been estimated by algorithm from the true positive findings [Shakeel et al., 2020]. It is expected that while the disease is suspected, a findings analysis would also be positive. The recall is the measurement of the model that identifies true positives accurately. Thus, remember how many patients have been recognized as having lung disease amongst all the people who genuinely have lung cancer.

$$TPR = \frac{TP}{TP + FN}$$ *Equation 3.18: Sensitivity calculation*

3.7.4   SPECIFICITY / SELECTIVITY / TRUE NEGATIVE RATE (TNR)

It measures the capacity of a test to precisely get negative results for individuals that will not have the infection that is being tested, which is often known as the "true negative rate." High-specificity testing would suitably discover almost all people who do not suffer from the disease and would not produce numerous false-positive results.

It is expected that when the disease is just not present, the results of the test would have been negative [Shakeel et al., 2020]. As per the formula, the number of true negative is divided by the summation of the number of true negative and the number of false positive.

$$TNR = \frac{TN}{TN + FP}$$ *Equation 3.19: Specificity calculation*

3.7.5   FALL OUT / FALSE POSITIVE RATE (FPR)

The relationship between a possible positive occurrence of the disease and the presumption that the disease does not exist is thus positive in terms of the outcome of this test.

$$FPR = \frac{FP}{FP + TN}$$ *Equation 3.20: FPR calculation*

3.7.6   MIS RATE / FALSE NEGATIVE RATE (FNR)

A relationship between the probability of negative disease findings and the possibility of negative disease testing results.

$$FNR = \frac{FN}{FN + TP}$$ *Equation 3.21: FNR calculation*

3.7.7   FALSE DISCOVERY RATE (FDR)

$$FDR = \frac{FP}{FP + TP}$$ *Equation 3.22: FDR calculation*

3.7.8   FALSE OMISSION RATE (FOR)

$$FOR = \frac{FN}{FN + TN}$$ *Equation 3.23: FOR calculation*

3.7.9   F1-SCORE / F-MEASURE/ SØRENSEN–DICE COEFFICIENT / DICE SIMILARITY COEFFICIENT (DSC)

In measuring an algorithm's performance, such a metric is often called the F Score or F Measure, which takes into consideration both precision and recall. The estimation of recall and precision with one situation plays a vital role, such that the issue can be eliminated by identifying models in low recall and high precision or vice versa [Dong et al., 2020]. The relational means of precision and recall are mathematically described as follows:

$$F1 - Score = 2 \times \frac{\Pr ecision \times \mathrm{Re}\, call}{\Pr ecision + \mathrm{Re}\, call}$$

$$F1 - Score = \frac{2TP}{2TP + FN + FP}$$ *Equation 3.24: F1-Score calculation*

The F1 score is the weighted average of precision and recall. False positives and false negatives into account have to be taken into account. There are low false-positives and low false-negatives that correctly identify true threats such that false alarms do not interrupt them. It is expected that two very different warnings and the performance should be subsequently calculated.

3.7.10   ACCURACY

It would be the most frequently accessed and maybe the most important option in the classification task assessing the performance of the algorithm. The ratio of specifically categorized data items to the overall set of observations may be described in the formula [Shakeel et al., 2020]. In certain circumstances accuracy, considering the widespread usability, is not the highest performance metric, particularly when the target classes of the data are unequalled.

$$Accuracy = \frac{TP + TN}{TP + TN + FP + FN}$$ *Equation 3.25: Accuracy calculation*

3.7.11   CONFUSION MATRIX

Such a matrix is one of the most logical and concise statistics in which a machine-learning model attains accuracy and appropriateness. The principal application lies in challenges of classification, in which the outcome may consist of two or more classes [Kavitha et al., 2019].

3.7.12 ROC-AUC SCORE

It is computed through the ROC curve that expresses the relation between the true-positive rate (formerly the sensitivity or recall) and false-positive rate. The ROC/AUC score is determined using the ROC curve (1- Specificity). The Area Under the ROC Curve or ROC-AUC for binary classification indicates how strong a model is in positive and negative target classes [Riti et al., 2016; Kavitha et al., 2019]. In particular, where the value of positive and negative groups is similar the success criterion for the ROC-AUC score could be beneficial.

3.8 CLASSIFICATION

The authors analyzed 1010 malignant as well as benign cases who want to pick a model training set, then test the pattern for accuracy, and have frequently sought both supervised and unsupervised learning. However, there is no question of collinearity between the variables as shown in Figures 3.5(a), 3.5(b), 3.5(c), and 3.5(d) to predict cancerous or non-cancerous cells [Vinod Kumar et al. 2021]. The correlation between the predictor sector, entropy, and the standard deviation is high [Hussein Sarfaraz et al., 2019]. Kurtosis and skewness are also closely related. This collinearity also means that such forecast variables are eliminated.

3.8.1 BIDIRECTIONAL LONG SHORT-TERM MEMORY NETWORKS (BLSTM)

Since the mid-twentieth century, the release of the long short-term memory (LSTM) [Zhang et al., 2018] would have rendered artificial neural networks [Kumar et al., 2012] (ANNs), a remarkably recurrent neural network (RNN). Through its straight ties in the groupings, RNN may become dependent on time. To demonstrate this, feed-forward networks should implement the use of static pattern mapping, e.g. MLP.

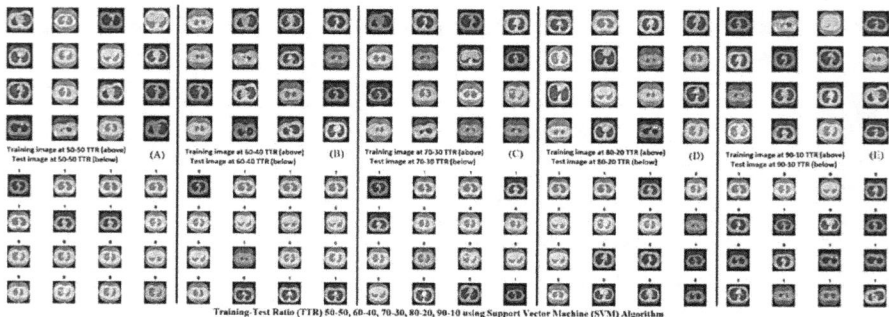

FIGURE 3.5(a)   Benign as zero and malign as one in a resulting CT image were identified after compilation of classifier SVM algorithms from 50% to 90% TTR respectively.

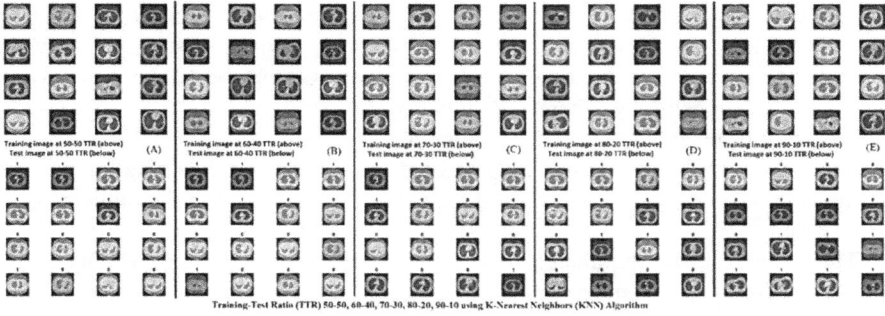

FIGURE 3.5(b) Benign as zero and malign as one in a resulting CT image were identified after compilation of classifier KNN algorithms from 50% to 90% TTR respectively.

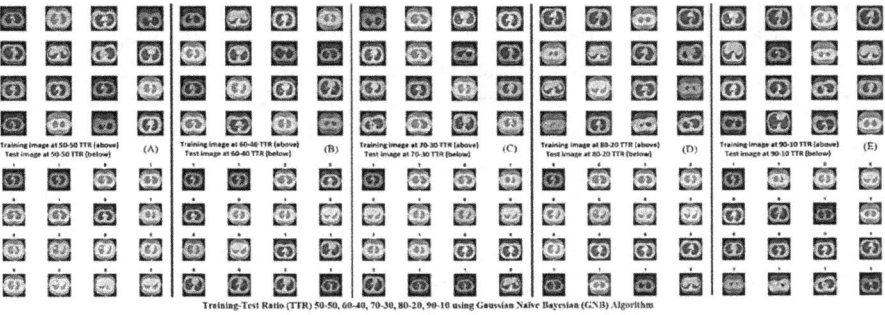

FIGURE 3.5(c) Benign as zero and malign as one in a resulting CT image were identified after compilation of classifier GNB algorithm from 50% to 90% TTR respectively.

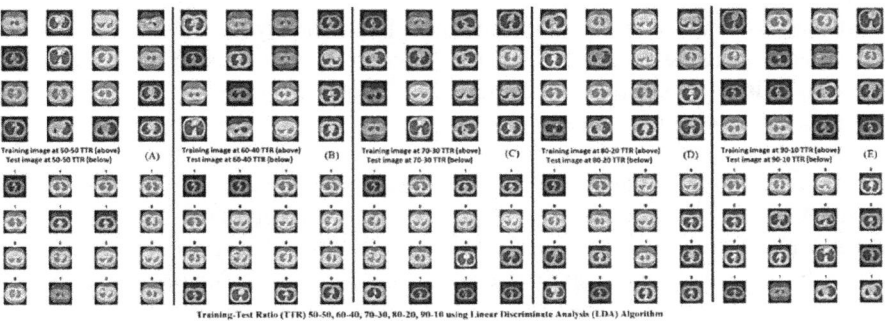

FIGURE 3.5(d) Benign as zero and malign as one in a resulting CT image were identified after compilation of classifier LDA algorithm from 50% to 90% TTR respectively.

Moreover, traditional RNNs [Wang Ran et al., 2018] became responsive, particularly in descending gradient preparation, to this extremely genuine issue, including the issue of the disappearing gradient. This issue arises because the gradient disappears or explodes exponentially. Therefore, after several steps, traditional RNNs "forget" interest points fewer than ten times after missing gradients. The impact may be unusually restricted towards the latter research methods, as well as the linkage and connection of progressively segregated preceding layers being very complicated and problematic.

The issue expertly illuminated by LSTM [Zhang et al., 2018] outlines itself as a special repeating arrangement that can see like differentiable capacity cells. A mistake inside the LSTM block, which can be maintained for long periods, which can be maintained at an intersection for a long period. LSTMs can oversee time-laps of up to 10,000 steps [Salehi et al., 2018] and for exceptionally long periods. LSTM [Salman et al., 2020] discharges a surprising capacity to learn from other abilities in repeating sets, e.g. precise timing, exact replication esteem, expansion, and now and then increase [Albu et al. 2019]. Analysts have used an angle to learn measurements of LSTMs from back-propagation time [Mhaske Diksha et al., 2019].

It relates to the unidirectional classic recurring network architectures. It implies that the input sequence to such a network is interpreted in the correct direction: it behaves in the right direction. The input background is collected in the previous context of a network. These are problems, therefore, in which only a prior context will not properly learn that problem. It may also be enough to obtain a potential context, for example, a clinical scenario to recognize the local, possibly disturbing portion of an input sequence by providing information about what is yet to come. A future context could only be provided exactly when at the point of calculation, the input sequences are often fully given. The principle of two-way recurrent neural networks allows certain past contexts and future contexts to be accessible in recurrent neural networks [Razzak et al., 2018]. Consider a classic RNN just with one hidden-layer recurring. Now, a second recurrent hidden-layer is introduced and linked with the input-layer as well as to the output layer, though specifically not by the other hidden-layer. Second, frequently in the forward direction, it introduces an input sequence to an input layer. Just the first computing of hidden layer and then all activations are processed with each process [Mobile et al., 2017]. Third, the input sequence is displayed backward. Just the second hidden layer works, and it also keeps all of its operations. After this, the output layer generates the output-sequence by incorporating the historical knowledge that is the first hidden layer, and future information that is the second hidden layer at every stage. The second layer contains a second, hidden layer. Now the backward hidden layer gradient calculation operates in almost the same way at a reversal time.

Later the concept of two-way layers was implemented analogously in LSTM networks [Mhaske Diksha et al., 2019]. The bidirectional long-term storage networks (BLSTMs) have become more prominent with BLSTMs for incredible manual classification accuracy. LSTMs as well as other techniques such as the hidden Markov model (HMM), thus choose BLSTMs, for the classification task set out in this chapter.

3.9   EXPERIMENTS AND RESULTS

The following requisites are given for BLSTM and SVM using the GPU frameworks MATLAB R2018b, Intel ® CoreTM i5-7200U 7th-general CPU @3.1 GHZ, Intel(R) HD Graphics 620 RAM 1 GB adapter, 8 GB DDR4 RAM. Various requirements are kept in mind for the classifier's results as shown in Figures 3.5(a), 3.5(b), 3.5(c), and 3.5(d).

The machine trained the system of 395 CT image fragments to detect pulmonary cancer and a dataset is obtained with the LIDC-IDRI following the transformation (dot) DCM of the file in (dot) PNG format. Using the MATLAB script, this system can detect and stagger the lung tumor. Following the tumor identification process, here the images of the cancerous and non-cancerous cells are represented by Figure 3.5(a), 3.5(b), 3.5(c), and 3.5(d), respectively, with zero and one.

During the training of images of the CAD system, the corresponding masks are being used to train the machine, and during the test process, researchers use an image of this map for output processing. Then, as per the distributions of the region of interest, in this case, the lung parenchyma, implement the mask as indicated in these figures.

The experimental dataset used here is the LIDC-IDRI [Pradhan et al., 2020] that covers diagnosing thorax CT scans and predictive lesion scanning for lung cancer. This is an excellent database available for improving the techniques for lung cancer classification and diagnosis [Kumar et al. 2021].

A manual segmentation has been given to an actual truth of its lung parenchyma, and researchers performed a pre-processing phase till experimenting on this data package to cut out images to eliminate any details that were not part of the field of research. This step was completed.

The image volume for a training test variation is used in the group of 50–50, 60–40, 70–30, 80–20, and consequently 90–10. The confusion matrix of BLSTM and LSTM is displayed in Table 3.2 and Table 3.3 with all the differences in the classification of these statistics. The predictive form can be categorized into four groups: True Positive (TP), True Negative (TN), False Positive (FP), and False Negative (FN). The percentage of positive outcomes is accurate and the forecast is supposed to be positive. The number of genuinely negative outcomes predicted to be positive is False Positive [Shakeel et al., 2020]. The number of positive results is a false negative, which is supposed to be negative. The number of negative outcomes is valid negatively, which is supposed to be negative.

The author elaborates the training and testing sets in which BLSTM was trained for 70%, TP for the value 6610, TN for the value 74 and 2137 and 292 for both FP and FN, which indicates that 6902 of the total exhibits had a malignant value and 2211 conveyors had a benign value as in Table 3.2 and 3.3.

Calculate the sensitivity and specificity, resulting respectively in improved performance at 95.77% and 96.65%. Whereas if LSTM testing is trained at 70%, TP and TN are 6578 and 2103 accordingly, FP is 108, whereas FN is 324, which implies 6902 images are malignant and 2211 are benign, as described in Table 3.3. Sensitivity and specificity, respectively, are 95.77% and 96.65%.

TABLE 3.2

Confusion matrix with advanced classification metrics using BLSTM at TTR 70–30 (%)

		Actual class			
		Positive	Negative		
Predicted class	Positive	6610(TP)	74 (FP)	Precision = TP/ (TP+FP) Pos. Pred Value = 98.89%	Total Test Positive
	Negative	292 (FN)	2137 (TN)	TN/(FN+TN) Neg. Pred Value = 62.14%	
		TP/(TP+FN) Sensitivity = 95.77%	TN/(TN+FP) Specificity = 96.65%	(TP+TN)/ (TP+TN+FN+FP) Accuracy = 95.98%	Total Test Negative
		Total Disease = 6902	Total Normal = 2211		

TABLE 3.3

Confusion matrix with advanced classification metrics using LSTM at TTR 70–30 (%)

		Actual class			
		Positive	Negative		
Predicted class	Positive	6578 (TP)	108 (FP)	Precision = TP/ (TP+FP) Pos. Pred Value = 98.39%	Total Test Positive
	Negative	324 (FN)	2103 (TN)	TN/(FN+TN) Neg. Pred Value = 86.65%	
		TP/(TP+FN) Sensitivity = 95.77%	TN/(TN+FP) Specificity = 96.65%	(TP+TN)/ (TP+TN+FN+FP) Accuracy = 95.98%	Total Test Negative
		Total Disease = 6902	Total Normal = 2211		

Cross-validation cannot be used for many purposes in the research work carried out. First, the need for cross-validation approaches takes too long in an artificial neural network, considering the huge quantity of data. Second, mutual evaluation picked the correct role of the training effectiveness, thereby producing the right model for rating.

However, such an approach is not even an efficient and reliable means for testing diagnostic systems with computer assistance. The researchers plan to be as thorough as feasible.

The suggested approach is being used several times in this respect, rather than cross-verification, and also the training and testing information has always been chosen completely randomly. Research works have been carried out ten times in this analysis. The findings demonstrate that the final results have been applied ten times.

The results of these studies are described in the statistics, and 70% of the training data and 30% of the data were included here. The selection of training and test data is based on its analysis. Returning to Figures 3.4(a), 3.4(b), 3.4(c), and 3.4(d), the following conclusions can be drawn. As shown in Figures 3.7 and 3.8, the accuracy is 93.79% of BLSTM at a training ratio of 70%, since this recall is 96.18%.

FALSE ALARM RATE

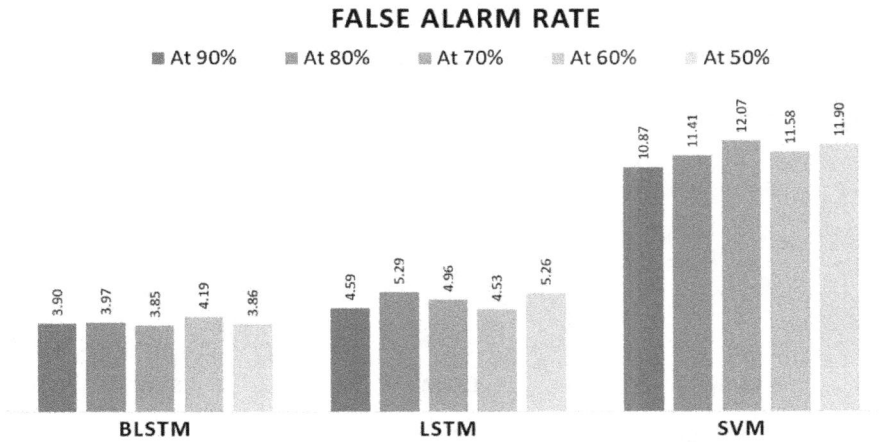

FIGURE 3.6 False alarm rate at 50% to 90% TTR using BLSTM, LSTM, and SVM.

PRECISION

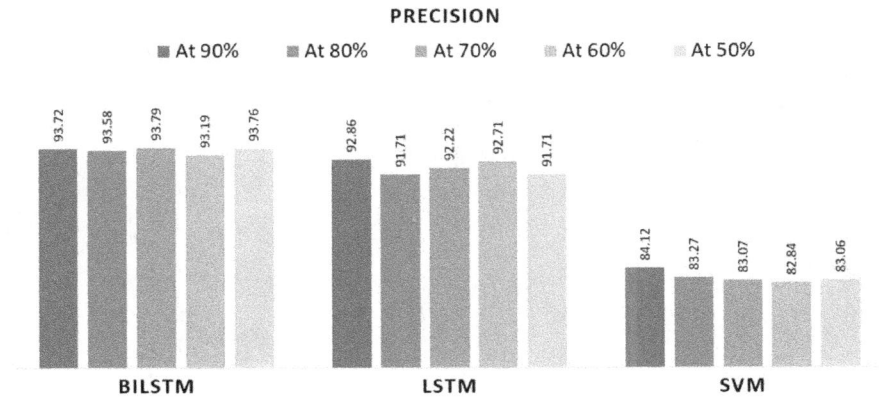

FIGURE 3.7 Precision at 50% to 90% TTR using BLSTM, LSTM, and SVM.

This strategy enables the accuracy to be accurately identified as 96.15% of BLSTM at 70% training, 95.45% of LSTM at 60% training, and 89.13% of SVM at 90% training, as described in Figure 3.9. Hence BLSTM is shown to be better than the other comparatively. The choice of BLSTM is, therefore, more plausible.

SVM is comparably fast. Overall, the SVM consumes less time (0.11 seconds with 80% training data), while BLSTM spends the longest time during the functional extraction process (20.48 seconds with 60% training data). The SVM is thus appropriate for evaluating as shown in Figure 3.11. According to Figures 3.7, 3.8, 3.9, 3.12, and 3.13, one of the important aspects to improve performance is the training

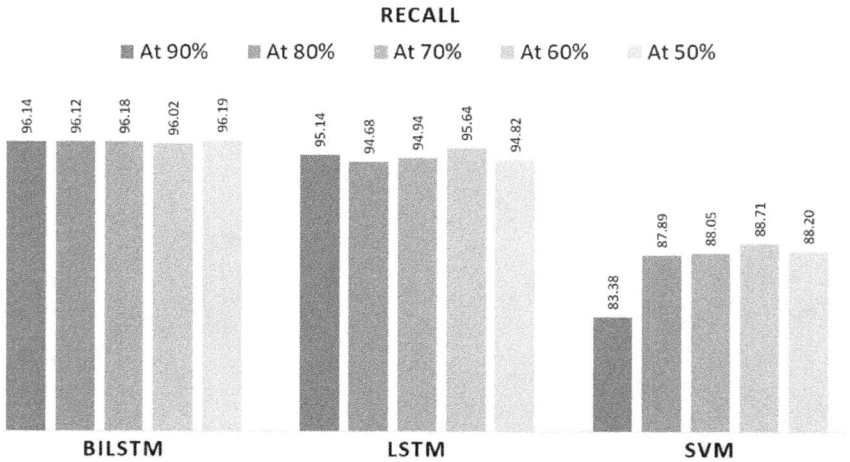

FIGURE 3.8   Sensitivity at 50% to 90% TTR using BLSTM, LSTM, and SVM.

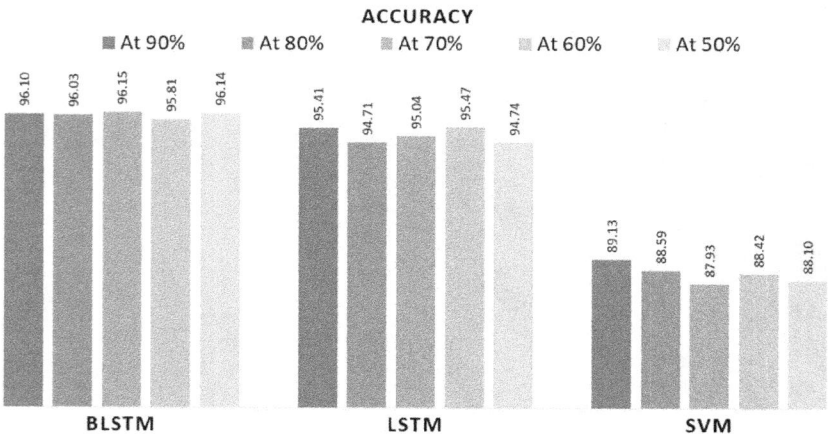

FIGURE 3.9   Accuracy at 50% to 90% TTR using BLSTM, LSTM, and SVM.

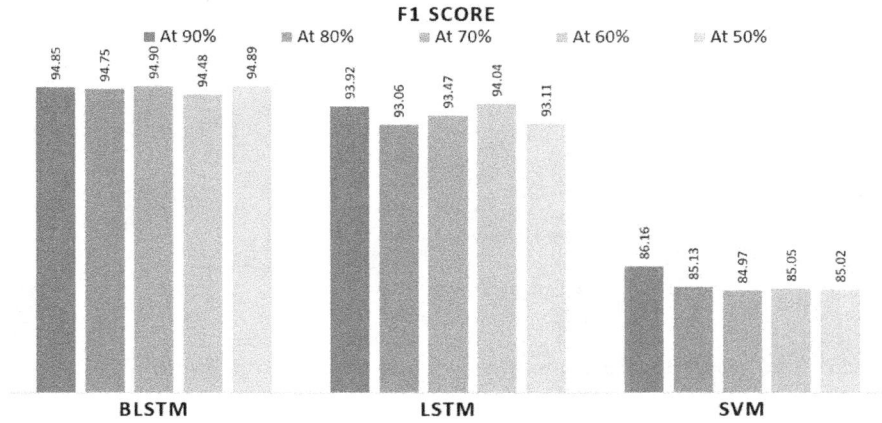

FIGURE 3.10 F1-score at 50% to 90% TTR using BLSTM, LSTM, and SVM.

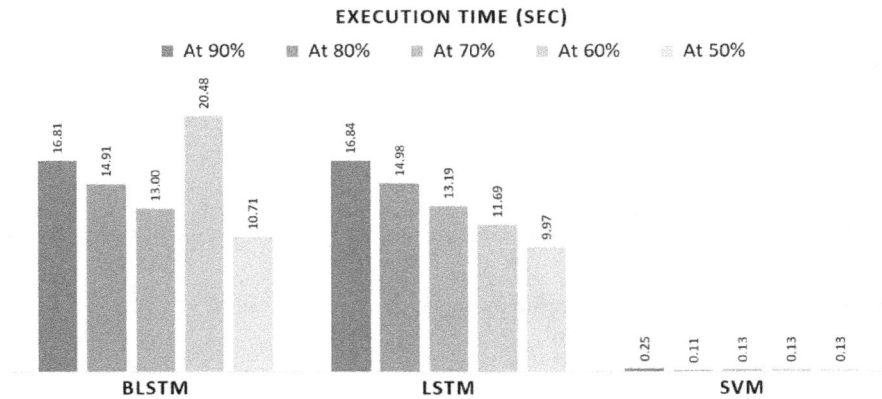

FIGURE 3.11 Execution time at 50% to 90% TTR using BLSTM, LSTM, and SVM.

algorithm. Within that analysis, the BLSTM was chosen because it makes it more sensitive and accurate than the SVM and the LSTM.

For this review, the SVM properly distinguished 6902 patients from 2211 individuals. In 89.13% of sufferers, SVM correctly identified cancer versus patients in non-cancer categories (healthy individuals and patients with other lung diseases). Breathed automated breathing evaluation with SVM enables patients with lung cancer to be discriminated against in safe objects and wrinkled of patients with various disease states. It may provide a level of discrimination between patients with cancer and a healthy population, as described in Figure 3.12.

The total number of 9113 of 2211 patients classified as 6902 is correctly categorized by LSTM. Cancer has been positive relative to non-cancer patients in 95.47% of LSTM. There is a classification of the level of segregation between victims of cancer and healthy society.

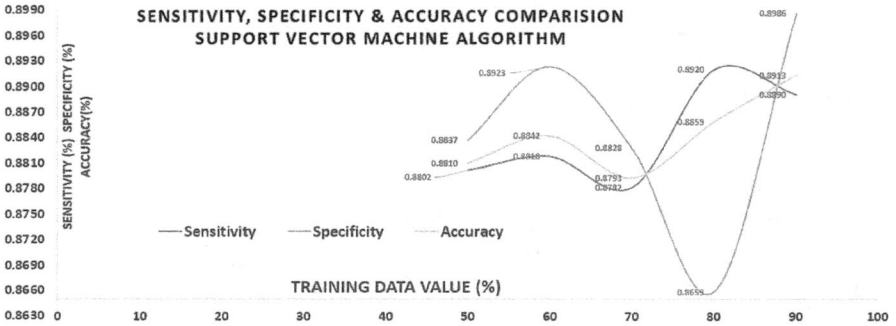

FIGURE 3.12 Analysis among sensitivity, specificity accuracy using SVM.

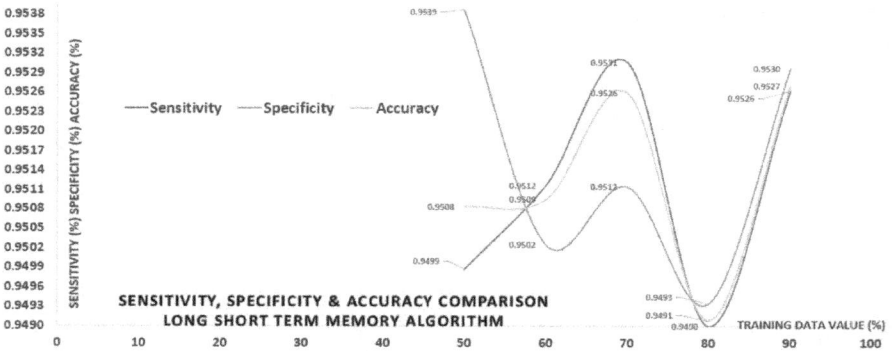

FIGURE 3.13 Analysis among sensitivity, specificity accuracy using LSTM.

There were a total of 6379 training samples and 2734 testing samples, which are at 70%–30% Training Test Ratio. The 118 samples are 6902 nodules and 2211 non-nodules. Table 3.2 displays nodules as well as non-nodules correctly and misclassified. The percentage of sensitivity and specificity obtained was 95.77 and 96.65. As shown in Figure 3.14, ROC curves reflect the total machine output for various test-to-test ratios. ROC was measured to verify how BLSTM differentiates between nodules and not nodules correctly. ROC serves as an effective tool to test the ability to discriminate against classifier performance. The curve of ROC is drawn from FPR with TPR. Figure 3.14 displays the BLSTM classifier ROC curve.

The learning processes were done properly along with their pre-processing level, lung segments, and extractions.

Following the procedure with the highest precision and sensitivity, accuracy, and others, the findings were estimated to be less erroneous. This involves more analysis on such strategies as segmenting datasets for the largest amount. High metabolism of tumors and better CT images were seen. For this imagery diseases, such as lung cancer, breast cancer, brain tumors, and cardiovascular disease may indeed be

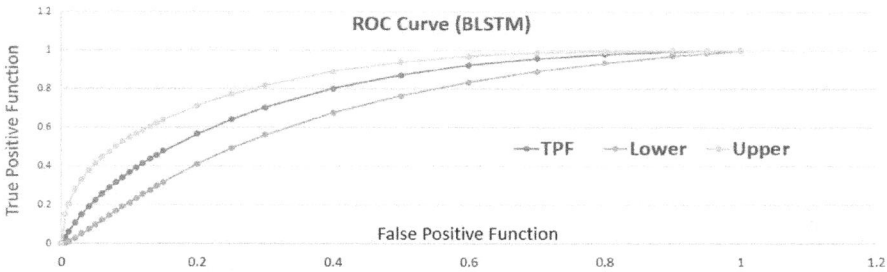

FIGURE 3.14 ROC curve analysis using BLSTM.

observed using the proposed approach. This procedure can provide sufferers with a much more detailed set of slices.

Accordingly, the care treatment has documented only such cuts, not all angles. It seems this conception has been proposed by the authors, which will provide advantages such as limited time to stay on images by minimizing the use of recorded slices, increasing effectiveness, less power usage, reducing financial costs, saving time and money, scanning expenses, etc. Everyone goes way easier to handle.

3.10 FUTURE ENHANCEMENT

A watershed marker segmentation has only been used by the researcher in the proposed method and a few other segmenting methods including comparisons have been suggested, which would improve the feature efficiency in the future. Also, several other filters and methods for image improvement had to be looked at. The accuracy of extra tumor properties is improved by increasing the data size given. Additional classifications, such as AlexNet, GoogLeNet, and others, based on artificial intelligence, should be applied.

3.11 CONCLUSION

The proposed detection of lung cancer identifies the tumor in the lung. The CT image is pre-processed and then segmented using a marker-controlled watershed. The learning processes were done properly along with their pre-processing level, lung segments, and extractions. Following the procedure with the highest precision and sensitivity, accuracy, and others, the findings were estimated to be less erroneous. This involves more analysis on such strategies as segmenting datasets for the largest amount. For the extraction of features, the segmented image is used. The tumor is discovered inside the lung with features extracted. For medical diagnostics, both the supervised and the unsupervised classifier are used. By the use of BLSTM, the maximum accuracy rate is 96.14%, with moderate use of LSTM 95.47% and another 89.13% using SVM. This system thus leads the radiologist to predict the tumor stage and enhance accuracy. The researchers have examined machine learning techniques that give health professionals and radiologists diagnostic equipment to support them to select their own victims' options for treatment focused on histological characteristics. Integrating artificial intelligence approaches with medical assistance can be helpful for lung cancers and this analysis discusses such advantages and existing deficiencies over the entire care cycle.

Conflict of interest: No conflict of interest.

REFERENCES

Alahmari S.S. and D. Cherezov, "Radiomics improves pulmonary nodule malignancy prediction in lung cancer screening." *IEEE Access* 6: 77796–77806 (2018).

Albu, Adriana, Radu-Emil Precup, and Teodor-Adrian Teban. "Results and challenges of artificial neural networks used for decision-making and control in medical applications." *Facta Universitatis, Series: Mechanical Engineering* 17, no. 3: 285–308 (2019).

Altaf, Fouzia, Syed M.S. Islam, Naveed Akhtar, and Naeem Khalid Janjua. "Going deep in medical image analysis: Concepts, methods, challenges, and future directions." *IEEE Access* 7: 99540–99572 (2019).

Asan, Onur, Alparslan Emrah Bayrak, and Avishek Choudhury. "Artificial intelligence and human trust in healthcare: Focus on clinicians." *Journal of Medical Internet Research* 22, no. 6: e15154 (2020).

Chen, Ying, Yerong Wang, Fei Hu, and Ding Wang. "A lung dense deep convolution neural network for robust lung parenchyma segmentation." *IEEE Access* (2020).

Cui, Sunan, Yi Luo, Huan-Hsin Tseng, Randall K. Ten Haken, and Issam El Naqa. "Artificial neural network with composite architectures for prediction of local control in radiotherapy." *IEEE Transactions on Radiation and Plasma Medical Sciences* 3, no. 2: 242–249 (2018).

Dartmouth-Hitchcock Medical Centre, "A new machine learning model can classify lung cancer slides at the pathologist level." *Sch of Comp. Dub Inst of Tech* (2019).

Deep Prakash, K., et al. "Early detection of lung cancer using the SVM classifier in biomedical image processing." *IEEE International Conference on Power, Control, Signals, and Instrumentation Engineering* (ICPCSI-2017), (2017).

Dong, Xing, Xu Dan, AoYawen, Xu Haibo, Li Huan, TuMengqi, Chen Linglong, and Ruan Zhao. "Identifying sarcopenia in advanced non-small cell lung cancer patients using skeletal muscle CT radiomics and machine learning." *Thoracic Cancer* 11, no. 9: 2650–2659 (2020).

Echegaray, Sebastian, Viswam Nair, Michael Kadoch, Ann Leung, Daniel Rubin, Olivier Gevaert, and Sandy Napel. "A rapid segmentation-insensitive 'digital biopsy' method for radiomic feature extraction: method and pilot study using CT images of non–small cell lung cancer." *Tomography* 2, no. 4: 283 (2016).

Gupta, Yubraj, Ramesh Kumar Lama, Sang-Woong Lee, and Goo-Rak Kwon. "An MRI brain disease classification system using PDFB-CT and GLCM with kernel-SVM for medical decision support." *Multimedia Tools and Applications* 79, no. 43: 32195–32224 (2020).

Herrmann, Markus D., David A. Clunie, AndriyFedorov, Sean W. Doyle, Steven Pieper, Veronica Klepeis, Long P. Le et al. "Implementing the DICOM standard for digital pathology." *Journal of Pathology Informatics* 9 (2018).

Hussein, Sarfaraz, PujanKandel, Candice W. Bolan, Michael B. Wallace, and UlasBagci. "Lung and pancreatic tumor characterization in the deep learning era: novel supervised and unsupervised learning approaches." *IEEE Transactions on Medical Imaging* 38, no. 8: 1777–1787 (2019).

Johora, F. Tuj, M. Jony, and Parvin Khatun. "A new strategy to detect lung cancer on CT images." *International Research Journal of Engineering and Technology (IRJET)* 5, no. 12: 27–32 (2018).

Kavitha, M.S., J. Shanthini, and R. Sabitha. "ECM-CSD: an efficient classification model for cancer stage diagnosis in CT lung images using FCM and SVM techniques." *Journal of Medical Systems* 43, no. 3: 73 (2019).

Kim, B.C. et al. "Deep feature learning for pulmonary nodule classification in a lung CT." In: 2016 4th International Winter Conference on Brain-Computer Interface (BCI), pp. 1–3. IEEE, (2016).

Kim, Catherine S., and Melenda D. Jeter. "Radiation therapy, early-stage non-small cell lung cancer." In StatPearls [Internet]. StatPearls Publishing, (2018).

Kumar, Vinod and Brijesh Bakariya, "Classification of malignant lung cancer using deep learning", *Journal of Medical Engineering & Technology*, 45, no. 2: 85–93, DOI: 10.1080/03091902.2020.1853837 (2021).

Kumar, Vinod, and Dr. Kanwal Garg. "Neural network-based approach for detection of abnormal regions of lung cancer in X-ray image." *International Journal of Engineering Research & Technology*, ISSN: 2278-0181 (2012).

Kumar, Vinod, Ashu Gupta, Rattan Rana, and Kanwal Garg. "Lung cancer detection from X-ray image using statistical features." *International Journal of Computing* 4, no. 6: 178–181 (2015).

Li, Lingling, Yuan Wu, Yi Yang, Lian Li, and Bin Wu. "A new strategy to detect lung cancer on CT images." In 2018 IEEE 3rd International Conference on Image, Vision, and Computing (ICIVC), pp. 716–722. IEEE, (2018).

Liu, Liang, Jianjiao Ni, and Xinhong He. "Upregulation of the long noncoding RNA SNHG3 promotes lung adenocarcinoma proliferation." *Disease Markers*, 2018, 5736716, https://doi.org/10.1155/2018/5736716 (2018).

Loyman, Mark, and Hayit Greenspan. "Semi-supervised lung nodule retrieval." arXiv preprint arXiv:2005.01805 (2020).

Luo, Yi, Daniel McShan, Dipankar Ray, Martha Matuszak, Shruti Jolly, Theodore Lawrence, Feng-Ming Kong, Randall Ten Haken, and Issam El Naqa. "Development of a fully cross-validated Bayesian network approach for local control prediction in lung cancer." *IEEE Transactions on Radiation and Plasma Medical Sciences* 3, no. 2: 232–241 (2018).

Manikandan, T., and N. Bharathi. "Lung cancer detection using fuzzy auto-seed cluster means morphological segmentation and SVM classifier." *Journal of Medical Systems* 40, no. 7: 181 (2016).

Mathur, Prashant, Krishnan Sathish Kumar, Meesha Chaturvedi, Priyanka Das, Kondalli Lakshmi Narayana Sudarshan, Stephen Santhappan, Vinodh Nallasamy et al. "Cancer statistics, 2020: Report from National Cancer Registry Programme, India." *JCO Global Oncology* 6: 1063–1075 (2020).

Mhaske, Diksha, Kannan Rajeswari, and Ruchita Tekade. "Deep learning algorithm for classification and prediction of lung cancer using CT scan images." In 2019 5th International Conference On Computing, Communication, Control, And Automation (ICCUBEA), pp. 1–5. IEEE, (2019).

Mobile, Aryan, Supratik Moulik, and Hien Van Nguyen. "Lung cancer screening using adaptive memory-augmented recurrent networks." arXiv preprint arXiv:1710.05719 (2017).

Murillo BR "Health of things algorithms for malignancy level classification of lung nodules." *IEEE Access.* https://doi.org/10.1109/ACCESS.2817614, (2018).

Mya Tun KM et al. "Implementation of lung cancer nodule feature extraction using digital image processing." *IJSETR* 03(09):1610–1618, (2014).

Nie, Allen, Ashley Zehnder, Rodney L. Page, Arturo L. Pineda, Manuel A. Rivas, Carlos D. Bustamante, and James Zou. "Deep tag: Inferring all-cause diagnoses from clinical notes in the under-resourced medical domain." arXiv preprint arXiv:1806.10722 (2018).

Patel, Vaibhavi, Samkit Shah, Harshal Trivedi, and Urja Naik. "An analysis of lung tumor classification using SVM and ANN with GLCM features." In Proceedings of First International Conference on Computing, Communications, and Cyber-Security (IC4S 2019), pp. 273–284. Springer, Singapore (2020).

Pelc, Norbert J., and Adam Wang. "CT statistical and iterative reconstructions and post processing." In *Computed Tomography*, pp. 45–59. Springer, Cham (2020).

Pradhan, Kanchan, and Priyanka Chawla. "Medical Internet of Things using machine learning algorithms for lung cancer detection." *Journal of Management Analytics* 7, no. 4: 591–623 (2020).

Rabbani, Mohamad, Jonathan Kanevsky, Kamran Kafi, Florent Chandelier, and Francis J. Giles. "Role of artificial intelligence in the care of patients with non-small cell lung cancer." *European Journal of Clinical Investigation* 48, no. 4: e12901 (2018).

Razzak, Muhammad Imran, Saeeda Naz, and Ahmad Zaib. "Deep learning for medical image processing: Overview, challenges, and the future." In *Classification in BioApps*, pp. 323–350. Springer, Cham (2018).

Riti, Yosefina Finsensia, Hanung Adi Nugroho, Sunu Wibirama, Budi Windarta, and Lina Choridah. "Feature extraction for lesion margin characteristic classification from CT Scan lungs image." In 2016 1st International Conference on Information Technology, Information Systems and Electrical Engineering (ICITISEE), pp. 54–58. IEEE, (2016).

Rodrigues, Murillo B., Raul Victor M. Da Nóbrega, Shara Shami A. Alves, Pedro Pedrosa Rebouças Filho, Joao Batista F. Duarte, Arun K. Sangaiah, and Victor Hugo C. De Albuquerque. "Health of things algorithms for malignancy level classification of lung nodules." *IEEE Access* 6: 18592–18601 (2018).

Salehi, Bahare, Paolo Zucca, Mehdi Sharifi-Rad, Raffaele Pezzani, Sadegh Rajabi, William N. Setzer, Elena Maria Varoni, Marcello Iriti, Farzad Kobarfard, and Javad Sharifi-Rad. "Phototherapeutics in cancer invasion and metastasis." *Phototherapy Research* 32, no. 8: 1425–1449 (2018).

Salman, Osamah Khaled Musleh, Bekir Aksoy, and Koray Özsoy. "Using deep learning techniques in detecting lung cancer." In *Deep Learning for Cancer Diagnosis*, pp. 135–146. Springer, Singapore (2020).

Taher, Fatma, Naoufel Werghi, and Hussain Al-Ahmad. "Rule-based classification of sputum images for early lung cancer detection." 2015 IEEE International Conference on Electronics, Circuits, and Systems (ICECS). IEEE, (2015).

Shah, Anwar, Javed Iqbal Bangash, Abdul Waheed Khan, Imran Ahmed, Abdullah Khan, Asfandyar Khan, and Arshad Khan. "Comparative analysis of median filter and its variants for removal of impulse noise from gray scale images." *Journal of King Saud University-Computer and Information Sciences* (2020).

Shakeel, P. Mohamed, Amr Tolba, Zafer Al-Makhadmeh, and Mustafa Musa Jaber. "Automatic detection of lung cancer from biomedical dataset using discrete AdaBoost optimized ensemble learning generalized neural networks." *Neural Computing and Applications* 32, no. 3: 777–790 (2020).

Sujitha, R. and V. Seenivasagam. "Classification of lung cancer stages with machine learning over big data healthcare framework." *Journal of Ambient Intelligence and Humanized Computing* 12: 5639–5649 (2021).

Tabish, Tanveer A., Md Zahidul I. Pranjol, Hasan Hayat, Alma AM Rahat, Trefa M. Abdullah, Jacqueline L. Whatmore, and Shaowei Zhang. "In vitro toxic effects of reduced graphene oxide Nano sheets on lung cancer cells." *Nanotechnology* 28, no. 50: 504001 (2017).

Taher, Fatma, Naoufel Werghi, and Hussain Al-Ahmad. "Rule-based classification of sputum images for early lung cancer detection." IEEE International Conference on Electronics, Circuits, and Systems (ICECS), pp. 29–32. IEEE, (2016).

Tripathi, Priyanshu, Shweta Tyagi, and Madhwendra Nath. "A comparative analysis of segmentation techniques for lung cancer detection." *Pattern Recognition and Image Analysis* 29, no. 1: 167–173 (2019).

Wang, Ran, Xiaokun Liang, Xuanyu Zhu, and Yaoqin Xie. "A feasibility of respiration prediction based on deep Bi-LSTM for real-time tumor tracking." *IEEE Access* 6: 51262–51268 (2018).

Wang, Shidan, Alyssa Chen, Lin Yang, Ling Cai, Yang Xie, Junya Fujimoto, Adi Gazdar, and Guanghua Xiao. "Comprehensive analysis of lung cancer pathology images to discover tumor shape and boundary features that predict survival outcome." *Scientific Reports* 8, no. 1: 1–9 (2018).

Wang, Shidan, Donghan M. Yang, Ruichen Rong, Xiaowei Zhan, Junya Fujimoto, Hongyu Liu, John Minna, Ignacio Ivan Wistuba, Yang Xie, and Guanghua Xiao. "Artificial intelligence in lung cancer pathology image analysis." *Cancers* 11, no. 11: 1673 (2019).

Wu, Jian, Chunfeng Lian, Su Ruan, Thomas R. Mazur, Sasa Mutic, Mark A. Anastasio, Perry W. Grigsby, Pierre Vera, and Hua Li. "Treatment outcome prediction for cancer patients based on radionics and belief function theory." *IEEE Transactions on Radiation and Plasma Medical Sciences* 3, no. 2: 216–224 (2018).

Wu, Jianrong, and Tianyi Qian. "A survey of pulmonary nodule detection, segmentation, and classification in computed tomography with deep learning techniques." *Journal of Medical Artificial Intelligence* 2: 2–8 (2019).

Yadav, Archana, and Ranjana Badre. "Lung carcinoma detection techniques: A survey." In 2020 12th International Conference on Computational Intelligence and Communication Networks (CICN), pp. 63–69. IEEE, (2020).

Yang, Shuo, Ran Wei, Jingzhi Guo, and Lida Xu. "Semantic inference on clinical documents: combining machine learning algorithms with an inference engine for effective clinical diagnosis and treatment." *IEEE Access* 5: 3529–3546 (2017).

Yiwen, X., H. Ahmed, et al., "Deep learning predicts lung cancer treatment response from serial medical imaging." *Clinical Cancer Research,* 25, no. 11, 3266-3275, https://doi.org/10.1158/1078-0432.CCR-18-2495 (2019).

Zhang, Junjie, Yong Xia, Hengfei Cui, and Yanning Zhang. "Pulmonary nodule detection in medical images: A survey." *Biomedical Signal Processing and Control* 43: 138–147 (2018).

4 Deep Learning in Content-Based Medical Image Retrieval

Harpal Singh, Priyanka Kaushal,
Meenakshi Garg, and Gaurav Dhiman

CONTENTS

4.1 INTRODUCTION

In recent years, rapid development in digital computers, multimedia and store networks has led to vast image repositories and multimedia libraries. These advances in data storage and the dissemination of information often benefit from clinical and medical testing. In hospitals with image diagnostics and analysis, vast amounts of image data are produced and the output of medical image collections has since been significantly increased. Therefore, an appropriate medical image retrieval technology must be developed in order for doctors to browse these vast databases. A number of algorithms for automatic medical image processing have been suggested in the literature to promote the creation and maintenance of such vast medical image databases

(Jiji 2014; Ponciano 2013; Quellec 2011; Rahman 2011; Zhang 2016; Garg et al. 2019; Krizhevsky et al. 2012).

An efficient method for diagnosing and treating a range of diseases, as well as an efficient instrument (Liu 2007; Dhiman and Kumar 2017; Dhiman and Kumar 2018a; Dhiman and Kumar 2018b) for managing a high volume of data, can be implemented through the use of content-based medical imaging systems (CBMIR). Content-based image recovery (CBIR) is a technology for computer vision that can look for huge datasets. The research concentrates on the characteristics of the picture, including colour, texture, form or other features derived from the picture. CBIR system reliability depends primarily on the functions chosen (Yan 2016; Dhiman and Kumar 2019). The images are first shown in a high-dimensional vacuum. The picture similarities contained in the database and the query frame are then determined by various distance metrices in the functional space. In terms of characteristics and selection of a resemblance measurement, the most important components of CBIR systems are imagery data displays. While several scholars have thoroughly studied these areas (Wan 2014), the main challenge for CBIR systems is the reduction of the "semantic gap." The data is lost due to the image characteristics, i.e. from semantine high levels to low levels (Kumar 2014). Machine learning research has dramatically evolved and deep learning is an achievement. Deep learning contains several algorithms that model high-level data abstractions using deep non-linear architectures (Bengio 2012). Deep learning imitates the human brain that has a deep architecture and knowledge in the human brain through several layers of transition. Thus deep learning techniques (deep network) enable the machine to learn complex characteristics in raw photos so that data from many abstractions can be automatically learned by exploring deep architecture. Deep-seed learning methods, such as image and video classifications (Zeiler 2014; Karpathy 2014; Wu 2016; Hinton 2012; Zhou 2013; Babenko 2015) visual tracking and linguistic processing are successfully implemented in many applications (Lin 2015). Deep learning methods for CBIR have been used in recent studies (Bay 2006; Yang 2007; Wu 2011) but deep learning methods for CBMIR work are less closely studied.

A methodology for deep learning (see Figure 4.1), i.e. the CNN, has been developed for various body and body imaging methods. Representations for learning feature. The 3D volumetric image usually consists of 2D body organ slices, obtained in medical imaging. This article focuses on the set of these two-dimensional divisions; classes were drawn up worldwide, i.e. photograms from different parts of the body were divided into different classes with their part of the body. The supervision is therefore very poor and time consuming to determine and the annotation required during the training phase is therefore reduced. This type of annotation is useful for medical imaging because annotations typically require professional advice and costly information. The CNN model is trained in the initial classification process and the feature representations for CBMIR are learned. A thorough review of the proposed method for the efficiency of recovery is provided for a series of images from a variety of imaging modes. There are three main contributions to this thesis: a multimodal data collection covering a wide variety of medical imaging focus areas is thoroughly compiled. A profound learning system is the model and instruction in a number of medical pictures. A strong medical image recall framework for a wide range of multimodal datasets is used for the teachings purpose.

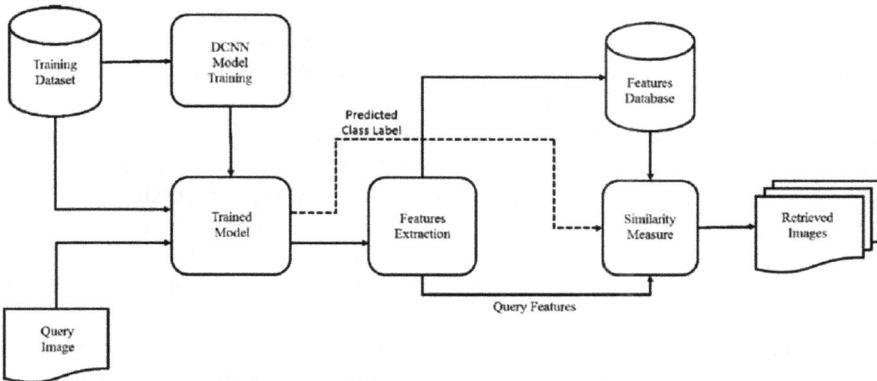

FIGURE 4.1 The proposed system for medical image recovery focused on content using a profound neural network.

These advances also benefit from clinical and diagnostic studies in the field of digital storage and information. Diagnostic and analytical imaging facilities in hospitals produce huge volumes of image data, thereby significantly enhancing medical image collection production. Therefore, an efficient medical image recovery system should be developed in order to help clinicians browse these broad datasets. A broad range of algorithms for automatic medical image processing are available in this literature to help build and maintain large-scale databases (Mansoori 2009; Alfanindya 2013; Dhiman 2017). In order to diagnose and treat efficiently a range of diseases effectively and efficiently (Dhiman 2018) to manage high quantities of data the Content-Based Medical Imaging System (CBMIR) can also be used. Without such a method, access, management and extraction from such vast collections of relevant information are very complicated. Medical pictures are processed on the basis of textual information such as tags and manual annotations because they require staff, clinical experience and time. Medical image recovery systems are therefore expected to automatically identify and recover images based on the functional displays derived from the images themselves. In support of clinical decision making, this will complement programs, analysis, and clinical trials to locate the knowledge from broad repositories. Content-based image recovery (CBIR) is a technology for a computer vision that can look for huge datasets. The research concentrates on the characteristics of the picture, including colour, texture, form or other features derived from the picture. The efficiency of a CBIR system depends primarily on these functions (Dhiman 2018; Kaur 2020). The images are first shown in a high-dimensional vacuum. In terms of characteristics and selection of a resemblance measurement, the most important components of CBIR systems are imagery data displays. Those are the most serious issues in CBIR systems, although many researchers have thoroughly studied them (Dhiman 2019), but the "semantic lacuna" is reduced from high to low semanticities because of their picture features (Dhiman 2020). Machine learning has fundamentally changed and profound learning is an accomplishment. Deep learning has a wide variety of algorithms to model high-level data abstractions with the use of profound and multi-linear methods (Garg 2017). Deep learning emulates the deep architectural human brain and processes knowledge across many levels of human brain transformation. This enables the framework (deep network) to automatically acquire data features on multiple abstraction levels across deeper algorithms and obtain representations of features without using manufactured features. Recent research reports that in many applications such as picture and seriousness, visual monitoring, language comprehension and natural language processing have been successfully applied in deep education methods. Latest experiments have used deep learning methods for CBIR tasks but deeper learning methods for CBMIR tasks have been given less consideration. This chapter examines the application of this article to the CBMIR challenge, based on the achievements of deep learning methods to bridge the distance. A deep learning technique, i.e. convolutionary neural system (CNN) is adapted for different representations of imagery and body organs. The 3D volumetric image usually consists of 2D body organ slices, obtained in medical imaging. This article focuses on the set of these two-dimensional divisions; classes were drawn up worldwide, i.e. photograms from different parts of the body were divided into different classes with their part of the body. Therefore supervision is very poor and takes very little time to mark, thus reducing the annotations during the training phase. This type of annotation is useful for medical imaging because annotations typically require professional advice and costly information. The CNN model is trained in the initial classification process and the feature representations for CBMIR are learned. A thorough review of the proposed method for the efficiency of recovery is provided for a series of images from a variety of imaging modes. The work presented has three important contributions.

1. A multimodal dataset that covers a wide variety of focus areas of medical imaging is extensively compiled.
2. The deep learning system is modeling and training on the medical image spectrum.
3. A highly effective medical image procurement method is showing the studied features and works with a huge multimodal dataset array.

Content-based image recovery (CBIR) is a technique used to get identical images based on extracted features from the picture database. Content-based access to medical pictures to facilitate clinical decision-making has long been suggested to ease clinical data management and scenarios incorporation into picture archiving and communication systems for content-based access methods (PACS). At least two problems in contrast with regular CBIR systems should be taken into account in the medical CBIR for content-based image retrieval in order to manage the complex medical images.

The first challenge is medical image extraction. The precision of the recovery method can be enhanced by a good feature representation. Those features are therefore subjective and incomplete within the conventional CBIR system, which can be extracted manually. In our method, we use CNN to derive objective and detailed features from medical images.

The second concern is the picture features with high dimensions. To improve search performance, large numbers of feature vector dimensions have traditionally been utilized to compare the similarity, and hundreds and even thousands of feature vector dimensions are available. The dimension curse was typically encountered in the study and arrangement of these high-dimensional feature vectors.

We have therefore built and used this high-dimensional data base to sweep through the dimensionality curse that we presented in SPIE Medical Imaging. This high-dimensional database was also used. To overcome automatic extraction features in our CBIR study, in this presentation and in this combined high-dimensional database indexing technique we implemented CBIR feature extraction methods based upon deep learning (DL) and reports.

A medical imagery database requires services that are not required in traditional text databases. This refers to:

- **Indexing non-textual**: Besides textual indexing, there needs to be a way for non-textual indexing with relations between the two data forms. For instance, by examining organ and tumour outlines, geometrical information can be obtained. However, physiological knowledge will come from sources such as laboratory findings and other parts of the medical record. Ties between the two groups would be important if a specific diagnostic or therapeutic decision is to consider the clinical consequences.
- **Made-to-measure scheme:** The clinical researcher will need tools to edit, change, and adapt to new requests, which allow end-user-designed ad hoc custom schema for recovery and search. Queries of different users in an image database might have very different query language demands. For instance, a doctor may ask complicated questions about an image that is related to the functionality and/or structure of organs in the image. Another way to access images that all represent a single morphological feature would be to provide a database for use in teaching.
- **The momentum**: Any query of a collection of evolving and complex medical images needs to be created by the user. A CT-sample patient may subsequently be found to have metastases of the liver to delineate a primary lung tumour. Thus, a doctor accessing the database needs to integrate in the data base new knowledge about hepatic metastases and is able to develop specific questions about the state of the patient, both past and current, related to the new information.
- **Modules of resemblance:** As the consumer creates potential theories to explore a database, it implies that new formalities are possible to search through a database to collect "interesting photos." Thus, tools that "show me one like that which is bigger," or "show me one between the two," can give the user a powerful way to learn new concepts and information. This approach to "change the field of view" is considered an essential feature of a medical database of imaging.
- **Modules of reference**: For purposes of comparing it to one under investigation, a further question structure would show ten "normal" instances to decide if this falls within normal limits. Special tools are required to make the shape flexible.
- **Smart demands:** Iconic queries are queries using user-generated examples. This may be drawings of essential features or prototypes. The user can create

personalized semantics by using icons and prototypes associations. Generic schemes are required to provide a starting point in schema evolution such that the consumer is able to classify which relationships and similarity measures are suitable for the problem under investigation (e.g. "this is what I want to describe as a left ventricular aneurysm"). Users are expected to support the iconic requests and the custom schema to a number of objects at various abstraction levels. For the full use of a medical imaging database, a pictorial query language would be necessary.

- **Language of explanation:** New image descriptors can contribute to new information and new disease staging types. For instance, a broad midline neuroblastoma that encloses the coeliac axis is unworkable, whereas a midline-long neuroblastoma that does not enclose the coeliac axis is operational. Both of them are pathologically referred to as "stage 3." New instruments need to be created to evolve new definitions that differentiate between these two possibilities.
- **Registration multi-modality:** Precise image registration from various imaging modalities brings new expertise to the decision-making process, but there are still no resources available to apply this technique quickly.
- **Manipulation of the frame:** As intrinsic working contexts, imagery databases could integrate many of the techniques that already exist for the image management: zooming, column, rotation, contrast enhancement, regional interest contours.

4.2 RELATED WORK

Existing work on our research is discussed briefly in this section.

4.2.1 IMAGE RETRIEVAL BASED ON CONTENT (CBIR)

Figure 4.1 shows a simple CBIR device block diagram. CBIR images are recovered from the large databases (Zylak 2000). Every system's first phase is offline and the second is online. Two phases of any system usually occur. Features from large image collections (used for device trainings) are extracted during the offline process to build a local functional database (Thompson 2020).

Over the past few decades, different purpose descriptors for images reflecting form and colour (Nishiura 2020) as well as textures were created worldwide (Tang 2020). Invariant function transition (SIFT), speed-up function interfaces have also been developed using certain local descriptors, such as SIFT and SURF (Jung 2020) and Wordbag (BoW) (Dhiman 2017; Soni 2020; Gomathi 2020). Local level definitions were created. The image recovery function (Mansoori 2009) was used in different experiments, but cannot fully overcome the semantic breakdown. We have recently reviewed semantic holes in the machine-based learning techniques (Dhiman 2017) and features at local level, based on hath or portable code (Dhiman 2018). A new method has been created for reducing the semantical gap through advanced machine learning methods, which have given a possibility to fill this gap directly by learning pictures without using manual features.

4.2.2 Medical Image Recovery Based on Content (CBMIR)

The extensive distribution of medical image collections in PCS hospitals is increasing (Dhiman 2019). Implementing an efficient medical image recovery system is therefore important for the handling of such large medical databases. In addition, a certain CBMIR method allows doctors to objectively decide on a particular disease or injury in addition to this database management task. A global system of functional extraction was unsuitable for medical images because clinically useful information in small image regions was highly localized. The Visual Words System Bag (BoVW) based approach was presented to the brain magnetic resonance image using SIFT characteristics in diagnosing Alzheimer's disease (MRI). They proposed LG-CHF as a vector function for matching images (Laguerre Circular Harmonic Functions Coefficients). For skin lesion images with reduced vector, classification and regression tree a content-based recovery system has been proposed. The effect of the medical CBIR result on the doctor's decision using a CBIR mammography algorithm with different characteristics, distance measurements and pertinency feedback has been suggested. Two wavelet adjustment CBIR techniques were utilized to define every image of a query and to estimate the best wavelet filter in a different wavelet. To optimize recovery efficiency, a regression function was used. A biomedical image regeneration approach was proposed that focuses on the classification of supervised learning. As a base and SVM support machine it used image filtering and semblance fusion to predict the query type.

4.2.3 Deep Learning

Deep learning is a subfield of machine learning that utilizes algorithms to model high level abstractions in data using a deep architecture, both linear and non-linear, with multiple transformation levels. The deep history of learning began in 1965, but only recently has it made significant progress with enhanced machine capacity, nonlinearity that enables deeper connectivity (Garg 2019), (Dhiman 2017), (Soni 2020) and enhanced networking practices (Dhiman 2019). Deep learning consists of artificial neural networks that simulate the functioning of the human brain. Neural feed networks consisting of several hidden layers are powerful examples of deep architectural models. A powerful type of neural network training remains the popular back propagation algorithm of the 1980s. In the classification framework for interstitial lung disease was presented in the convolutional neural network (CNN). Its dataset included seven groups, seven of which were ILD and tissue-healthy trends. By characterizing lung habits they obtained an efficiency of classification of 85.5%. A Boltzmann Machine Based Approach (CT) study has been proposed for a convolutions classification, combining generative and discrimination in representation learning (Zylak 2000). There have been two separate datasets: one for the description of lung texture and one for the recognition of airways. Vocals were listed (Nishiura 2020) in brain tissue classes with a multiscale CNN for automatic MR image segmentation. Depending on the patch size used, the network was educated in multiple image patches of various kernel sizes. A two-stage, profound-learning system for body organ recognition was introduced (Tang 2020). The first phase was followed by training on CNN to extract bigotry and insightful patches from local patches. The second stage of the procedure changed the network to eliminate unequal patches, with 12 groups including 2D CT and MR slices (CAD). The investigation involved three facets of the CNN: multiple CNN architectures, a dataset scale and a pre-trained medical picture model transfer analysis. Researchers may use various safety schemes or protocols to create stable contact to transmit various medical pictures over the network (Soni 2020; Limbasiya 2018).

4.3 METHODOLOGY

This study introduces a new classification system for the processing of images from the medical database. Figure 4.2 provides a detailed description of the classification method proposed. The DCNN model underlying is designed to learn the filter kernel via an abstract display of data in each layer. The pooling layer is characterised by standard sample maps of convolutionary layers often produced by defining local maximums in one area. Pooling also gives the translation invariance and limits the number of neurons for further treatment. Each neuron has a thicker relationship in fully connected layers than the convolutionary layers. The DCNN part is referred to as the extraction function before fully connected layers and is then referred to as the classifier. The following phases provide the proposed method with a comprehensive work.

4.3.1 PHASE 1

The first part of the classification phase includes the training of a deep convolutionary neural network in classifying medical images using a supervised method of learning. The study was conducted with 2D images, which means that every image must be classified into a class that eventually formats a multifaceted image classification problem. There are usually two modules in the picture classification algorithms: extract

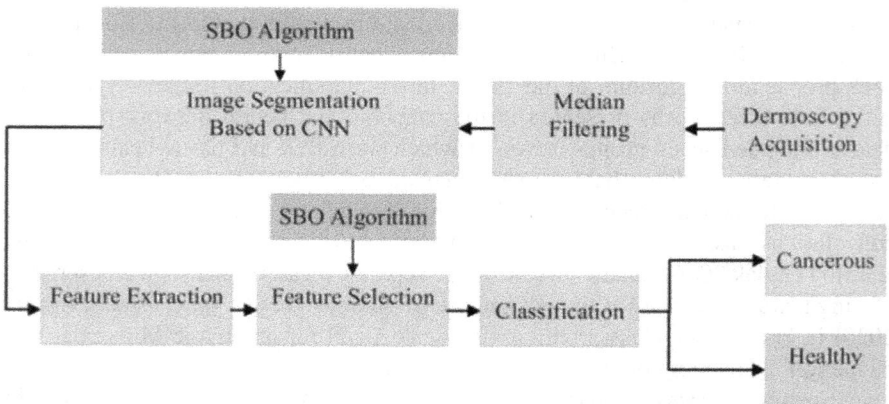

FIGURE 4.2 Architecture of CBMIR.

and grading module. DCNN learns the hierarchy of deep convolution and the classification of images in a full learning system. Instead of making prevailing assumptions, as is the case with handcrafted functions, a profound learning algorithm learning features mid, abstract, and low-level. The following are the architecture and training models.

4.3.1.1 The DCNN Model Architecture

For preparation eight layers were used, five were convolutionary and three were connected completely, as shown in Figure 4.3. There were eight layers of the models. CVL and FCL representing the convolutionary and fully connected layers, e.g. CVL1 represents the first convolutionary layer. A softmax function with 24 outputs was added to the output of the last fully connected layer (FCL3), which offers class distribution. The probability vector size 1 is then 24, where each vector element corresponds to the class of dataset. Compared to the CNN model, grey images of dimensions 224 and 224 are accepted as inputs. In CVL1 the input from 64 kernels of 11 to 11 with steps of 4 pixels will be screened. Stride is the distance from the kernel map's reciprocal neuron fields. The first convolutionary layer contribution is translated into a nonlinearity and the max layer is subsequently synthesised for pooling neurons. The rectified linear unit (ReLU) has no linearity for the output of any convolutionary or fully connected layer. The ReLUs network can train more than once than its equivalent of tanh units and has deeper difficulties with the loss of gradients. The CVL2 uses the CVL1 as a non-linear ReLU and max spatial pooling input with a size of 198 kernels each, between 5 and 5. The CVL3 consists of 392 5×5 kernels and gets input from pooled outputs of the second layer. Size 3, per 3, of both convolutionary layers, are 256 kernels, CVL4 and CVL5. Neurons are equivalent in all fully connected stages, i.e. 409. After the first, second and fifth convolution layers, a pooling layer was used. It comprises a grid of 2-point pooling units, each of which sums up the neurons in the 3–3 neighbourhood in the position of the pooling unit. It is difficult to overlap the model with the over-expansion operation during training. After both the first and the second fully connected regulatory drop out layers, overcrowding with a 1 pp probability of neuron dropout is prevented, where p is the likelihood of neurons

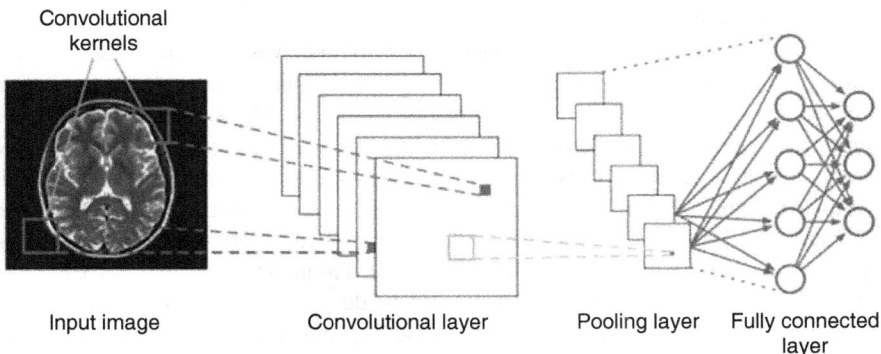

Convolutional kernels

Input image   Convolutional layer   Pooling layer   Fully connected
                                                                          layer

FIGURE 4.3   Training of image with DCNN.

preservation. The dropped neurons are not both forward and reverse, i.e. all links coming into and out of dropped neurons are eliminated during a training process. After the exercise, the falling neurons are replaced with their original weights for the next stage. The neurons are used during the testing period. However, a preserved chance factor p tests these in order to correspond to the predicted neuronal benefits in the testing process. Finally, 4096 inputs and 24 outputs are supplied to the log-max in the output FCL3. Log-softmax for an N-dimensional vector is a soft max function that calculates a vector over [0.1] and has a total value of 1. This assigns the distribution of probability to a class (Garg 2019).

4.3.1.2   Details of Training

In order to practise for all images in data bases, the middle was cut down using 224 = 224 moves. The model was learned by reverse propagation of the Stochastic Gradient (SGD). The learning rate was extremely low at 0.0001 with age 30 SGD. It has been maximally optimised. Negative Log Likelihood (NLL) has been used in SGD testing as an objective function or criterion. For N-classification issues, NLL is also used. Gaussian, zero and 0.01 were used to initialise weights for all layers. In the three completely connected layers '1' and '0,' the biases for CVL2, CVL4, and CVL5 have already been known. The study rate has continuously been maintained and the learning rate decreased for all storm gradients, including 0.0001, but in turn the duration of the training increased. Another argument for continuous learning is that our data have already been developed with a very low learning rate, and it clearly requires additional staff and more resources to reduce it. A 0.0422 training error for up to 30 planning times was used to customise the model. SGD is the most widely used neural network algorithm and is highly effective in the analysis of discriminatory linear classifications under a convex-loss function, such as SVM or logistic regression. The two main benefits of SGD includes reliability and ease of deployment including coordination options for networks, such as number of iterations, learning rate, decay in rates, etc. The need for hyperparameters such as numerous times or iterations and control parameters include some of the disadvantages of SGD. For each sample of xi and yj training SGD updates the parameters. The SGD parameter is modified. The most common neural networking algorithm is back propagation in combination with certain optimization strategies, such as the SGD. All back propagation parameters determine the gradients.

When the objective function is reduced, the SGD receives measured gradients and updates them as inputs. Returns include the known objective for the determination of the input losses, i.e. the actual class mark. The chain rule is used to measure the gradients in relation to the loss function for each layer. Each layer consists of three important back stretching phases, that is, the forward transition, the reverse transition and the derivative of operation, where the layer includes parameters as shown in Figure 4.3. There is no parameter available for such layers, such as the pooling layer, and the derivative should not be specified. Parameters and derivatives of the cost function system w.r.t. parameters would be defined if the layer has parameters An input is provided to each layer and an output is produced. The loss E is taken from the output layer and spread backwards from each layer. Many of these layers exist in each layer and the vector parameters should be familiar with all the layer parameters.

4.3.2   PHASE 2

Reprinted features may be obtained from the last three fully connected model layers, e.g., FSC1 – FCL 3, as soon as the DCNN model is refined and trained for identification of multimodal medical images. For creating a database, the features F1i, F2i and F3i are then extracted from fully connected 1–3 layers of that specific image by feed from a trained DCNN classification model, respectively. Similarly, F1i, which is a function database extracted from FCL1 and F2i and F3i, reflects FCL2 and FCL3 databases where i=1 to P and P equals the number of training samples.

In any query formulation, similar images to those in a query image can be found using

$$d(a,b) = \sqrt{\sum_{i=1}^{P}(a_i - b_i)^2} \qquad\qquad (4.3.1)$$

When ai and bi are query- and database-image functions, by comparing feature representations extract for the query-image, using Euclidian distance metric. Finally, comparative analysis of the efficiency of feature recovery from FCL1 to FCL3 are carried out.

4.3.3   PHASE 3: OPTIMIZATION

4.3.3.1   Spotted Hyena Optimizer (SHO)

In this subsection, SHO algorithm (Dhiman 2018) are mathematical models, i.e. search, encircle, hunt, and attack prey.

4.3.3.1.1   Encircling Prey

The current best candidate solution is seen as the target prey to mathematically form the social hierarchy of spotted hyenas. Now the other quest agents update their best approach.

$$\overrightarrow{D_h} = |\ \overrightarrow{B}.\overrightarrow{P_p}(x) - \overrightarrow{P}(x)\ | \qquad\qquad (4.3.2)$$

$$\overrightarrow{P}(x+1) = \overrightarrow{P_p}(x) - \overrightarrow{E}.\overrightarrow{D_h} \qquad\qquad (4.3.3)$$

Where Dh defines the distance of the prey from the spotted hyena, x indicates current isolation, B and E are coefficient vectors, Pp indicates the location vector of the prey, P is the location vector of the spotted hyena. The vectors B and E are as follows:

$$\overrightarrow{B} = 2.r\overrightarrow{d_1} \qquad\qquad (4.3.4)$$

$$\overrightarrow{E} = 2\overrightarrow{h}.r\overrightarrow{d_2} - \overrightarrow{h} \qquad\qquad (4.3.5)$$

$$\overrightarrow{h} = 5 - (Iteration \times (5\,/\,Max_{Iteration})) \qquad\qquad (4.3.6)$$

Where, Iteration=1,2,3,...., $Max_{Iteration}$

h is reduced linearly from 5 to 0 over the course of full iterations to balance discovery and operation (MaxIteration). In addition, with an increase in iteration value, this process facilitates greater exploitation. Rd1 and rd2 are, however, random vectors within [0, 1].

4.3.3.1.2 Hunting

It is believed that the best search agent knows the location of prey in order to mathematically identify the hunting behaviour of spotted hyenas. The other search officers form a cluster of trustworthy friends to the best search officer and have been able to offer the best options so far to update their position. In this mechanism, the following equations are proposed:

$$\vec{D}_h = |\vec{B}.\vec{P}_h - \vec{P}_k|$$ (4.3.7)

$$\vec{P}_k = \vec{P}_h - \vec{E}.\vec{D}_h$$ (4.3.8)

$$\vec{C}_h = \vec{P}_k + \vec{P}_{k+1} + ... + \vec{P}_{k+N}$$ (4.3.9)

Where Ph determines the first best spotted hyena location, and Pk indicates the other spotted hyena position. The number of hyenas contained here as measured as follows, is indicated by N:

$$N = count_{nos}(\vec{P}_h, \vec{P}_{h+1}, \vec{P}_{h+2}, ..., (\vec{P}_h + \vec{M}))$$ (4.3.10)

Where M is a random vector in the area of [0,5, 1], and nos defines the number of solutions and includes all the solutions used, after the addition of M, which is very similar to being the best solution in a given search field.

4.3.3.1.3 Attacking

The value of vector h is reduced in order to mathematically model the attack on a prey. Vector E variance is also minimized by changing the value in Vector h from 5 to 0 during the iteration process. The following is the mathematical formula for attacking the prey:

$$\vec{P}(x+1) = \frac{\vec{C}_h}{N}$$ (4.3.11)

When P (x+1) saves the best solution and changes other search agents' location in accordance with the best search agent's position. The SHO algorithm enables its search agents to update its location and attack it.

4.3.3.1.4 Searching

Often detected hyenas pursue the prey according to the location of the cluster of detected hyenas residing in the C_h vector. They move to search and strike for their prey from each other. Thus, vector E is used to push the search agents to go far from the prey through random values that are greater than 1 or less than 1. The SHO algorithm can be searched globally using this mechanism.

Archive and grid mechanisms are used in the multi-target Spotted Hyena Optimizer (MOSHO) algorithm (Dhiman 2018). The group selection approach is used for better exploration and exploitation after upgrading the search agents. The parameters of the CNN pattern are calculated by this algorithm.

4.4 DATASET

For supervised learning from publicly available medical courses, the dataset used for the proposed CBMIR challenge has been collected. During this study, the chest X-ray images of 50 COVID-19 patients were taken from a GitHub Open Source repository supplied by Joseph Cohen (Open database of COVID-19 cases with chest X-ray or CT images n.d.). The collection includes X-ray/CT pictures of the thoracic lung, mostly ARDS patients, COVID-19, MERS patients, pneumonia, extreme acute respiratory system and ARDS patients. This archive includes (SARS). In addition, 50 regular images of the chest X-ray were used from the "Pneumonia" Kaggle repository (Chest X-Ray Images n.d.). Our experiments were conducted using a chest X-ray image dataset of 50 average patients. All pictures in this dataset are restated to a 280×280 pixel scale. Figures 4.4 and 4.5 show photographs of COVID-19 and daily patients.

4.5 EXPERIMENTAL RESULTS

To build and train the proposed deep learning system, a common and widespread, profound learning method is used. The simulations were conducted on a laptop with a clock speed of 2.40 GHz and a RAM of 6.00 GB with a laptop Dell Inspiron 14.04 with an Intel core i3 Processor. Classification and recovery results have been assessed for the proposed method. Figures 4.6–4.7 show the convergence analysis of various layers of DCNN.

4.6 PERFORMANCE METRICS

The performance of the framework proposed for CBMIR was checked with the most widely used CBIR systems, namely precision and recall. The results are shown in Table 4.1.

- $Precision = \dfrac{True\_Positive}{\left(True\_Positive + False\_positive\right)}$

- $Recall = \dfrac{True\_positive}{\left(True\_positive + False\_negative\right)}$

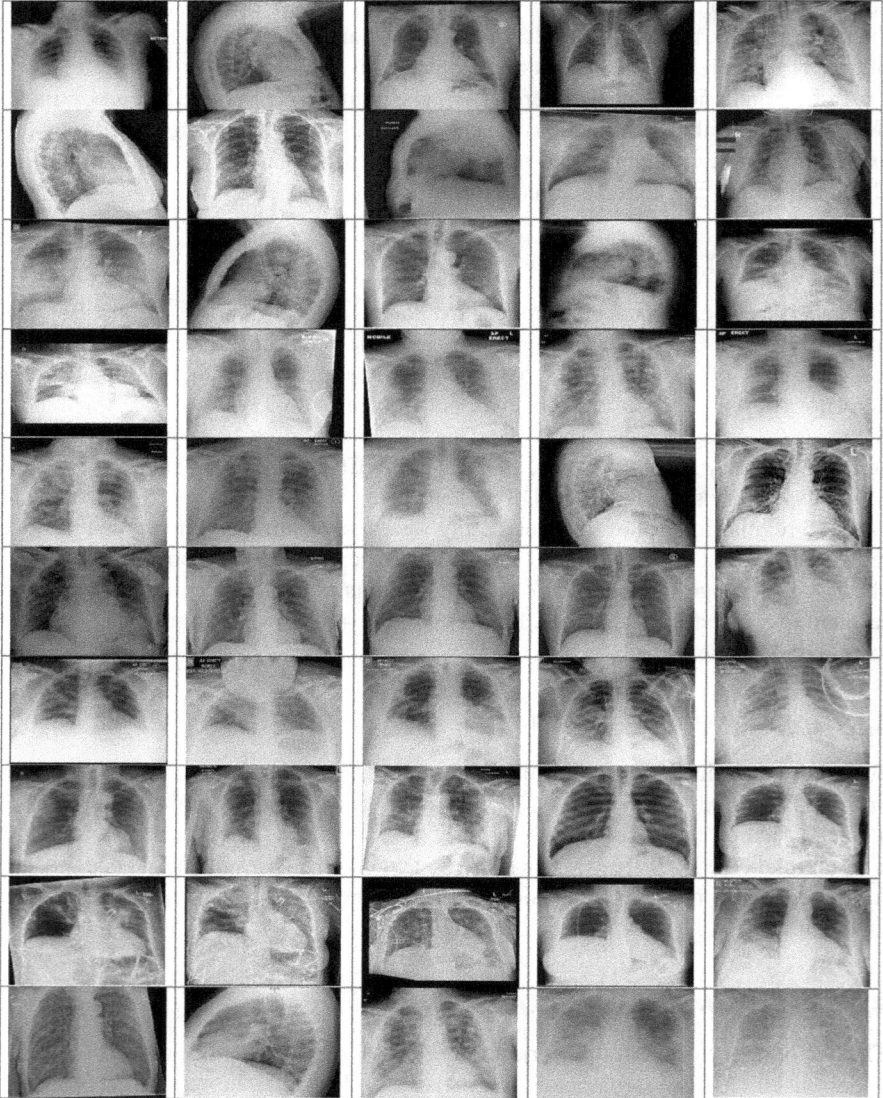

FIGURE 4.4   X-ray pictures of patients suffering from Coronavirus (COVID-19).

4.7   CONCLUSION

This article proposes a deep learning system for medical image retrieval based on content by training a profound convolutionary neural network for the purpose of classification. The first is to get a prediction of the query class of the query picture by the qualified network and then find related pictures in that particular class. The second approach is to search for all applicable images in the entire database without

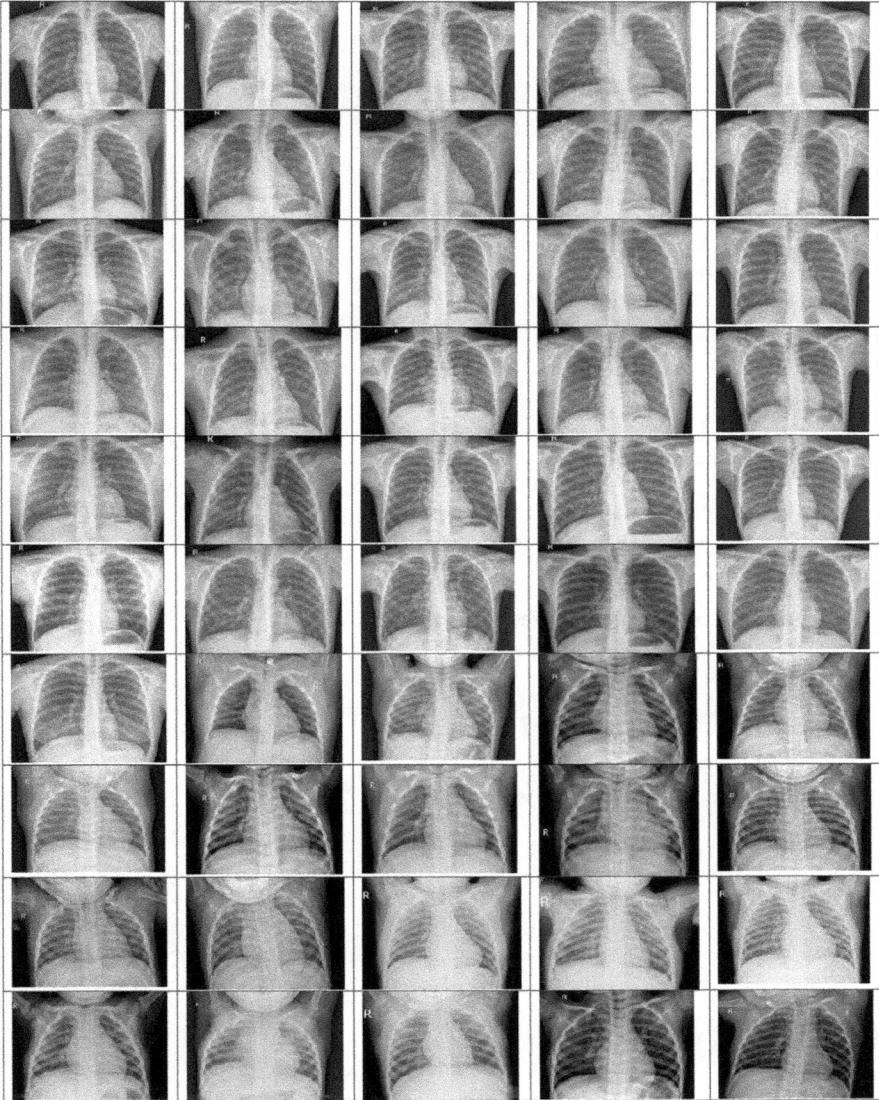

FIGURE 4.5 The normal patient X-ray images.

adding details about the query image type. The deep learning-based technique Deep Neural Network (DNN) classifier is used to classify medical images for categorizing the images. For tuning the parameters of deep learning models, the Spotted Hyena Optimizer (SHO) is used. By learning discriminatory characteristics from pictures directly, the proposed solution reduces semantic distance. The network has succeeded in training medical images with a 99.77 per cent average rating accuracy. In order to

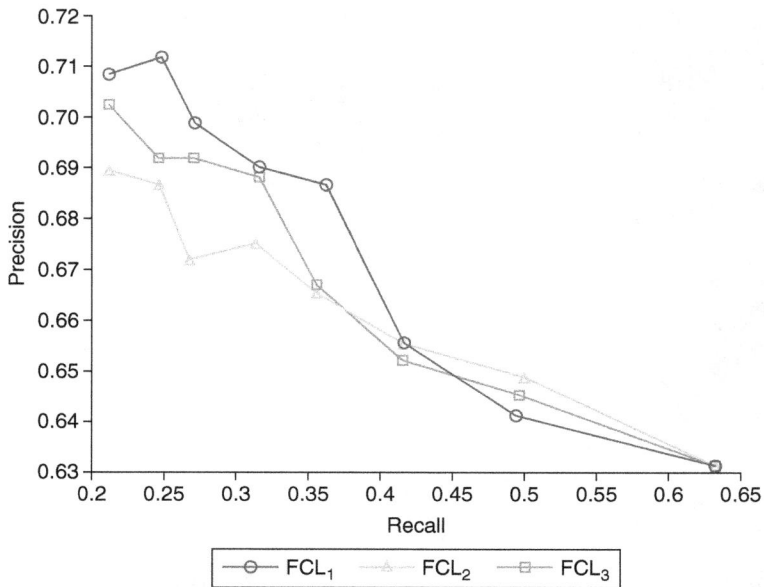

FIGURE 4.6   Class prediction accuracy vs CBMIR recall.

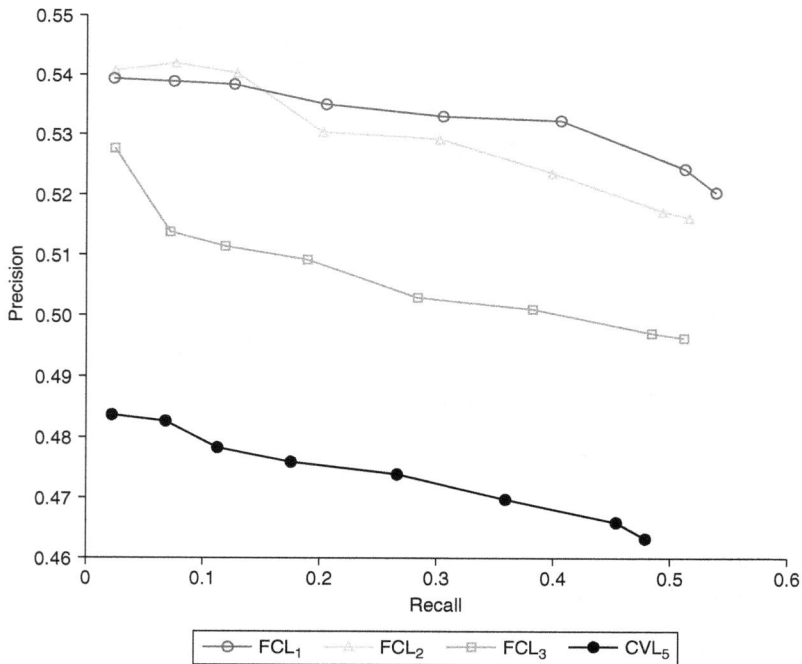

FIGURE 4.7   CBMIR accuracy vs class prediction recall.

TABLE 4.1
Experimental results on performance metrics

Method	No. of images	Training (images)	Testing (images)	Precision (%)	Recall (%)
On whole pictures, DCNN trained (proposed)	1000	50	1000	99.76	99.77
DCNN trained on local picture patches in two phases [24]	1000	50	1000	98.43	97.28

remove the recovery functionality, the last three fully connected network layers were used. Widespread measurements such as accuracy and reminders have been used to assess the efficiency of the proposed system for recovery of medical images, which helps the doctors to analyse the disease better.

REFERENCES

Alfanindya, A., Hashim, N. and Eswaran, C., 2013, June. Content based image retrieval and classification using speeded-up robust features (SURF) and grouped bag-of-visual-words (GBoVW). In *2013 International Conference on Technology, Informatics, Management, Engineering and Environment* pp. 77–82. IEEE.

Babenko, A. and Lempitsky, V., 2015. Aggregating local deep features for image retrieval. In *Proceedings of the IEEE International Conference on Computer Vision* (pp. 1269–1277).

Bay, H., Tuytelaars, T. and Van Gool, L., 2006, May. Surf: Speeded up robust features. In *European Conference on Computer Vision* (pp. 404–417). Springer, Berlin, Heidelberg.

Bengio, Y., Courville, A.C. and Vincent, P., 2012. Unsupervised feature learning and deep learning: A review and new perspectives. *CoRR, abs/1206.5538, 1*, p.2012.

Chest X-Ray Images (Pneumonia) www.kaggle.com/paultimothymooney/chest-xray-pneumonia.

Dhiman, G. and Kumar, V., 2017. Spotted hyena optimizer: a novel bio-inspired based metaheuristic technique for engineering applications. *Advances in Engineering Software, 114*, pp.48–70.

Dhiman, G. and Kumar, V., 2018. Multi-objective spotted hyena optimizer: a multi-objective optimization algorithm for engineering problems. *Knowledge-Based Systems, 150*, pp.175–197.

Dhiman, G. and Kumar, V., 2019. Seagull optimization algorithm: Theory and its applications for large-scale industrial engineering problems. *Knowledge-Based Systems, 165*, pp.169–196.

Dhiman, G. and Kumar, V., 2017. Spotted hyena optimizer: a novel bio-inspired based metaheuristic technique for engineering applications. *Advances in Engineering Software, 114*, pp.48–70.

Dhiman, G. and Kumar, V., 2018a. Emperor penguin optimizer: a bio-inspired algorithm for engineering problems. *Knowledge-Based Systems, 159*, pp.20–50.

Dhiman, G. and Kumar, V., 2018b. Multi-objective spotted hyena optimizer: a multi-objective optimization algorithm for engineering problems. *Knowledge-Based Systems, 150,* pp.175–197.

Dhiman, G., 2019. *Multi-objective metaheuristic approaches for data clustering in engineering application (s)* (Doctoral dissertation).

Dhiman, G., Soni, M., Pandey, H.M., Slowik, A. and Kaur, H., 2020. A novel hybrid hypervolume indicator and reference vector adaptation strategies based evolutionary algorithm for many-objective optimization. *Engineering with Computers,* pp.1–19.

Garg, M. and Malhotra, M., 2017. Retrieval of images on the basis of content: a survey. *International Journal of Engineering Development and Research, 5,* pp.757–760.

Garg, M., Malhotra, M. and Singh, H., 2019. A novel CBIR-based system using texture fused LBP variants and GLCM features. *International Journal of Innovative Technology and Exploring Engineering, 9*(2), pp.1247–1257.

Garg, M., Malhotra, M. and Singh, H., 2019. Comparison of deep learning techniques on content based image retrieval. *Modern Physics Letters A,* p.1950285.

Gomathi, S., Kohli, R., Soni, M., Dhiman, G. and Nair, R., 2020. Pattern analysis: predicting COVID-19 pandemic in India using AutoML. *World Journal of Engineering.*

Garg, M., Singh, H. and Malhotra, M., 2019. Fuzzy-NN approach with statistical features for description and classification of efficient image retrieval. *Modern Physics Letters A, 34*(03), p.1950022.

Hinton, G., Deng, L., Yu, D., Dahl, G.E., Mohamed, A.R., Jaitly, N., Senior, A., Vanhoucke, V., Nguyen, P., Sainath, T.N. and Kingsbury, B., 2012. Deep neural networks for acoustic modeling in speech recognition: The shared views of four research groups. *IEEE Signal Processing Magazine, 29*(6), pp.82–97. https://biodifferences.com/difference-between-x-ray-and-ct-scan.html.

Jung, S.M., Akhmetzhanov, A.R., Hayashi, K., Linton, N.M., Yang, Y., Yuan, B., Kobayashi, T., Kinoshita, R. and Nishiura, H., 2020. Real-time estimation of the risk of death from novel coronavirus (COVID-19) infection: inference using exported cases. *Journal of Clinical Medicine, 9*(2), p.523.

Jiji, G.W. and Raj, P.S.J.D., 2014. Content-based image retrieval in dermatology using intelligent technique. *IET Image Processing, 9*(4), pp.306–317.

Kaur, S., Awasthi, L.K., Sangal, A.L. and Dhiman, G., 2020. Tunicate swarm algorithm: a new bio-inspired based metaheuristic paradigm for global optimization. *Engineering Applications Artificial Intelligence, 90,* p.103541.

Karpathy, A., Toderici, G., Shetty, S., Leung, T., Sukthankar, R. and Fei-Fei, L., 2014. Large-scale video classification with convolutional neural networks. In *Proceedings of the IEEE conference on Computer Vision and Pattern Recognition* (pp. 1725–1732).

Krizhevsky, A., Sutskever, I. and Hinton, G.E., 2012. Imagenet classification with deep convolutional neural networks. *Advances in Neural Information Processing Systems, 25,* pp.1097–1105.

Kumar, K.K. and Gopal, T.V., 2014, March. A novel approach to self order feature reweighting in CBIR to reduce semantic gap using Relevance Feedback. In *2014 International Conference on Circuits, Power and Computing Technologies [ICCPCT-2014]* (pp. 1437–1442). IEEE.

Limbasiya, T., Soni, M. and Mishra, S.K., 2018. Advanced formal authentication protocol using smart cards for network applicants. *Computers & Electrical Engineering, 66,* pp.50–63.

Lin, K., Yang, H.F., Hsiao, J.H. and Chen, C.S., 2015. Deep learning of binary hash codes for fast image retrieval. In *Proceedings of the IEEE Conference on Computer Vision and Pattern Recognition Workshops* (pp. 27–35).

Liu, Y., Zhang, D., Lu, G. and Ma, W.Y., 2007. A survey of content-based image retrieval with high-level semantics. *Pattern Recognition*, *40*(1), pp.262–282.

Mansoori, Z. and Jamzad, M., 2009, May. Content based image retrieval using the knowledge of texture, color and binary tree structure. In *2009 Canadian Conference on Electrical and Computer Engineering* (pp. 999–1003). IEEE.

Nishiura, H., Kobayashi, T., Yang, Y., Hayashi, K., Miyama, T., Kinoshita, R., Linton, N.M., Jung, S.M., Yuan, B., Suzuki, A. and Akhmetzhanov, A.R., 2020. The rate of under as certainment of novel coronavirus (2019-nCoV) infection: estimation using Japanese passengers data on evacuation flights.

Open database of COVID-19 cases with chest X-ray or CT images https://github.com/ieee8023/covid-chestxray-dataset.

Ponciano-Silva, M., Souza, J.P., Bugatti, P.H., Bedo, M.V., Kaster, D.S., Braga, R.T., Bellucci, A.D., Azevedo-Marques, P.M., Traina, C. and Traina, A.J., 2013, June. Does a CBIR system really impact decisions of physicians in a clinical environment. In *Proceedings of the 26th IEEE International Symposium on Computer-Based Medical Systems* (pp. 41–46). IEEE.

Quellec, G., Lamard, M., Cazuguel, G., Cochener, B. and Roux, C., 2011. Fast wavelet-based image characterization for highly adaptive image retrieval. *IEEE Transactions on Image Processing*, *21*(4), pp.1613–1623.

Rahman, M.M., Antani, S.K. and Thoma, G.R., 2011. A learning-based similarity fusion and filtering approach for biomedical image retrieval using SVM classification and relevance feedback. *IEEE Transactions on Information Technology in Biomedicine*, *15*(4), pp.640–646.

Soni, M., Chauhan, S., Bajpai, B. and Puri, T., 2020, September. An approach to enhance fall detection using machine learning classifier. In *2020 12th International Conference on Computational Intelligence and Communication Networks (CICN)* (pp. 229–233). IEEE.

Soni, M., Gomathi, S. and Adhyaru, Y.B.K., 2020, July. Natural language processing for the job portal enhancement. In *2020 7th International Conference on Smart Structures and Systems (ICSSS)* (pp. 1–4). IEEE.

Thompson, R.N., 2020. Novel coronavirus outbreak in Wuhan, China, 2020: intense surveillance is vital for preventing sustained transmission in new locations. *Journal of Clinical Medicine*, *9*(2), p.498.

Tang, B., Wang, X., Li, Q., Bragazzi, N.L., Tang, S., Xiao, Y. and Wu, J., 2020. Estimation of the transmission risk of the 2019-nCoV and its implication for public health interventions. *Journal of Clinical Medicine*, *9*(2), p.462.

Wan, J., Wang, D., Hoi, S.C.H., Wu, P., Zhu, J., Zhang, Y. and Li, J., 2014, November. Deep learning for content-based image retrieval: A comprehensive study. In *Proceedings of the 22nd ACM International Conference on Multimedia* (pp. 157–166).

Wu, G., Lu, W., Gao, G., Zhao, C. and Liu, J., 2016. Regional deep learning model for visual tracking. *Neurocomputing*, *175*, pp.310–323.

Wu, L. and Hoi, S.C., 2011. Enhancing bag-of-words models with semantics-preserving metric learning. *IEEE MultiMedia*, *18*(1), pp.24–37.

Yan, Z., Zhan, Y., Peng, Z., Liao, S., Shinagawa, Y., Zhang, S., Metaxas, D.N. and Zhou, X.S., 2016. Multi-instance deep learning: Discover discriminative local anatomies for bodypart recognition. *IEEE Transactions on Medical Imaging*, *35*(5), pp.1332–1343.

Yang, J., Jiang, Y.G., Hauptmann, A.G. and Ngo, C.W., 2007, September. Evaluating bag-of-visual- Words representations in scene classification. In *Proceedings of the International Workshop on Workshop on Multimedia Information Retrieval* (pp. 197–206).

Zhang, F., Song, Y., Cai, W., Hauptmann, A.G., Liu, S., Pujol, S., Kikinis, R., Fulham, M.J., Feng, D.D. and Chen, M., 2016. Dictionary pruning with visual word significance for medical image retrieval. *Neurocomputing*, *177*, pp.75–88.

Zhou, S., Chen, Q. and Wang, X., 2013. Active deep learning method for semi-supervised sentiment classification. *Neurocomputing*, *120*, pp.536–546.

Zylak, C.M., Standen, J.R., Barnes, G.R. and Zylak, C.J., 2000. Pneumomediastinum revisited. *Radiographics*, *20*(4), pp.1043–1057.

Zeiler, M.D. and Fergus, R., 2014, September. Visualizing and understanding convolutional networks. In *European Conference on Computer Vision* (pp. 818–833). Springer, Cham.

5 Implication of Image Pre-Processing in Object Detection Using Machine Learning Techniques

Neeru Mago, Jagmohan Mago, Sonia Mago, and Rajeev Kumar Dang

CONTENTS

DOI: 10.1201/9781003153405-5

5.1 INTRODUCTION

An image is a collection of numbers (pixels) arranged in a two-dimensional (2D) array that ranges from 0 to 255. It can be defined by a mathematical function $f(x, y)$ where x and y are horizontal and vertical co-ordinates respectively. Pre-processing is the most important task which improves image's quality. The main objective of pre-processing is to improve an image or enhance some important image features, that may have suppressed, for further processing. So, we performed some pre-operation on such images like resizing it, removing noise, segmentation and morphology. Therefore, image pre-processing is necessary for picture enhancement, removing noise, rescaling and converting it into machine understandable format.

5.2 LITERATURE REVIEW

Li et al. (2016) proposed a real-time video copy detection system. This system operates using the Hadoop platform. After careful observation, it was concluded that Hadoop platform was far better than other systems. Shah (2015) elaborated that there is an urgent need of performing big video data analytics in the modern world. Moreover, it has been observed that one has to face several issues in performing large-scale video data using Hadoop and even a solution was offered after a careful analysis of the problem.

Natarajan et al. (2015) introduced a quiet actual time traffic analytics system. Road accidents from live video streams can be automatically deducted by it. The system will notify all the nearby hospitals and highways rescue teams in case of any accident. The system cannot only detect road congestion but it can also broadcast alternative route information automatically. Parsola et al. (2017) suggested a more efficient system that can be used for the detection of moving objects and collection of their co-ordination. These collected coordinates can then be utilized for the localizing of post-event investigation. Kumar et al. (2016) processed frames to identify moving objects. Gay color information is extracted by the suggested system to improve object segmentation for background subtraction. Then the moving objects are discovered carefully by it. This approach requires MATLAB for its better execution.

Jamkhandikar et al. (2014) recommended a framework that is actually effective for the accurate detection of an object using color feature. The system includes several stages to track an object automatically. Moreover, this system prefers to use the Euclidian filter as it has the advantages of being used in moving images. As gray images need less time to proceed so the color images are changed into gray ones. Ghaware et al. (2016) demonstrated how Hadoop technology has been used for the analysis of huge amount of video data and for the detection of the object. For the detection of scene changes in video, block-based algorithms are used. If there is any change in the scene, that change will automatically be stored in the server. After this, the videos, those are stored in the server, are divided into a number of chunks and then these chunks are sent to different nodes for analytical purposes with the help of Hadoop's MapReduce technology. The object tracking algorithm, using a novel Bayesian Kalman filter with simplified Gaussian mixture are used to detect objects. Dhinakaran et al. (2015) brought out a system called HDVFS that made object tracking more convenient and effective as it offers a MapReduce framework. During the process it becomes easier to detect the stipulated image object and its behavior from the unstructured video in the cloud depositary. Therefore, this system proves to be capable of storing an extremely large size file like the Amazon 53 storage bucket.

Saxena et al. (2016) proposed various concepts that have been used for the execution of parallel image processing such as parallel computing and currently available parallel architecture, tools and techniques. Behind this approach, the chief motive is to find out the problems concerned with parallel computing of various image processing applications. Ganesh and Appavu (2015) emphasized the decomposition of the large-scale video frameworks into simple lightweight aspects of a video surveillance algorithm. In order to analyze large-scale video, a blog tracking based video surveillance algorithm has been used. The process of traffic hours videos, those are captured in local stores sites, is cleaned by using Hadoop MapReduce functions. Aradhye et al. (2016) has a main focus on that object which is directed as an intruder. Thus this system helps in security purposes and the continuous monitoring of the vicinity is simplified by this approach. Three algorithms are used to serve the purpose namely mean square difference, color histograms, and a structural similarity index. Hadoop helps in reducing the analysis time. If there is any deviation in the scene in the starting, ending or total time duration of unethical part of the video system, then a red mark is indicated.

Agrawal and Sivagami (2015) made the use of a more sound method in order to detect objects from the database even if it is lying in an unstructured format. The relational database cannot process the same. Parallel processing has been used for not only changing the techniques such as implementation online approximation techniques using mean filters but for object recognition also. Kumar et al. (2015) had utilized three algorithms to identify the object in the network. These are color histogram, mean square difference and structural difference. BKF-SGM-IMS tracks the same objects by using Hadoop, which in turn reduces the time and storage space. Moreover, at the starting and ending time of the video, results are illuminated in the form of intrusion video duration.

5.3 PREPROCESSING METHODS

The method that is of primary importance in influencing automatic detection of defects is pre-processing. There are some preprocessing methods available in literature (Anitha & Radha, 2010; Ravi & Ashokkumar, 2017) like contrast adjustment, intensity adjustment, histogram equalization, morphological operation, etc.

5.3.1 CONTRAST ADJUSTMENT

When there is no clear-cut difference between black and white images, it means that it lacks contrast and we have to adjust this problem. Absolute pixel intensities are not as sensitive to contrast as the human eye is. It helps in getting a better image as shown in Figure 5.1.

5.3.2 INTENSITY ADJUSTMENT

This technique maps an image's intensity values to a new range. These techniques are used so that an image could be improved. But here, improvement is defined in two ways objective and subjective. Objective improvement increases the signal-to-noise ratio and subjective improvement makes certain features easier to see by modifying the colors for intensities as shown in Figure 5.2.

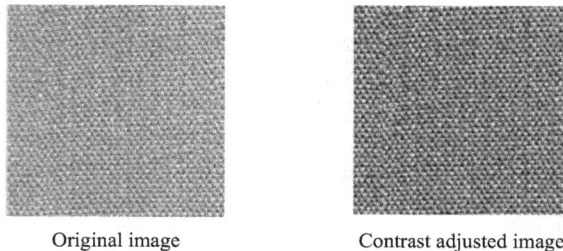

Original image   Contrast adjusted image

FIGURE 5.1   Image after contrast adjustment.

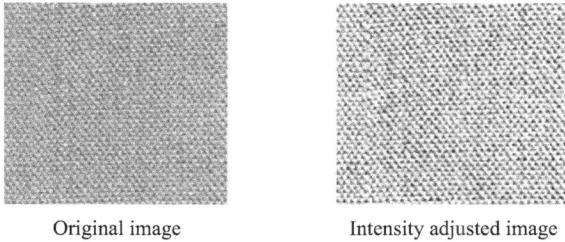

Original image   Intensity adjusted image

FIGURE 5.2   Image after intensity adjustment.

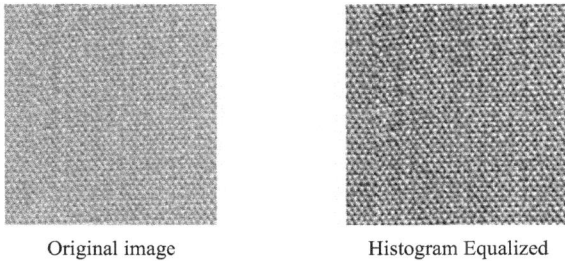

Original image   Histogram Equalized

FIGURE 5.3   Image after histogram equalized.

5.3.3 HISTOGRAM EQUALIZATION

The intensity distribution of an image is graphically represented by a histogram. In order to consider the entire range of intensities, equalization distributes the occurrence of pixel intensities equally. By using this method the global contrast of images can be usually increased. This process helps in distributing the intensities on the histogram in a better way as shown in Figure 5.3.

5.3.4 MORPHOLOGICAL OPERATION

Each texture image is implemented with morphological reconstruction by thresholding. The regional maxima computation of the corresponding H-maxima transformation is the extended maxima transformation. Consequently, a binary image is produced by it. In order to evaluate the characteristics and the location of every object, a connected-component labelling operation is performed. The regional maxima of the H-Transform is computed by the extended maxima transform as shown in Figure 5.4.

5.4   IMAGE ENHANCEMENT TECHNIQUES

Image enhancement is considered as one of the most vital pre-processing steps that is used in a number of computer vision applications like medical image processing, computer vision, radar and satellite image processing, remote sensing, underwater image processing, human face detection and recognition, and many more. Numerous

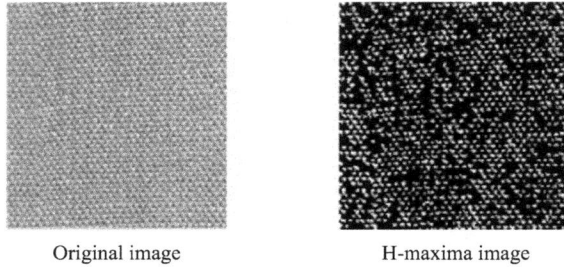

Original image   H-maxima image

FIGURE 5.4   Image after morphological operation.

image enhancement algorithms have been developed. Each and every algorithm is distinguished from one another in respect of processing speed, computational complexity, and quality of image and so on. The primary motive of image enhancement technique is to bring forth the hidden details of an image. Moreover, it also helps in increasing the contrast in a low contrast image. The image enhancement techniques operate by changing the intensity of the pixels of an image and hence provide a better and more transparent image in the form of output. Besides this, these techniques also help in creating an improved input for various image processing systems. A plethora of advanced image enhancement techniques are found in literature. But the most significant and the most commonly used techniques are discussed below:

5.4.1   INTENSITY ENHANCEMENT (IE)

This is a technique (Al-Ameen et al., 2015) that is used in the adjustment of an image by making an image intensity values to a new range. Not only this, it also provides an image having better quality. But it must be kept in mind that it is highly important to identify the upper pixel value as well as the lower pixel value limits before applying the image enhancement technique on an image. Moreover, these are the values limits over which the test image has to be regulated. The objective of the intensity enhancement approach is to magnify the image intensity. This all is done by converting every gray scale value into an updated cumulative distribution function (CDF). Further, this function is attained from the intensity histogram.

5.4.2   HISTOGRAM EQUALIZATION (HE)

This is one of the easiest techniques that can be implemented in the spatial domain. Histogram equalization (Jaya & Gopikakumari, 2013) improves the contrast level of an image by changing its image intensity. Besides, it is also known as a transformation technique that helps in improving the image contrast nature by proper redistribution of image gray level in a consistent way. Histogram equalization can be explained as follows:

$$P\left(r_k\right) = n_k / n \qquad (5.1)$$

where $n_k = 0,1,2....,L-1$ and r_k has been taken as the kth gray level and n is represented as the total number of pixels those are available in an image with gray level r_k.

The contrast of the image is increased successfully by the algorithm when there is an expansion in the dynamic range of the same. This method not only helps in enhancing the contrast of the image greatly but it also puts forward some artifacts, washed out appearance, impulse visual sense in video system and loss of image details (Jaya & Gopikakumari, 2013). Furthermore, this technique does not preserve the mean brightness of the image, which results in the extreme brightening of some local areas.

5.4.3 Adaptive Histogram Equalization (AHE)

This technique was developed (Jaya & Gopikakumari, 2013) in order to overturn the demerits of the histogram equalization method, which has been already discussed in the above paragraph. This technique functions by partitioning the input image into numerous sub blocks or tiles. After that the histogram of the contained pixels is calculated for every given tile. The final step involves the application of histogram equalization method for the center pixel with the help of the CDF of that tile on the basis of the pixel values in the neighborhood. Both the adaptive histogram equalization and ordinary histogram equalization perform the same derivation function of every tile in order to get the transformation function. Nevertheless, it is a local operation and therefore the regions with varied grayscale ranges can be enhanced simultaneously. Hence, it is proved that the cons of the histogram equalization process can be overcome with the help of this technique. Moreover, the filtering process or the interpolation method can be used in order to decrease the artifacts that were introduced due to the division of an image into tiles. Still, the adaptive histogram equalization method too has some cons to its credit. First, it overly magnifies the background noise in the homogeneous regions. Second, the computational costs are rather expensive (Saenpaen et al., 2018) because it is not at all a straightforward and simple task to use a perfect block size as it can increase all parts of the image.

5.4.4 Contrast Limited Adaptive Histogram Equalization (CLAHE)

A new concept of clip limit is used by the contrast limited adaptive histogram equalization (Irmak & Ertas, 2016) in order to deal with the noise problems of AHE. This method helps in putting up limits on the amplification of an image. It is done by clipping the histogram at a predefined value before the computation of the cumulative distribution function (CDF). In addition, the clip limit solely banks on the normalization of histogram and the region size of the neighborhood (Saenpaen et al., 2018). After that, the pixel values that are lying over the clip limit are redistributed among the histogram values situated under the clip limit. The pixel values will be pushed over the clip limit again by this redistribution process. This all results in an effective clip limit that is rather bigger than the one decided in prior. Hence, this process of redistribution goes on repetitively until all the exceeded pixel values reach below the clip limit. Once more, an interpolation function is performed so that the final CLAHE image could be amassed.

The following mathematical expression (Jaya & Gopikakumari, 2013) in terms of uniform distribution can be used in order to express the standard CLAHE with transform degree levels:

$$g = \left[g_{max} - g_{min} \right] * P(f) + g_{min} \tag{5.2}$$

where, g_{max} is considered as maximum pixel value and g_{min} is recognized as the minimum pixel value. *P(f)* is taken as the cumulative probability distribution whereas g is regarded as the computed pixel value.

5.4.5 Brightness Preserving Bi-histogram Equalization (BBHE)

In this technique, first of all the histogram of the input image is divided into two partitions on the basis of its mean (Rivera et al., 2012). The first part contains the range from minimum gray level to mean and the second part involves ranges from mean to maximum gray level. Then, the two histograms are equalized independently. The histogram containing the range from 0 to K-1 is split into two parts having different intensity X_T. Consequently, two histograms are produced by this separation. Where the first histogram contains the range of 0 to X_T the second one contains the range of X_{T+1} to X_{K-1}. It has been made evident both mathematically and experimentally that this technique has no match in order to sustain the original brightness to a greater level.

5.4.6 Dualistic Sub-Image Histogram Equalization (DSHIE)

We have already seen how the BBHE method is used to decompose the original image into two sub-images and to equalize the histograms of the sub-images in a separate way. But, on the other hand, the dualistic sub image histogram equalization (DSHIE) does not divide the image on the basis of mean value, rather it divides the image into two parts on the basis of a CDF of 0.5 (Rivera et al., 2012). This is done in order to enable the output image produce the highest entropy.

5.4.7 Minimum Mean Brightness Error Bi-Histogram Equalization (MMBEBHE)

Minimum mean brightness error bi-histogram equalization method (Rivera et al., 2012) is considered to be better than the BBHE method. The major reason behind its priority over the previous method is that it can assure the perseverance of the brightness far much better than BBHE and DSIHE. Another major difference between the previous methods and the MMBEBHE method is that the latter one proposes a search for a threshold level. This threshold level is further used to decompose the image *I* into two sub-images. The first image is represented by *I[0,1t]* and the second image by *I[1t+1, K-1]*. As a result, it helps in differentiating between the minimum brightness difference of the input image and the output image, which was not considered by the previous methods.

5.4.8 BACKGROUND BRIGHTNESS PRESERVING HISTOGRAM EQUALIZATION (BBPHE)

It has been observed that plain images contain the background levels with much higher density as compared to the other levels. Not only this, in plain images, the total density of background levels can be more than half of the total pixels. That is why, in BBPHE (Rivera et al., 2012), the input image is decomposed into sub-images on the basis of background levels and non-background levels range. Then, the next step involves the equalization of every sub-image in an independent way and then combining into the final output image. Therefore, this process alleviates the problem of over-enhancement by stretching the background levels within the original range. Moreover, background brightness preserving histogram equalization technique is capable of expanding the comparatively low density gray levels of other sub-images into a wide range. This technique achieves this function by normalization.

5.5 PERFORMANCE METRICS FOR IMAGE ENHANCEMENT TECHNIQUES

There are various performance metrics available for image enhancement techniques as discussed below:

5.5.1 MEAN SQUARED ERROR (MSE)

$$MSE = \frac{1}{MN} \sum_{j=1}^{M} \sum_{k=1}^{N} \left(I_{j,k} - O_{j,k} \right)^2 \qquad (5.3)$$

In the above given expression, M and N represents the rows and the columns of an image respectively. Here $I_{j,k}$ stands for the original input image, and $O_{j,k}$ represents the corresponding output image. In order to reduce the mean squared error (Yadav et al., 2016), there should be a close pixel intensity between the input and output images.

5.5.2 PEAK SIGNAL TO NOISE RATIO (PSNR)

$$PSNR = 10 \log_{10} \frac{255^2}{MSE} \qquad (5.4)$$

There is a condition that the peak signal to noise ratio (Kober, 2006) should be as abundant as possible because it will ensure that the content of the signal in the output is great whereas the noise is low. Moreover, the voice value of the signal is taken as maximum, which is 255 because it is peak signal to noise ratio and the gray scale images range from 0 to 255.

5.5.3 Absolute Mean Brightness Error (AMBE)

The brightness reservation is fixed by absolute mean brightness error method (Jaya & Gopikakumari, 2013). This system is based on a dissimilar brightness level between the original and enhanced image. For the improved reservation of image brightness, the value of AMBE has to be lowered. The following expression will demonstrate it clearly.

$$AMBE = \left| E(I) - E(O) \right| \tag{5.5}$$

In this equation, $E(I)$ represents the input image average intensity and $E(O)$ stands for the enhanced image average intensity.

5.5.4 Contrast

Contrast stands for the difference in brightness and the color of an image that make an object perceptible.

$$CONTRAST = \left(I_{max} - I_{min} \right) / (I_{max} + I_{min}) \tag{5.6}$$

In this expression I_{max} and I_{min} denotes the highest and lowest luminance.

5.5.5 Entropy

$$ENTROPY = -\Sigma\, h(O) \times log_2 h(O)$$

In this expression, $h(O)$ stands for the normalized histogram of output image. For the measurement of the content of the image, entropy (Jintasuttisak & Intajag, 2014) has been used. Moreover, higher lues show those images which contain affluent details.

5.5.6 Normalized Absolute Error (NAE)

$$NAE = \frac{\sum_{j=1}^{M} \sum_{k=1}^{N} \left| I_{j,k} - O_{j,k} \right|}{\sum_{j=1}^{M} \sum_{k=1}^{N} \left| I_{j,k} \right|} \tag{5.8}$$

where, $I_{j,k}$ represents the original input image and $O_{j,k}$ stands for the corresponding output image. Therefore, the normalized absolute error (Kober, 2006) should be minimal. The difference between the input image and output images is also provided by the NAE.

5.5.7   Measure of Enhancement (EME)

The measure of enhancement method is primarily known as an image enhancement method that is employed to conduct a test between original and the enhanced image. For instance, take an input image $I(n,m)$. Then partition this image into two different blocks called k_1 and k_2 of window size $w_{k,l}(i,j)$. The measurement of enhancement can be shown as follows by using all of the parameters mentioned above.

$$EME = \frac{1}{k_1 k_2} \sum_{l=1}^{k_1} \sum_{k=1}^{k_2} 20 \mathrm{In} \left(\frac{I^w_{max;k,l}}{I^w_{min;k,l}} \right) \tag{5.9}$$

In this equation, $I^w_{max;k,l}$ stands for the maximum value of an image $I(n,m)$ within $w_{k,l}$ block and $I^w_{min;k,l}$ represents the minimum value of an image $I(n,m)$ within $w_{k,l}$ block.

5.5.8   Measure of Enhancement by Entropy (EMEE)

Measure of enhancement by entropy is a method that operates with the help of entropy in order to enhance the image. For example, take an image $I(n,m)$ and then divide this image into two different blocks such as $k1$ and $k2$ of window size $w_{k,l}(i,j)$, α is a constant scalar value. The following expression will illustrate EMEE clearly:

$$EMEE = \frac{1}{k_1 k_2} \sum_{l=1}^{k_2} \sum_{k=1}^{k_1} \alpha \left(\frac{I^w_{max;k,l}}{I^w_{min;k,l}} \right)^{\alpha} \mathrm{In} \left(\frac{I^w_{max;k,l}}{I^w_{min;k,l}} \right) \tag{5.10}$$

Therefore, the EMEE is the entropy of the contrast ratio for each block $wk,,l$ scaled by α, averaged over the entire image.

5.5.9   Contrast Improvement Index (CII)

The contrast improvement index (Jaya & Gopikakumari, 2013) has proved to be a ground-breaking function that assists in comparing the performance of enhancement techniques. It operates by measuring the ratio of local contrast of enhanced image and the input image by following the below expression:

$$CII = A_O / A_I \tag{5.11}$$

In this expression, A_O stands for the average local contrast level of an enhanced image with a 3 × 3 window size. On the other hand, A_I stands for the average local contrast level of an input image with 3 × 3 window size. The more the value of CII, the more improvement can be shown over image contrast nature.

5.5.10 TENEGRAD MEASUREMENT (TM)

The Tenegrad measurement (Jaya & Gopikakumari, 2013) is often represented as the maximization of the gradient measure. For instance, take an input image I. Here, the gradient value $I(x,y)$ is used for the estimation of the Tenegrad value and the Sobel filter derivative is implemented for the estimation of partial derivatives with convolution kernels I_x and I_y. Hence, the gradient magnitude of TM can be expressed as follows:

$$T(x,y) = \sqrt[2]{I_x \times I(x,y)^2 + I_y \times I(x,y)^2} \qquad (5.12)$$

Hence, Tenegrad measures can be estimated with the help of the following expression:

$$TM = \sum_x \sum_y T(x,y)^2 \qquad (5.13)$$

When the value of TM will be higher, the image quality will also be higher. Thus, this function will ensure that the image structural information is preserved.

5.6 IMAGE SEGMENTATION TECHNIQUES

Image segmentation is the process by which a digital image is partitioned into several segments in computer vision. The fundamental motive behind this segmentation is to abbreviate and change the composition of an image in a new way so that it becomes convenient to analyze only the significant and relevant part of that image. Moreover, image segmentation is one of the vital operations to be performed in several applications in the vision area such as object detection and classification. It splits an image into foreground and background as well as into regions or categories. Consequently, these divisions reciprocate either to various objects or parts of objects. There are numerous image segmentation techniques present in literature as discussed below:

5.6.1 THRESHOLDING

This technique (Abd Elaziz et al., 2019) is considered to be the most simplest and convenient technique for image segmentation. By using this technique, the image pixels are partitioned according to their intensity level. These techniques are implemented only over the images that have lighter objects than background. One can select these techniques either manually or automatically. But it must be kept in mind that this selection will be solely based on prior knowledge of image features. There are basically three types of thresholding:

5.6.1.1 Global Thresholding

Any applicable threshold value/T is valid for its operation. However, this proposed value T will remain consistent for the entire image. The peaks of images histograms are necessary in order to calculate threshold values. The output image $O(x,y)$ based on the T value can be achieved from the original image $I(x,y)$ in the following form:

$$O(x,y) = \begin{cases} 1, if\ I(x,y) > T \\ 0, if\ I(x,y) \leq T \end{cases}$$ (5.14)

5.6.1.2 Variable Thresholding

This type of thresholding enable the value of T to change over the image. This type of thresholding is further divided into two subtypes:

- **Local threshold**: Here, the T value relies upon the neighborhood of x and y.
- **Adaptive threshold**: In this type, the value of T is a function of x and y.

5.6.1.3 Multiple Thresholding

This kind of thresholding provides multiple threshold values like $T0$ and $T1$ which can be used to calculate the output image in the following way:

$$O(x,y) = \begin{cases} m, if\ I(x,y) > T1 \\ n, if\ I(x,y) \leq T1 \\ o, if\ I(x,y) \leq T0 \end{cases}$$ (5.15)

Otsu's, Eridas and Quadratic Integral Ratio (QIR) [29] are some of the algorithms that can be utilized to perform segmentation using thresholding techniques.

5.6.2 Edge Detection Based Techniques

Edge detection based techniques are one of the most refined techniques for the processing of images. These techniques bring a tremendous change in the intensity value of an image. This is because a single intensity value is not capable of supplying beneficial information about edges. Moreover, edge detection techniques are designed so that they can find the edges in both the cases; where the first derivative of intensity is greater than a particular threshold or where the second derivative has zero crossings (Abd Elaziz et al., 2019). In these methods, the first step is to locate all the edges and then the second step is to connect them together so that the object boundaries could be formed in order to divide the necessary regions. The Sobel operator, canny operator, Prewitt and Robert's operator are some of the most fundamental edge detection techniques. The primary objective of these techniques is to produce a binary image at the end. All of these techniques are structural techniques that rely upon discontinuity detection (Kang et al., 2009).

Furthermore, in the beginning, the edge detection technique pre-processes the images into binary form and then filters those images with the help of the Gaussian 5×5 filter so that the noise can be removed from the images. Then the next step involves the estimation of gradient directions and magnitude of the filtered image according to the definition of the first order derivative, which can be formulated in the following way:

If G_x and G_y denote the gradient directions which correspond to x and y axis, then gradient direction continuous with respect to x-axis orientation can be described as follow:

$$G_x = I(x+1) - I(x) \tag{5.16}$$

where I represents an input image, $I(x+1)$ and $I(x)$ indicate the next pixel and current pixel positions. In the same way, the gradient direction continuous according to y-axis orientation can be described as follow:

$$G_y = I(y+1) - I(y) \tag{5.17}$$

And the gradient magnitude M is given by:

$$M = \sqrt{G_x^2 + G_y^2} \tag{5.18}$$

All the non-maximum pixel amplitudes are suppressed according to the gradient outcomes, while retaining the local maxima pixels in the image.

5.6.3 Region Based Techniques

These are the segmentation techniques that operate by splitting the image into various regions that have identical features. This technique is further divided into two sub-types.

5.6.3.1 Region Growing Methods

These methods divide the images into numerous regions according to the growth of the cells, which are also known as pixels. One can choose these seeds either manually, with the help of previous knowledge, or automatically, with the help of a particular application. Moreover, the connectivity between the pixels can dominate the growth of seeds. With the help of prior knowledge of a problem, this can be stopped.

5.6.3.2 Region Splitting and Merging Methods

In these techniques two most fundamental elements like splitting and merging are used in order to segment an image into numerous regions. Splitting is responsible for continuously dividing an image into regions consisting of identical features, whereas merging stands for the combination of the neighborhood similar regions.

5.6.4 Clustering Based Techniques

Through this technique, the images are divided into the clusters containing pixels of same peculiarities. With the help of clustering, the population or data points are partitioned into numerous groups. This is done because the data points in the same group are more identical as compared to the other data points of other groups. That is why these groups are called clusters. *K-means* (Bali & Singh, 2015) is one of the most frequently opted clustering algorithm. This technique involves the calculation of every center and then assignment of every pixel to its nearest center. The main goal behind this function is to increase the intra-cluster similarity and decrease the inter-cluster equality.

5.6.5 WATERSHED BASED TECHNIQUES

The notion of topological interpretation is utilized by the watershed based technique. Here, the intensity presents the basins that have a hole in their minima, which results in water spilling. The nearby basins are merged together when water approaches the border of the basin. Therefore, the borders of the region of segmentation are indispensable in order to maintain the distance between basin dams. Hence, these dams are built with the help of dilation or enlargement. The gradient of image is taken as a topographic surface by the watershed techniques. And the pixels, which have greater gradients, are known as boundaries and these boundaries are continuous.

5.6.6 PARTIAL DIFFERENTIAL EQUATION BASED TECHNIQUES

Partial differential equation (PDE) based techniques are marked as the quickest methods of image segmentation. These methods are best suited in the applications that are time critical. These methods involved mainly two PDE methods. The first is known as a non-linear isotropic diffusion filter, which is used to increase the edges, and the second one is convex non-quadratic variation restoration, which is utilized in order to avoid noise in the image. The resulting edges and boundaries in the PDE method are obscured and close operators are used to shift them. In addition, the fourth order PDE method is implemented so that the noise from the images could be lessened whereas the second order PDE method is utilized for the better identification of the edges and boundaries of the image.

5.6.7 ARTIFICIAL NEURAL NETWORK BASED TECHNIQUES

The learning strategies of human brain are replicated with the help of artificial neural network-based segmentation techniques. Consequently, it helps in decision-making. In the recent times, this technique is frequently opted for the segmentation of medical images. The main purpose of this method is to detach the necessary image from the background. A neural network is created of a huge number of connected nodes and every connection has a peculiar weight. This technique contains two fundamental steps. The first is the extraction of features and the second is segmentation by neural network. Moreover, in this method the network "learns" by changing the interconnections, weights between the layers and makes the related output simple for a set of input data (Min et al., 2015).

5.6.8 Fuzzy Logic Based Techniques

Fuzzy logic based techniques (Bali & Singh, 2015) are the segmentation methods based on the fuzzy membership function's masking methods, which are based on some rules and it ensures better results (Kang et al., 2009). Here, every pixel is parted into fuzzy sets. It is also possible that every pixel can be partially a part of many sets and regions of image (Min et al., 2015). This technique is more effective and convenient as compared to the other techniques (Kang et al., 2009).

5.6.9 Genetic Algorithm Based Techniques

The genetic algorithm based technique (Bali & Singh, 2015) was created from the evolution theory. This theory consists of three main functions. The first function is selection and the second and third are crossover and mutation respectively. These methods are best used in pattern recognition applications.

5.6.10 Saliency Based Technique

One of the fastest progresses in the computer vision applications is the saliency based technique (Shokoufandeh et al., 1999). This technique assists in the detection of the conspicuous object in the image which plays a relatively very important role in these days. Moreover, while finding a resolution for segmentation problem, saliency object detection techniques separate the prominent foreground objects from the background. It does not split an image into the regions of similar properties, which is generally done in the above mentioned segmentation techniques. Hence, it is the new method for image segmentation.

5.7 EXPERIMENTAL RESULTS

Let's take an example of an image captured by high-definition camera and apply noise removal techniques for face detection. The results are shown in Figures 5.5(a) to 5.5(e).

Algorithm for Face Detection Using Image Processing Techniques

```
a = imread('face.jpg');
imshow(a);
I = rgb2gray(a);
J = imnoise(I,'gaussian',0,0.025);
imshow(J);
title('Portion of the Image with Added Gaussian Noise');
K = wiener2(J,[5 5]);
figure
```

```
imshow(K);
title('Portion of the Image with Noise Removed by Wiener Filter');
det = vision.CascadeObjectDetector;
det.MergeThreshold=10;
bb = step(det,a);
outp = insertObjectAnnotation(a,'rectangle',bb,'detectaction');
imshow(out);
```

As we can see from Figures 5.5(a) to 5.5(e), first of all, we converted the original RGB image into a gray scale image, which is full of noise. Next we applied the noise removal technique using the Weiner filter to remove noise from the gray scale image. Finally, we applied the face detector algorithm that detects faces in the image and converted back the detected faces image into an RGB image.

FIGURE 5.5(a)   Original image in RGB.

FIGURE 5.5(b)   Grey scale image with noise.

FIGURE 5.5(c)   Grey scale image after noise removal.

FIGURE 5.5(d)   Face detection in gray scale.

FIGURE 5.5(e)   Face detection in RGB image.

5.8 CONCLUSION

In this chapter, various image pre-processing methods, image enhancement, performance metrics for image enhancement and image segmentation techniques are discussed. Further, an algorithm is designed for face detection in MATLAB 2018a. It can be concluded that image pre-processing is an important task before feeding images to any intelligent system. Pre-processing mainly involves enhancement techniques, which are used for removing noise that may occur due to blurriness, night view, dim light, occlusion, shadows, etc. Therefore, when we design a system for surveillance or detection of objects in a video or any other image recognition task, the first step is image pre-processing so that high performance of the system can be achieved.

REFERENCES

Abd Elaziz, M., Bhattacharyya, S., & Lu, S. (2019). Swarm selection method for multilevel thresholding image segmentation. *Expert Systems with Applications*, *138*, 112818.

Agrawal, A. K., & Sivagami, M, P. S. (2015). Object recognition in Hadoop using HIPI. *International Journal of Scientific & Technology Research*, *4*(7), 115–118.

Al-Ameen, Z., Sulong, G., Rehman, A., Al-Dhelaan, A., Saba, T., & Al-Rodhaan, M. (2015). An innovative technique for contrast enhancement of computed tomography images using normalized gamma-corrected contrast-limited adaptive histogram equalization. *EURASIP Journal on Advances in Signal Processing*, *2015*(1), 1–12.

Anitha, S., & Radha, V. (2010). Comparison of image preprocessing techniques for textile texture images. *International Journal of Engineering Science and Technology*, *2*(12), 7619–7625.

Aradhye, A. A., Inamdar, H. S., Patil, G. S., & Jagdale, Y. B. (2016). Object detection avenue for video surveillance using Hadoop. International Research Journal of Engineering and Technology (IRJET), 3(5), 497–499.

Bali, A., & Singh, S. N. (2015). A review on the strategies and techniques of image segmentation. *2015 Fifth International Conference on Advanced Computing & Communication Technologies*, 113–120.

Dhinakaran, K., Silviya Nancy, J., & Duraimurugan, N. (2015). Video analytics using HDVFS in cloud environment. *ARPN J Eng Appl Sci*, *10*, 13.

Ganesh, R. B., & Appavu, S. (2015). An intelligent video surveillance framework with big data management for Indian road traffic system. *International Journal of Computer Applications*, *123*(10), 12–19.

Ghaware, S., Harke, V., Jadhav, S., & Khanuja, H. (2016). *Approaches for Video Surveillance Using Hadoop. 4863*(February), 47–50.

Irmak, E., & Ertas, A. H. (2016). A review of robust image enhancement algorithms and their applications. *2016 IEEE Smart Energy Grid Engineering (SEGE)*, 371–375.

Jamkhandikar, D., Mytri, V. D., & Shahapure, P. (2014). Object detection and tracking in real time video based on color. *International Journal of Engineering Research and Development*, *10*, 33–37.

Jaya, V. L., & Gopikakumari, R. (2013). IEM: a new image enhancement metric for contrast and sharpness measurements. *International Journal of Computer Applications*, *79*(9), 1–9.

Jintasuttisak, T., & Intajag, S. (2014). Color retinal image enhancement by Rayleigh contrast-limited adaptive histogram equalization. *2014 14th International Conference on Control, Automation and Systems (ICCAS 2014)*, 692–697.

Kang, W.-X., Yang, Q.-Q., & Liang, R.-P. (2009). The comparative research on image segmentation algorithms. *2009 First International Workshop on Education Technology and Computer Science*, *2*, 703–707.

Kober, V. (2006). Robust and efficient algorithm of image enhancement. *IEEE Transactions on Consumer Electronics*, *52*(2), 655–659.

Kumar, A. K. P. B. K., Mahankud, A., & Mishra, P. K. (2016). Detection of objects in a video of traffic. *International Research Journal of Engineering and Technology*, 3(5), 676–678.

Kumar, N., Pathan, E., Yadav, L., & Ransubhe, V. (2015). Video surveillance over camera network using Hadoop. *International Journal of Innovative Research in Computer and Communication Engineering*, 4(1), 1111–1115.

Li, J., Wu, Q., Lian, X., & Sun, J. (2016). Real-time video copy detection based on Hadoop. *2016 Sixth International Conference on Information Science and Technology (ICIST)*, 492–497.

Min, H., Jia, W., Wang, X.-F., Zhao, Y., Hu, R.-X., Luo, Y.-T., Xue, F., & Lu, J.-T. (2015). An intensity-texture model based level set method for image segmentation. *Pattern Recognition*, *48*(4), 1547–1562.

Natarajan, V. A., Jothilakshmi, S., & Gudivada, V. N. (2015). Scalable traffic video analytics using Hadoop MapReduce. *ALLDATA 2015* (The First International Conference on Big Data, Small Data, Linked Data and Open Data, Copyright (c) IARIA, 2015. ISBN: 978-1-61208-445-9), 18.

Parsola, J., Gangodkar, D., & Mittal, A. (2017). Efficient storage and processing of video data for moving object detection using Hadoop/MapReduce. *Proceedings of the International Conference on Signal, Networks, Computing, and Systems*, 137–147.

Ravi, P., & Ashokkumar, A. (2017). Analysis of various image processing techniques. *Int J Adv Netw Appl*, *8*(5), 86–89.

Rivera, A. R., Ryu, B., & Chae, O. (2012). Content-aware dark image enhancement through channel division. *IEEE Transactions on Image Processing*, *21*(9), 3967–3980.

Saenpaen, J., Arwatchananukul, S., & Aunsri, N. (2018). A comparison of image enhancement methods for lumbar spine X-ray image. *2018 15th International Conference on Electrical Engineering/Electronics, Computer, Telecommunications and Information Technology (ECTI-CON)*, 798–801.

Saxena, S., Sharma, S., & Sharma, N. (2016). Parallel image processing techniques, benefits and limitations. *Research Journal of Applied Sciences, Engineering and Technology*, *12*(2), 223–238.

Shah, V. (2015). *Big video data analytics using Hadoop*. Jaunpur: IJARCSSE.

Shokoufandeh, A., Marsic, I., & Dickinson, S. J. (1999). View-based object recognition using saliency maps. *Image and Vision Computing*, *17*(5–6), 445–460.

Yadav, G., Maheshwari, S., & Agarwal, A. (2016). Multi-domain image enhancement of foggy images using contrast limited adaptive histogram equalization method. *Proceedings of the International Conference on Recent Cognizance in Wireless Communication & Image Processing*, 31–38.

6 Forecasting Time Series Data Using Artificial Neural Network
A Review

Sarbjit Singh, Kulwinder Singh Parmar,
Harpreet Kaur, and Jatinder Kaur

CONTENTS

6.1 INTRODUCTION

Forecasting is a highly challenging task. Various statistical models and soft computing techniques have been proposed by researchers and professional analysts to obtain more accurate forecasts. Among the various forecasting models, artificial neural networks (ANNs) are the most accurate, fast and reliable techniques. ANN finds its application in almost all the fields whether it is social, economic, engineering, atmospheric sciences or field of applied mathematics. The most salient features of ANN are that it is able to handle non-linearity, seasonality, noise and

DOI: 10.1201/9781003153405-6

robustness of data (Khashei and Bijari, 2010). The motive behind ANN based computation is to design such mathematical algorithms that impersonate works in parallel to information processing as done by the human brain (Basheer and Hajmeer, 2000). The most significant feature of ANN is that they have the capacity of approximating any continuous function with high precision (Zhang and Min Qi, 2005; Chen et al. 2003). ANN has the unique ability of generalizing. The unseen part of a sample can be correctly inferred with ANN. No prior assumptions about the model are required while using ANN unlike linear/regression models. This data driven approach makes ANN a magnificent tool in forecasting real-life complex problems, thus making it an appropriate tool for forecasting stock markets, environment pollution, river flow forecasting, weather, agriculture and many more areas of science and engineering where forecasting plays a pivotal role. In forecasting the multi-layer feed forward method is the widely used paradigm of ANN. The present chapter is an attempt to provide researchers in the field of forecasting a detailed study about the development of different types of neural networks and the methodology of modelling and forecasting different types of time series. This chapter also deals with the review of supervised learning techniques using ANNS to solve forecasting problems related to environment, stock market, hydrology and agriculture.

6.2 HISTORICAL DEVELOPMENTS IN THE FIELD OF ANN

The McCulloch-Pitts neuron model of computing element was designed by Warren McCulloch and Walter Pitts in 1943. In this model, each element is assigned the weighted sum of the inputs and then the threshold logic operation is applied (McCulloch and Pitts 1943). Several analytical computations were performed using combinations of these computing elements (Figure 6.1).The major shortcoming of this model was that the weights were fixed, which makes the model incapable of performing learning operations.

A learning scheme for adjustable connection weights was proposed by Donald Hebbin 1949, which became a vital learning rule in the field of neural networks (Hebb, 1949). An adaptive learning scheme was created by Marvin Minsky in 1954, which could adjust connection strengths automatically (Minsky and Papert, 1969). The famous perceptron model was proposed by Rosenblatt in 1958. In it weights

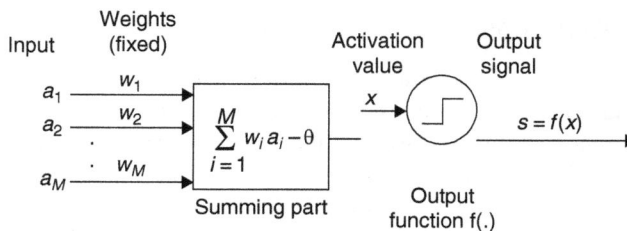

FIGURE 6.1 McCulloch-Pitts neuron model.

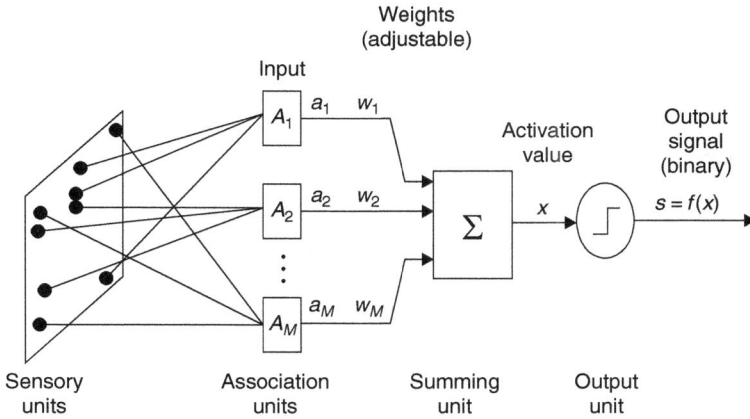

FIGURE 6.2 Rosenblatt's perceptron model of a neuron.

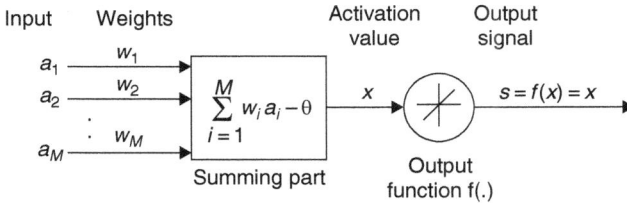

FIGURE 6.3 Widrow's Adaline model of a neuron.

were adjusted by using the perceptron learning law (Rosenblatt, 1958). The law is best suited for pattern classification problems using the multilayer perceptron (Figure 6.2).

The major disadvantage of the perceptron model was the inability to adjust the weights in reorganizing the classification task. This was overcome by Minsky and Papert in 1969 (Minsky and Papert, 1969). In 1960, the Adaline model for the Ae: adaptive element was proposed by Widrow and Hoff to adjust the weights (Widrow and Hoff, 1960) (Figure 6.3). Due to convergence of the algorithm, it was successfully applied in adaptive signal processing problems.

In 1964, Hu in his thesis used Widrow's adaptive network to forecast weather. Due to a lack of training algorithms the research was limited until 1986 when the back-propagation algorithm was introduced by Chauvin and Rumelhart (1995).

During the past 30 years, several remarkable contributions such as concepts of competitive learning, self-organization and simulated annealing influenced this field of artificial neural networks, which led to the development of new learning algorithms and problems in pattern recognition. At present, the methodologies of fuzzy logic and the neural networks are coupled to develop an adaptive framework to deal efficiently with real world problems.

6.3 METHODOLOGY

Artificial neural networks (ANNs) can be considered to contain a more simple structure of interconnected processing units than a biological neural network. ANNs comprise several interconnected processing units to undertake the stipulated task. ANN consists of three important layers, i.e. input, hidden and output layers. Input and output layers have nodes respectively corresponding to input and output variables (Figure 6.4).

Data is transferred between layers across weighted connections (Kaul et al., 2005). The input is given to the summing part of processing unit, which calculates the weighted sum to generate the output from this sum. Discreet or continuous input values can be given to processing unit to generate the output as well. This input to processing unit is either given independently or from outputs of other handling units. The power of connecting units is reflected in the weight value. Each unit in an ANN has a unique activation value at a given time, which is responsible for the activation state of the network. The output function unit generates overall output, which is determined by the activation value (Figure 6.5). Different types of activation

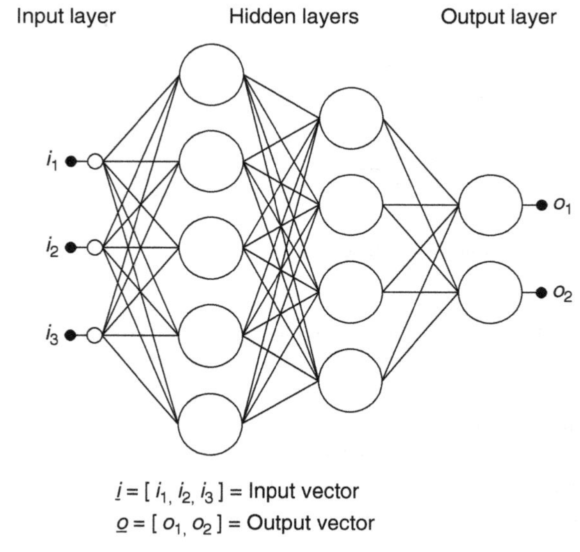

$\underline{i} = [\, i_1, i_2, i_3 \,] =$ Input vector

$\underline{o} = [\, o_1, o_2 \,] =$ Output vector

FIGURE 6.4   Basic architecture of an artificial neural network.

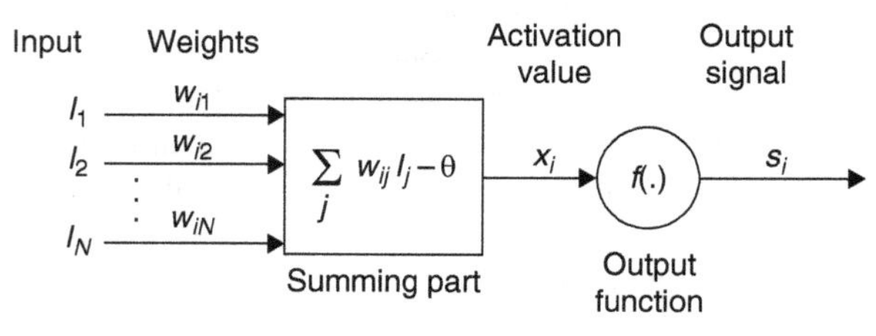

FIGURE 6.5   Processing unit of artificial neural network.

functions such as linear, step, ramp, sigmoid, hyperbolic tangent, and Gaussian functions can be used to generate activation values (Figure 6.6) and to determine the output path of network.

6.4 CLASSIFICATION OF ANN

Depending on the structure of the connections among the layers ANNs are classified as follows:

6.4.1 FEED-FORWARD NEURAL NETWORKS

A feed-forward neural network is the most basic form of an artificial neural network in which the flow of information is unidirectional and never backwards. The connections between nodes in a feed-forward ANN does not form a cycle (Figure 6.7).

6.4.2 RECURRENT NEURAL NETWORKS

In recurrent neural networks, the same operations are performed over and over the set of sequential input. The output generated from some previous stages serves as the input for the next stages of operation. The hidden layers do the important job of recollecting some information of input data that is being fed again and again. Due to this feature, recurrent neural networks have memory which increases its performance (Figure 6.8).

6.4.3 ELMAN BACK PROPAGATION NEURAL NETWORKS

An Elman neural network is a kind of feedback neural network, the hidden layers of which have been inter-connected output layer as well as to another layer, called the context layer with weights equal to one. The context layer acts as a delay operator to retain the memory so that the network has capacity to adapt the time-varying active features, which will add more stability (Figure 6.9).

6.4.4 TIME-DELAY FEED-FORWARD NEURAL NETWORK

The time-delay feed-forward neural network uses the time-delay feature to the network structure for building short-term memory to make the neural network dynamic. The time-delay feature is employed to the input layer of the neural network (Figure 6.10).

6.5 STRUCTURAL DESIGN OF ANN

The structural layout of ANN primarily consists of defining the required number of input, output and hidden layers along with the interaction scheme between the neurons (Gardner and Dorling 1998). The number of neurons in the input and output layer varies from problem to problem. But there is no specific criterion that would

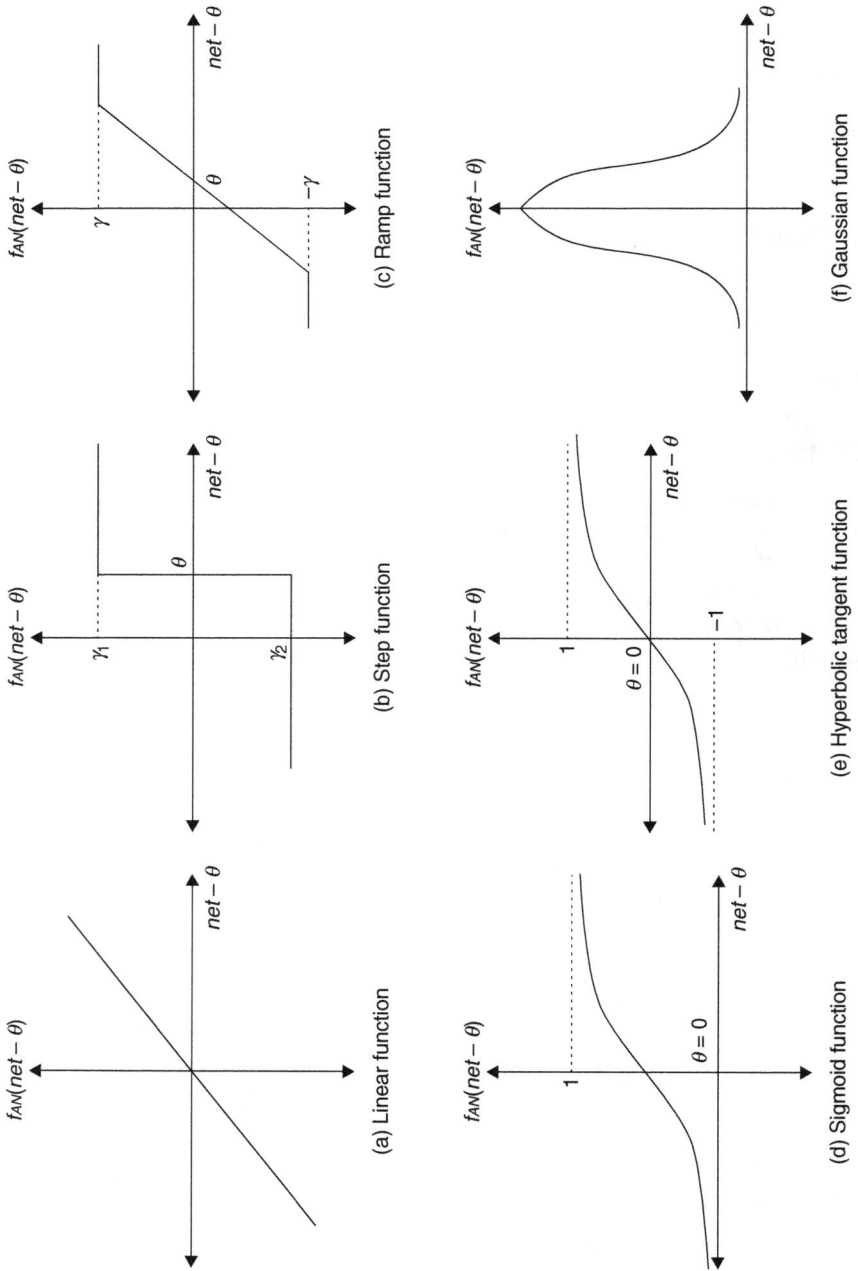

FIGURE 6.6 Different types of activation functions (a) linear function (b) step Function (c) Ramp function (d) sigmoid function (e) hyperbolic tangent function (f) Gaussian function.

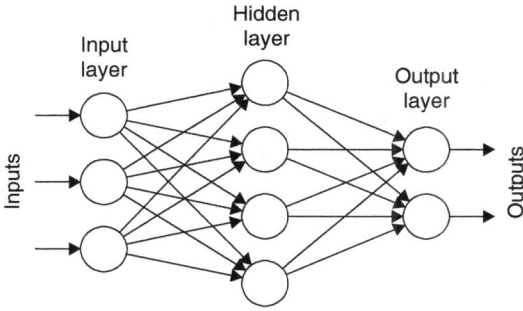

FIGURE 6.7   Architecture of feed-forward artificial neural network.

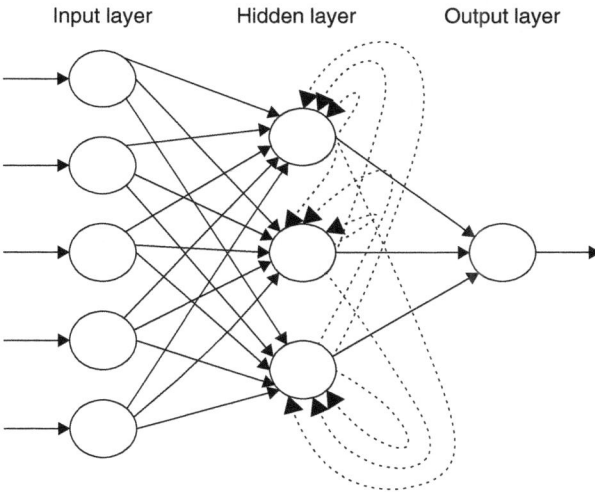

FIGURE 6.8   Architecture of recurrent artificial neural network.

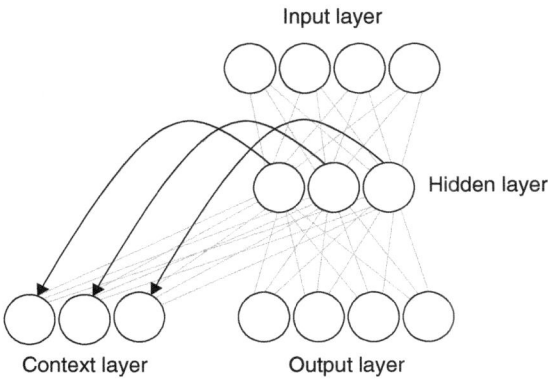

FIGURE 6.9   Architecture of Elman back propagation neural networks.

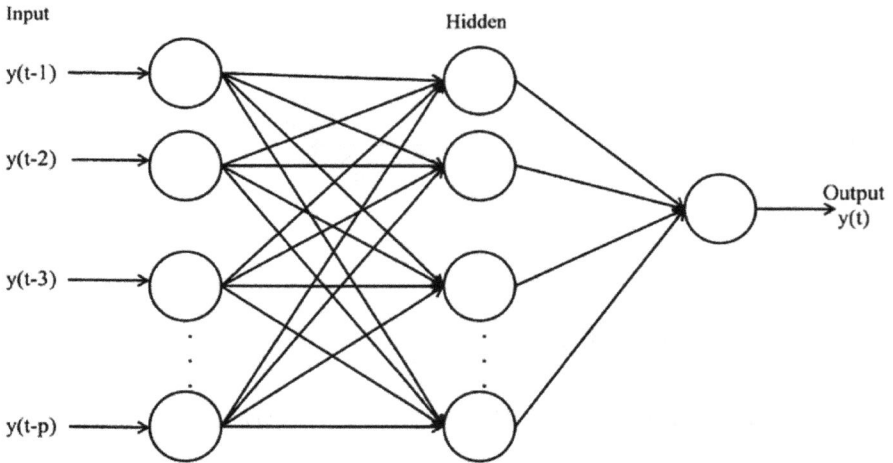

FIGURE 6.10 Architecture of time-delay feed-forward neural network.

determine how many neurons are required in each hidden layer. A lower number of neurons in the hidden layer results in a high training error (under fitting) while a larger count of neurons gave rise to a high validation error (over fitting). Both errors are not admissible. Based on the inputs the model predicts the output.

The processing units of each layer are joined by acyclic links. The relationship between the output (y_t) and the inputs ($y_{t-1}, \ldots . y_{t-p}$) is given by

$$y_t = w_0 + \sum_{j=1}^{q} w_j \, g \left(w_{0,j} + \sum_{j=1}^{p} w_{i,j} \, y_{t-i} \right) + \epsilon_t$$

Where $w_{i,j} \left(i = 0,1,2,\ldots p, j = 1,2,\ldots .q \right)$ and $w_j \left(j = 0,1,2,\ldots q \right)$ are the connection weights; p and q are number of input and hidden nodes respectively.

6.6   REVIEW OF APPLICATIONS, RESULTS AND DISCUSSIONS

Zhang (1998) found that ANN can handle linearity as well as non-linearity with equal desired accuracy. In continuation of this, Gorr in 1994 added that ANN can deal with seasonality as well as non-linearity, both simultaneously. When 88 seasonal time series from the famous M-competition were analysed by Sharda and Patil (1992) they came out with similar conclusions that prior de-seasoning the data is not required while using ANN. However, the data forecast on de-seasonalized is notably better than with the seasonal data (Nelson et al.1999). The changing seasonal patterns were identified by Franses and Draisma (1997). The ability of a neural network of self-organization and its flexible nature make it a suitable tool to be applied in stock markets where fluctuations are large and frequent (Bing et al. 2012). ANN thus has wide range of applications but here we are discussing broadly four areas of high utility of ANN.

6.6.1 RIVER FLOW FORECASTING

Since the conception of application of ANN in daily and hourly forecast of river flow by Kang et al. (1993) myriad studies (Abrahart and See 1998, Jain et al., 1999, Zealand et al., 1999) have used ANN in predicting different river flow parameters. Among the various models used the multi-layer perceptron model upgraded with back propagation (BP) algorithm is the most popular and widely used technique. The radial basis function also finds its large application in river flow forecasting (Dawson et al. (2002); Dibike and Solomatine (2001); Fernando and Jayawardena(1998)). For better and accurate forecasting on the non-stationary data ANN has been clubbed with wavelets and various hybrid models have been developed. Hybrid models of wavelet and artificial neural networks (WA–ANN) models were developed and applied to forecast one-day advanced stream flow forecasting in France and the US (Anctil and Tape 2004), monthly forecasting in Italy, Turkey (Kisi 2009), for daily forecasting of river flow (Kisi 2009). All these studies concluded that WA–ANN models perform better than ANN models for river flow forecasting.

6.6.2 ENVIRONMENTAL POLLUTION FORECASTING

To protect people from the harmful effects of pollution, it is necessary for health agencies to have some reliable advance predictions ahead regarding emission of pollutants and quality of air around them. The concentration of air pollution and its statistical relationship with climate is the basis for environmental pollution forecasting. Long multivariate linear regression models are being used for this purpose (Comrie 1997). However, the pollution-weather relationship is non-linear and complex. Thus, ANN models provide a better alternative to statistical models because of their computational efficiency and generalization ability. ANN is able to handle the non-linear complex pollution-weather relationship very aptly. Even a high dimensional data can be handled with equal precision. But the heterogeneity of data and wide range of influencing parameters makes environmental models hard to be ever fully investigated (Jakeman et al. 2006). In Santiago a 24-hour advance prediction of concentration of $PM_{2.5}$particles was achieved by Perez and Trier (2001). A three-layer feed forward neural network having 24 inputs and one output was used in this regard. Back in 1998 an extensive and comprehensive review on the application of artificial neural networks in the atmospheric sciences was conducted by Gardner and Dorling. The MLP structure is considered to be most suitable particularly in the study of atmospheric sciences. This technique was employed by Moseholm et al. (1996) while estimating the concentration levels of CO in the University District of North Seattle (a sheltered urban intersection). A comparison of the multivariate regression model and a three-layered neural network was done in Delhi to forecast (SO_2) concentrations by Chelani et al. (2002) to conclude that the latter gave more accurate and reliable results than the former.

6.6.3 STOCK MARKET FORECASTING

Stock market forecasting is a challenging task as it involves a lot of risk. Different methods have been used by various researchers to model and forecast stock market data. But due to the non-linear and complex nature of stock market data, artificial neural networks are found to be the most accurate and exact soft computing techniques. A review of application ANNs in various stock markets has been conducted by Li and Ma (2010) and it involves the application of ANNs to forecast future values of various stock markets and financial indices. The financial time series data was forecasted by Wang and Wang (2015) with the help of a coupled approach using the stochastic time effective function neural network and principal component analysis. The forecasts generated by the proposed hybrid approach are more accurate compared to the usual backpropagation neural network (BPNN) and PCA-BPNN. Niaki and Hoseinzade (2013) applied an artificial neural network (ANN) methodology to predict Standard & Poor's 500 (S&P 500) indices on a daily basis. They employed the design of experiments to select the most powerful features among 27 financial and economic variables, which were predicted using ANN. The performance of proposed technique was found to be better than the traditional buy-and-hold policy. Zahedi and Rounaghi (2015) employed the ANN model and principal component analysis (PCA) method to predict Tehran Stock Exchange prices and extracted 20 predominant features of 20 accounting variables. The accurate predictions came up in the form of new patterns of all the variables. Wang and Wang (2016) developed a novel technique of forecasting crude oil price fluctuations by combining MLP and ERNN with stochastic time effective function. The development of ERNN with distinguished feature of time-varying predictive system has the capacity to keep the record of recent events in order to forecast future values. The main objective of stochastic time effective function is to represent the robustness of recent information than the old information for the investors.

6.6.4 AGRICULTURE RELATED FORECASTING

Timely management of crops is of utmost priority in the field of agronomy and hence simulation can play key role in this field. Environmental factors such as temperature, rainfall, water stress, etc. have a strong influence on the growth and yield of many agricultural plants such as paddy, corn, soybean and wheat etc. (Khairunniza-Bejo et al. 2014). ANN can be effective tool in crop development modelling (Elizondo et al., 1994), soil–water retention estimations (Schaap and Bouten, 1996), pesticide and nutrient loss assessments (Yang et al., 1997) and disease prediction (Batchelor et al. 1997). In general, most of the agriculture models have either a mechanical or experimental approach (Poluektov and Topaj, 2001). The mechanical models are complex whereas empirical models have the limitation of not being able to predict beyond the data set range (Wang and Wang 2016). Since ANN is simple and more versatile, therefore it is more effective and an apt technique in agronomy. Pachepsky et al. (1996) compared ANN with regression models while calculating soil water content and found that the former gave better results than the latter. A similar comparison of ANN with traditional methods was carried out by Batchelor et al. (1997) in forecasting soybean rust to reach similar conclusions. Taking 204 sandy soil samples, Schaap and Bouten (1996) tried to model the drying water retention curve. They used a chain of NN models. Each time the number of input and output variables was increased for better forecasts. Though the number of linear and non-linear variants was large still ANN modelling gave excellent results for the water retention curve. ANN can be of great help in making fast and relevant decisions in the field of pesticides also. This was demonstrated by Yang et al. (1997). Accumulated daily rainfall, potential evapotranspiration, soil temperature and the number of days elapsed after pesticide application were taken as input variables. The outputs of the ANN model were the daily accumulated amounts of pesticide levels in the soil. Regardless of limited data, ANN gave better results in forecasting the concentration of pesticides in the ploughed field with RMSE and S.D. lower than 0.2 g/g. Continuing the research in the parallel direction Yang et al. (1997) applied the ANN model to the simulation of soil temperature at Ottawa (Canada). This time daily rainfall, potential evapotranspiration, maximum and minimum air temperature, and the day of the year were taken as the input variables. The results obtained from ANN models varied within the RMS difference range from *0.63* to *1.39 °C*, S.D. from *0.61* to *1.39 °C* and coefficients of determination (r^2) from *0.937* to *0.985*. The results showed that even the complex phenomenon in agriculture system can be made simple yet fast with the help of ANN. The fast back-propagation (FBP) network and a self-organizing radial basis function (RBF) network are the two most popular forms of ANN models. These two models were compared by done by Sharma et al. 2003 for simulation of subsurface drain outflow and nitrate-nitrogen concentration in tile effluent. The RBF (having 20 neurons in the hidden layer) model gave superior results to the FBP model (547 neurons in the hidden layer) in forecasting the concentration of nitrate-nitrogen in drain outflow due to the application of fertilizers/manure. This information can prevent the loss of valuable nitrogen fertilizer. Taking daily max. and min. air temperature, photoperiod, and days after planting of soybean as inputs of ANN, Elizondo et al. (1994) confirmed that this technique has high potential in predicting flowering and physiological maturity of the crop. The average relative error for date of flowering prediction and for date of physiological maturity prediction was +*0.143* days ($n=21$, R^2 =0.987) and +*2.19* days ($n=21$, R^2 = 0.950) respectively. The ease with which a neural network can be applied to predict various physiological properties particularly sensory colours of plants like tomato and peach was shown by Thai et al. (1991).

Table 6.1 represents the application of ANN methodology in single as well in joint mode, results and discussion of literature surveyed. Some frequently used performance measures of forecasting accuracy are MSE (mean squared error), RMSE (root mean squared error), MSE (mean squared error), MAE (mean absolute error), MdAE (median absolute error), MAPE (mean absolute percentage error), MdAPE (median absolute percentage error), sMAPE (symmetric mean absolute percentage error), sMdAPE (symmetric median absolute percentage error), MRAE (mean relative absolute error), MdRAE (median relative absolute error), GMRAE (geometric mean relative absolute error), RelMAE (relative mean absolute error), RelRMSE (relative root mean squared error), LMR (log mean squared error ratio), PB (percentage better), PB(MAE) (percentage better (MAE)), PB(MSE) (percentage better (MSE)).

TABLE 6.1
Discussion of the results of literature surveyed

S. No.	Article	Model	Input Data	Outcome
1.	Pachepsky et al. (1996)	ANN and statistical regression	Data consists of water contents at eight matric potentials for 130 Haplustoll and 100 AquicUstoll	ANN estimated water contents at selected matric potentials better than regression model
2.	Batchelor et al. (1997)	Three layered feed-forward neural network	Data of sequential weekly plantings of TK 5 soybean cultivar for the year 1980 and 1981, at the Asian Vegetable Research and Development Center in Taiwan was considered.	ANN outperforms than traditional methods
3.	SchaapandBouten. (1996)	Series of Neural Network models with varying number of input and output variables	Drying water retention curve containing 204 sandy soil samples from particle-size distribution (PSD), soil organic matter content (SOM), and bulk density (BD)	Neural networks are very useful tool in modelling drying water retention curve
4.	Yang et al. (1997)	Artificial neural network model	Concentration of pesticides in the ploughed field	Regardless of limited data, ANN gave better results in forecasting concentration of pesticides in the ploughed field with RMSE and S.D. lower than *0.2 g/g*.
5.	Yang et al. (1997)	Artificial neural network model	Data of soil temperature along with daily rainfall, potential evapotranspiration, max. and min. air temperature, and the day of the year were taken as the input variables.	Improved values near to 1 of coefficients of determination were obtained with ANN.

No.	Reference			
6.	Sharma et al. (2003)	Fast back-propagation network and a self-organizing radial basis function network.	Subsurface drain outflow and nitrate-nitrogen concentration in tile effluent	The RBF model gave superior results than the FBP model.
7.	Elizondo et al. (1994)	Artificial neural network model	To predict the data of flowering and physiological maturity for soybean	Neural network models efficiently predicted the flowering and physiological maturity of Soyabean
8.	Thai et al. (1991)	Feed-forward neural networks having sigmoidal transfer functions and statistical regression	Data of physical measurements of external colour for tomato and peach	Neural network requires less computation steps and therefore more suitable than baseline statistical methods
9.	Comrie, A. C. (1997)	Multiple regression models and neural networks	Ground-level Ozone pollution data of eight cities in US	Neural networks are more suitable for ozone forecasting than regression models
10.	Perez, P. and Trier, A. (2001)	A three-layer feed-forward neural network with 24 inputs and one output	Data of NO and NO_2 particle concentrations in the atmosphere of Santiago	More accuracy of multi-layer ANN as compared linear regressions
11.	Gardner, M. W. and Dorling, S. R. (1998)	ANN (Multilayer Perceptron)	Atmospheric Data	Multilayer ANN are useful alternatives to traditional statistical modelling techniques
12.	Moseholm al. (1996)	Data of concentration levels of carbon monoxide (CO)	Multilayer perceptron network	Neural networks are useful tool to understand the complex relationships between traffic, wind and CO concentrations than standard linear regression models as well as two dispersion models
13.	Chelani et al. (2002)	Data of SO_2 concentrations in three different sites in Delhi	Multivariate regression model and a three-layered neural network with Levenberg–Marquardt algorithm	More accurate and reliable results with ANN than multivariate regression model

(continued)

TABLE 6.1 Continued
Discussion of the results of literature surveyed

S. No.	Article	Model	Input Data	Outcome
14.	Abrahart, R.J., and See, L. (1998)	Three-year period of river flow data for two catchments: the Upper River Wye (Central Wales) and the River Ouse (Yorkshire) was taken for study	Neural n etworks and ARMA models	Improved levels of modelling performance with neural network than ARMA model
15.	Zealand et al. (1999)	Data of streamflow to a portion of the Winnipeg River system in Northwest Ontario, Canada	ANN and conventional models	Better performance by ANNs as compared to conventional models
16.	Fernando, D.A.K., and Jayawardena, A.W. (1998).	Rainfall runoff data	Radial basis function (RBF) network with orthogonal least-squares (OLS) algorithm, ANN model with back propagation (BP) algorithm, ARMAX model	ANNs with back propagation (BP) produce more accurate results than other methods
17.	Dibike, Y. B. and Solomatine, D.P. (2001)	Rainfall-runoff data	Multi-layer perceptron (MLP) network, radial basis function (RBF) Network	ANN outperforms the conceptual rainfall-runoff models
18	Dawson, C.W., Harpham, C., Wilby, R.L. and Chen, Y. (2002).	Rainfall-run off data for the River Yangtze, China	Multi-layer perceptron (MLP) network, radial basis function (RBF) network, regression model, ARIMA	Both types of neural network generate more accurate forecast than the traditional statistical techniques
19.	Anctil, F., and Tape, D.G. (2004)	Rainfall and runoff data in US and France	ANN, and Hybrid Wavelet decomposition with ANN	Better accuracy of joint wavelet decomposition with ANN
20.	Li and Ma (2010)	Data of various stock markets and financial indices	Artificial neural networks	ANNs are very successful in dealing with complex and non-linear stock market data

No.	Author	Data	Method	Findings
21.	Wang and Wang (2015)	Financial time series indices SSE, HS300, S&P 500 and DJIA	Hybrid approach using stochastic time effective function neural network (STNN) and principal component analysis (PCA)	Hybrid approach produce more accurate as compared to traditional backpropagation neural network (BPNN) and PCA-BPNN
22.	Niaki, S. T. A. and Hoseinzade, S. (2013)	Daily data of Standard & Poor's 500 (S&P 500) index	Artificial neural network (ANN) method	ANN successfully predict most powerful features among 27 financial and economic variables
23.	Zahedi and Rounaghi (2015)	Tehran Stock Exchange price data	ANN model and principal component analysis (PCA)	The accurate predictions with ANN-PCA in the form of new patterns of 20 accounting variables
24.	Wang and Wang (2016).	Crude oil price data	Multilayer perception and ERNN (Elman recurrent neural networks) with stochastic time effective function	ERNN is more useful than other models.

6.7 CONCLUSIONS

This chapter aimed to survey some significant contributions in the field of artificial neural networks to solve the prediction problems related to finance and economics, environment, hydrology and agriculture preferably. A detailed methodology of artificial neural networks with historical background has been discussed here. The results of survey reveal that ANNs give more accurate forecasts than traditional and baseline regression, ARMA and ARIMA models, etc. The major contribution of this chapter is to provide the basic terminology of ANN architecture and methodology useful for different forecasting problems, and survey the available sources of different type of data to define a new problem in this field for future research.

REFERENCES

Anctil, F., and Tape, D.G. (2004). "An exploration of artificial neural network rainfall-runoff forecasting combined with wavelet decomposition". *Journal of Environmental Engineering and Science*, 3(1): 121–128.

Abrahart, R.J., and See, L. (1998). "Neural network vs. ARMA modeling: constructing benchmark case studies of river flow prediction". In: *Proceedings of the 3rd International Conference on Geocomputation*. University of Bristol.

Basheer, I.A., and Hajmeer, M. (2000). "Artificial neural networks: fundamentals, computing, design, and application". *Journal of Microbiological Methods*, 43(1): 3–31.

Batchelor, W.D., Yang, X.B., and Tshanz, A.T. (1997). "Development of a neural network for soybean rust epidemics". *Transactions of the ASAE*, 40: 247–252.

Bing, Y., Hao, J.K., and Zhang, S.C. (2012). "Stock market prediction using artificial neural networks". *Advanced Engineering Forum*, 6: 1055–1060.

Chauvin, Y., and Rumelhart, D.E. (Eds.). (1995). *Backpropagation: Theory, Architectures, and Applications*. London: Psychology Press.

Chelani, A.B., Rao, C.C., Phadke, K.M., and Hasan, M.Z. (2002). "Prediction of sulphur dioxide concentration using artificial neural networks". *Environmental Modelling & Software*, 17(2): 159–166.

Chen, A.S., Leung, M.T., and Daouk, H. (2003). "Application of neural networks to an emerging financial market: forecasting and trading the Taiwan StockIndex". *Computers and Operations Research*, 30(6): 901–923.

Comrie, A.C. (1997). "Comparing neural networks and regression models for ozone forecasting". *Journal of the Air & Waste Management Association*, 47(6): 653–663.

Dawson, C.W., Harpham, C., Wilby, R.L., and Chen, Y. (2002). "Evaluation of artificial neural network techniques for flow forecasting in the River Yangtze, China". *Hydrology and Earth System Sciences*, 6: 619–626.

Dibike, Y., and Solomatine, D. (2001). "River flow forecasting using artificial neural networks". *Journal of Physics and Chemistry of the Earth, Part B: Hydrology, Oceans and Atmosphere* 26: 1–8.

Elizondo, D.A., McClendon, R.W., and Hoogenboom, G. (1994). "Neural network models for predicting flowering and physiological maturity of soybean". *Transactions of the ASAE*, 37: 981–988.

Fernando, D.A.K., and Jayawardena, A.W. (1998). "Runoff forecasting using RBF networks with OLS algorithm". *Journal of Hydrological Engineering*, 3: 203–209.

Franses, P.H., and Draisma, G. (1997). "Recognizing changing seasonal patterns using artificial neural networks". *Journal of Econometrics*, 81: 273–280.

Gardner, M.W., and Dorling, S.R. (1998). "Artificial neural networks (the multilayer perceptron):a review of applications in the atmospheric sciences". *Atmospheric Environment*, 32(14–15): 2627–2636.

Gorr, W.L. (1994). "Research prospective on neural network forecasting". *International Journal of Forecasting*, 10: 1–4.

Hebb, D.O. (1949). *The Organization of Behaviour: A Neuropsychological Theory.* New York: Wiley.

Jain, S.K., Das, A., and Srivastava, D.K. (1999). "Application of ANN for reservoir in flow prediction and operation". *Journal of Water Resources Planning and Management*, 125: 263–271.

Jakeman, A.J., Letcher, R.A., and Norton, J.P. (2006). "Ten iterative steps in development and evaluation of environmental models". *Environmental Modelling & Software*, 21(5): 602–614.

Kang, K.W., Kim, J.H., Park, C.Y., and Ham, K.J. (1993). "Evaluation of hydrological forecasting system based on neural network model". In: *Proceedings of the 25th Congress of the International Association for Hydraulic Research, Delft, Netherlands.* pp. 257–264.

Kaul, M., Hill, R.L., and Walthall, C. (2005). "Artificial neural networks for corn and soybean yield prediction". *Agricultural Systems*, 85(1): 1–18.

Khairunniza-Bejo, S., Mustaffha, S., and Ismail, W.I.W. (2014). "Application of artificial neural network in predicting crop yield: a review". *Journal of Food Science and Engineering*, 4(1): 1.

Khashei, M., and Bijari, M. (2010). "An artificial neural network model for time series forecasting". *Expert Systems with Applications*, 37(1): 479–489.

Kisi, O. (2009). "Daily pan evaporation modelling using multi-layer perceptrons and radial basis neural networks". *Hydrological Processes*, 23, 213–223.

Li, Y, Ma, W. (2010)."Applications of artificial neural networks in financial economics: a survey". In: *International Symposium on Computational Intelligence and Design. IEEE*, pp. 211–214.

McCulloch, W.S. and Pitts, W. (1943). "A logical calculus of the ideas immanent in nervous activity", *Bulletin of Mathematical. Biophysics*, 5: 115–133.

Minsky, M.L. and Papert, S.A. (1969). *Perceptrons*, Cambridge, MA: MIT Press.

Moseholm, L., Silva, J., and Larson, T. (1996). "Forecasting carbon monoxide concentrations near a sheltered intersection using video traffic surveillance and neural networks". *Transportation Research Part D: Transport and Environment,* 1(1): 15–28.

Niaki, S. T. A., and Hoseinzade, S. (2013). "Forecasting S&P 500 index using artificial neural networks and design of experiments". *Journal of Industrial Engineering International*, 9(1): 1.

Nelson, M., Hill, T., Remus, T., and Connor, M. (1999). "Time series forecasting using NNs: should the data be de-seasonalized first?" *Journal of Forecasting* 18: 359–367.

Pachepsky, Y.A., Timlin, D., and Varallyay, G. (1996). "Artificial neural networks to estimate soil water retention from easily measurable data". *Soil Science Society of American Journal* 60: 727–733.

Perez, P., and Trier, A. (2001). "Prediction of NO and NO2 concentrations near a street with heavy traffic in Santiago, Chile". *Atmospheric Environment*, 35(10): 1783–1789.

Poluektov, R.A., and Topaj, A.G. (2001). "Crop modeling: nostalgia about present or reminiscence about future". *Agronomy Journal* 93: 653–659.

Rosenblatt,F. (1958). "The perceptron: A probabilistic model for information storage and organization in the brain". *Psychological Review*, 65: 386–408.

Schaap, M.G., and Bouten, W. (1996). "Modelling water retention curves of sandy soils using neural networks". *Water Resources Research*, 32(10): 3033–3040.

Sharda, R., and Patil, R.B. (1992). "Connectionist approach to time series prediction: an empirical test". *Journal of Intelligent Manufacturing*, 3: 317–323.

Sharma, V., Negi, S.C., Rudra, R.P., and Yang, S. (2003). "Neural networks for predicting nitrate-nitrogen in drainage water". *Agricultural Water Management*, 63(3): 169–183.

Thai, C.N. and Shewfelt, R.L. (1991). "Modelling sensory colour quality of tomato and peach: neural networks and statistical regression". *Transactions of the ASAE*, 34(3): 950–955.

Wang, J., and Wang, J. (2015). "Forecasting stock market indexes using principle component analysis and stochastic time effective neural networks". *Neurocomputing*, 156: 68–78.

Wang, J., and Wang, J. (2016). "Forecasting energy market indices with recurrent neural networks: Case study of crude oil price fluctuations". *Energy*, 102: 365–374.

Widrow, B. and Hoff, M.E. (1960). "Adaptive switching circuits". *IRE WESCON Convention Record*, 4: 96–104.

Yang, C.C., Prasher, S.O., Mehuys, G.R., and Patni, N.K. (1997). "Application of artificial neural networks for simulation of soil temperature". *Transactions of the ASAE*, 40(3): 649–656.

Yang, C.C., Prasher, S.O., Sreekanth, S., Patni, N.K., and Masse, L. (1997). "An artificial neural network model for simulating pesticide concentrations in soil". *Transactions of the ASAE*, 40: 1285–1294.

Zahedi, J., and Rounaghi, M.M. (2015). "Application of artificial neural network models and principal component analysis method in predicting stock prices on Tehran Stock Exchange". *Physica A: Statistical Mechanics and its Applications*, 438: 178–187.

Zealand, C.M., Bum, D.H., and Simonovic, S.P. (1999). "Short term streamflow forecasting using artificial neural networks". *Journal of Hydrology*, 214: 32–48.

Zhang, G., P., B.E., and Hu, M.Y. (1998). "Forecasting with artificial neural networks: the state of the art". *International Journal of Forecasting*, 14: 35–62.

Zhang, G.P., and Qi, M. (2005). "Neural network forecasting for seasonal and trend timeseries". *European Journal of Operational Research*, 160(2): 501–514.

7 Internet of Things
An Emerging Paradigm for Social Safety and Security

Bhanu Sharma, Krishan Dutt Sharma, and Archana Mantri

CONTENTS

DOI: 10.1201/9781003153405-7

7.1 INTRODUCTION

Human intellect and curiosity keep us ever on our toes to thrive through interminable build outs. Humans ceaselessly probe the status quo of contemporary realizations to peep into the next possibilities. The Internet of Things (IoT) is a result of this human character that has realized the unimaginable. With IoT, humans have created a world where lifeless things have been animated to interact with one another and with humans. IoT has lent the modern world a revolutionary phenomenon where the whole world has become a huge brain with people, machines, things, and even animals that can connect and control one another (Asghari et al. 2019).

The term IoT normally refers to frameworks where web connectivity combines the computing capability and enables everyday objects to generate, imbibe, and exchange data with the help of sensors and that's even without human intervention. IoT being a technology complex allows several devices and machines to interact with humans, objects, and animals with unique identifiers (UIDs) to connect over a common platform where they can communicate and exchange data.

IoT uses an IP (Internet Protocol) address to connect many things like vehicles, home appliances, cell phones, and some wearables equipped with sensors and actuators. It enables these objects to mutually exchange data over a network. IoT, sometimes, is confused with the internet, but it is a much smarter system that does not require human involvement. It is competent in creating and analyzing human behaviors and even taking up the required actions.

IoT is a system of "Connected devices" that supersedes smart gadgets like laptops or smartphones. These connected devices can minutely record shared data and assist us to accomplish simple as well as complex tasks. Figure 7.1 shows the applications of IoT.

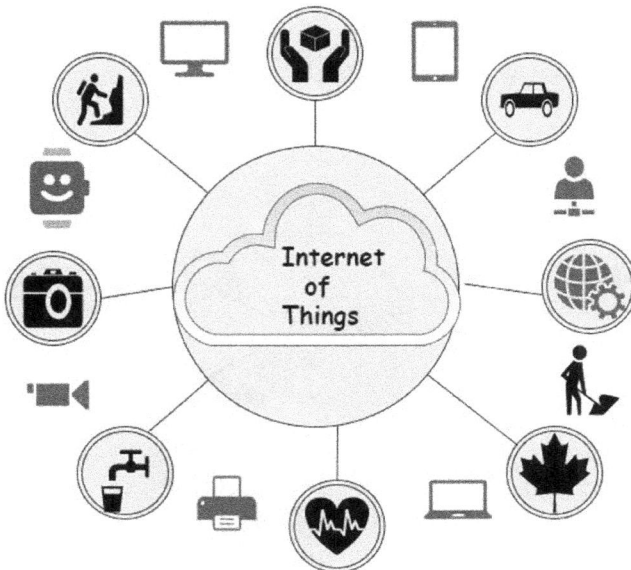

FIGURE 7.1 Internet of Things.

IoT has merged the digital world with the physical world and has scaffolded innovation to create newer applications. IoT has improved the quality of modern life with enhanced social and industrial productivity. Connected wearables, connected cars, smart appliances, smart homes, smart cities, and connected healthcare (Bhatt & Bhatt 2017), agriculture, etc. are some of the obvious manifestations of IoT that lead us to believe that we are heading towards a fully automated world (Brito et al. 2017).

7.2   THE IOT SYSTEM: TOOLS AND TECHNOLOGIES

The IoT system has the following core components: sensors and actuators.

The IoT system is a multi-layered technology complex with sensors and actuators as its key components. These sensors and actuators enable simple objects to exchange data over the internet, which they ultimately process to deliver intelligence and autonomous actions to leverage each other's functionality.

Sensors and actuators are indispensable to the IoT infrastructure and help it to monitor, control, and leverage operations within the IoT framework (García et al. 2017). Innovation has enhanced their application and expansion in such a way that they create powerful cloud-based software that provides analytical and intelligent solutions for people, things, and machines alike.

A sensor is a device that detects changes like air, light, temperature, humidity, movement, chemical substance, etc. in immediate surroundings and converts them into interpretable electrical impulses.

An actuator, on the contrary, functions oppositely to a sensor. It makes sense of the electrical signals created by sensors and delivers mechanical outputs. An actuator responds to physical change around it by a variety of actions like putting connected devices in motion, making them emit light or sound, changing their positions or angles, etc. Actuators can better be termed as 'movers.' Figure 7.2 shows the different types of sensors and actuators used in IoT.

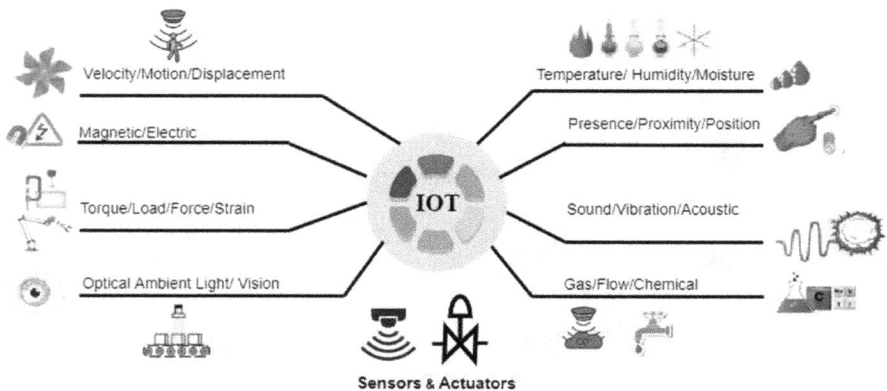

Sensors & Actuators

FIGURE 7.2   Sensors and actuators.

7.2.1 Sensors

Sensors complement the IoT framework since they can work as standalone or as assistive tools to smarten the functionality of machines. Sensors are termed as per the physical occurrence they are supposed to sense and quantify (García et al. 2017). An overview of some commonly used sensors and their functionality is worth discussing here to have a deep understanding of IoT systems.

7.2.1.1 Temperature Sensor

Measuring the thermal count of the air in the work environment is quite often required in various applications. Temperature sensors are mostly used in manufacturing plants, weather forecasts, warehouses, and agriculture to monitor soil temperature for balanced and high growth.

7.2.1.2 Thermistor

This is a resistor that can sense temperature. Its resistance varies with the temperature variation. It has many applications in electronics and in some other systems to prevent an excessive increase in current.

7.2.1.3 Thermocouples

An electric circuit with elements of two different conductors makes a thermocouple. It makes use of the phenomenon that electromotive force between its connectors is proportional to the difference in temperature. Thermocouples can also be used to sense temperature and sometimes as an alternate source of power delivering small voltage but a considerable amount of current. Thermocouples are mostly used at meteorology stations for the weather forecast.

7.2.1.4 Hair Tension Moisture Sensor

This is a traditional moisture sensor that uses the property of human or horsehair to expand due to moisture. As currently explored, it, however, works well with cotton or synthetic fibers too. It measures moisture with the help of a pointer on the scale that moves with the change in the hair's length. The device is cheap and damage resistant.

7.2.1.5 Light Sensors

Light sensors detect ambient light intensity and enable smartphones, smart TVs, computer screens to adjust brightness. Light sensors are quite common in electronics, but currently, they are increasingly being used in smart city applications too. Adaptive street and smart urban lighting are the modern applications of light sensors.

7.2.1.6 Smart Acoustic Sensors

These sensors detect noise levels in the environment and thus help to monitor and control noise pollution. They are widely used as smart city solutions.

7.2.1.7 Hydrophone

This device detects sounds in liquids or water bodies. It is a structural component of passive sonars. It is used to find fish in different aquatic environments.

7.2.1.8 Geophone

This sensor is used for environmental protection since it can monitor water levels and is used in flood warning systems. It also finds applications in industries.

7.2.1.9 Hydro-Pressure Sensor

This sensor measures the liquid or water level in a tank regardless of its shape or volume. It operates on the hydrostatic paradox that the hydrostatic pressure of the liquid is proportional to its height.

7.2.1.10 Optical Sensor

Optical sensors have an edge over the traditional mechanical level detectors since they detect water levels through the refraction of light by a prism. This sensor emits electromagnetic radiation and detects the presence of the target object. It has numerous applications in industry, transportation, smart vehicles, and robotics.

7.2.1.11 Motion Sensor

Motion sensors are increasingly used in many IoT smart building applications. These sensors help in monitoring private or public spaces for any criminal activities like intrusion or theft. Apart from this motion sensors leverage energy management solutions, smart cameras, and automated devices.

7.2.1.12 Passive Infrared Sensor (PIR)

These are electronic sensors used to detect motion and are generally used in automatic light systems, alarms, and ventilation systems.

7.2.1.13 Gyroscope Sensors

These sensors detect rotation and angular velocity and are used in various applications in defense, transport, security, and industry where rotation of objects is involved. These sensors are used in IoT devices for athletes to accurately measure and analyze body movements and to improve sports performance. Accelerometer and heading indicators are some other gyroscopes sensors that measure the rotation and angular velocity of the objects on which they are installed.

7.2.1.14 Electrochemical Breathalyzer

This sensor detects alcohol content in the blood. It analyzes breath accurately for alcohol intake and is resistant to other accompanied odors. It is used by traffic police to nab drunken drivers.

7.2.1.15 Electronic Nose

This is a set of detectors that detects the presence of chemical particles dissolved in the environment and useful in determining the chemical constitution of air.

7.2.1.16 Image Sensors

These sensors convert visuals into electrical signals and are installed on smart devices to view the surroundings. They are used in various applications in military, transportation, medical, and security agencies.

There are many other sensors used in IoT technology.

7.2.2 ACTUATORS

Actuators respond to their immediate environment and act upon the received input to enhance the functionality of the embedded machines or objects. They are frequently used in vehicles and machines and automation systems in industries. Actuators can be divided into four categories based on their design and purpose in IoT applications.

7.2.2.1 Linear Actuators

These are applied to achieve linear motion of objects.

7.2.2.2 Motors

They provide rotational motion to devices as well as their components.

7.2.2.3 Relays

These are electromagnet-based actuators and are used to control power buttons in many devices.

7.2.2.4 Solenoids

These are mostly applied in household instruments to enable mechanisms like locking or triggering. They are used as gas or water leak controllers in IoT-based monitoring systems.

7.3 CONNECTIVITY

To achieve interconnectivity among devices within the IoT system, there is a wealth of options available. Cellular, satellite, Bluetooth, NFC, Wi-Fi, LPWAN, Ethernet, and RFID are some of many possible options to connect different devices. Figure 7.3 shows the tools and technologies used for IoT.

7.3.1 LPWANs

LPWANs support the huge IoT networks involving industries and business organizations and facilitate long-range connectivity. They connect the majority of sensors and can support several IoT applications like asset tracking, occupancy detection, environmental monitoring, consumables monitoring, etc.

7.3.2 CELLULAR (3G/4G/5G)

Cellular networks support voice and video communications through broadband. They are most used in smart cars and other transportation applications like smart fleets or

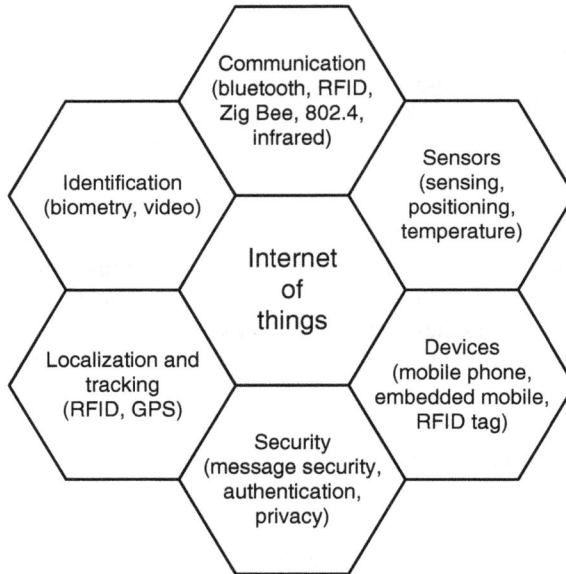

FIGURE 7.3   Tools and technologies for IoT.

logistics. 5G is the next-gen technology that will support autonomous vehicles. It will enable augmented reality-based real-time surveillance to support industrial automation, public safety, and healthcare.

7.3.3   ZIGBEE

Zigbee, mostly used as a mesh protocol, communicates sensor data to different sensor nodes and extends coverage. It consumes low power and facilitates short-range wireless connectivity. Zigbee is used in various smart home applications such as security, lighting, climate, and energy management, etc.

7.3.4   BLUETOOTH

Bluetooth is a WPAN short-range data exchange network. The latest version of Bluetooth Low Energy (BLE) is, however, most used in small IoT applications. It is employed to assist communication devices like smartphones, fitness wearables, and many home automation applications.

7.3.5   WI-FI

Wi-Fi is a well-known connectivity tool, most utilized in various connecting devices in homes and industries. It does not, however, support the IoT framework due to limited coverage and high-power consumption. Wi-Fi 6, with enhanced data throughput and higher bandwidth, is the next-gen Wi-Fi that promises a better connectivity experience.

7.3.6 RFID

This technology transmits data to the reader from an RFID tag using radio signals. RFID finds many applications in retail, security, and logistics.

7.4 IOT APPLICATIONS

IoT applications are increasingly being exploited for personal, public, or industrial benefits. Major areas include home automation, transportation and logistics, retail and marketing, security and surveillance, energy management, agriculture, and healthcare. The following sections deal with some popular IoT applications.

7.4.1 SMART HOMES

The smart home is a revolutionary concept realized by IoT. The IoT applications, have enabled us to exercise control over our household affairs without being physically present there. Figure 7.4 shows the Home automation using IoT. These applications can help us to monitor and control several household activities even while being away from the house.

One can control lights, fans, doors, air conditioning, water supply, and many other home instruments with a click on one's smartphone. Bitdefender box2, Logitech circle 2, Amazon echo plus (2nd gen), Nest thermostat (3rd gen), Awaire air-quality monitor, Nest protect smoke plus, Philips hue bulbs and lighting system, Singleque gesture control, Ring video doorbell pro, August smart lock pro, Ecovacs deebot N79S, Logitech harmony remote control, Blossom smart watering controller, etc. are some popular IoT smart home devices.

7.4.2 SMART CITIES

Population outburst has made city life difficult. There are problems with traffic, parking, power supplies, water distribution, surveillance, and security, etc. IoT, however, has come up with smart solutions to leverage these utilities and services to comfort the modern man (Mohanty 2016). Smart infrastructure, smart city air management, smart traffic management, smart parking, smart waste management, smart water, and power supply are the IoT applications that are making our cities smart. With smart IoT applications, many problems of city-life like pollution, transportation clogging, and inadequacy of power supplies, etc. can be solved easily. A smart city must be clean too. Applications like Smart Belly trash will assist the municipal corporation in tidying up the city. Figure 7.5 shows the glimpse of smart cities.

Vehicle parking is another rising problem in big cities. IoT, nevertheless, has developed applications to guide and assist citizens with their parking issues. There are smart applications available now that can support household electricity-related issues such as malfunctioning, installation, and meter tampering, etc.

FIGURE 7.4   Smart home.

7.4.3   WEARABLES

Wearables are the sensors and software embedded devices that assist users with their health, fitness, and entertainment-related requirements. These devices collect and process data and other related information about the users and help them monitor, plan, control, and enhance their many activities and experiences (Lavanya et al. 2019). Figure 7.6 shows the wearables used for different applications with the help of IoT. Commonly used wearables include smart watches, smart shoes, smart clothes, gesture control, NFC smart ring, fitness tracking bands, smart posture trainer, gaming simulators, bracelets, and smart bands for blinds, etc.

These wearables do not affect the users for they consume or release very little energy.

7.4.4   AGRICULTURE

Smart farming is a fast-developing IoT application that has come with a solution to various farming issues. Food consumption has increased with the population. This calls for immediate developments in farming methods and techniques. IoT has brought in smart solutions like FarmLogs and Cropio to address these issues. Figure 7.7 shows

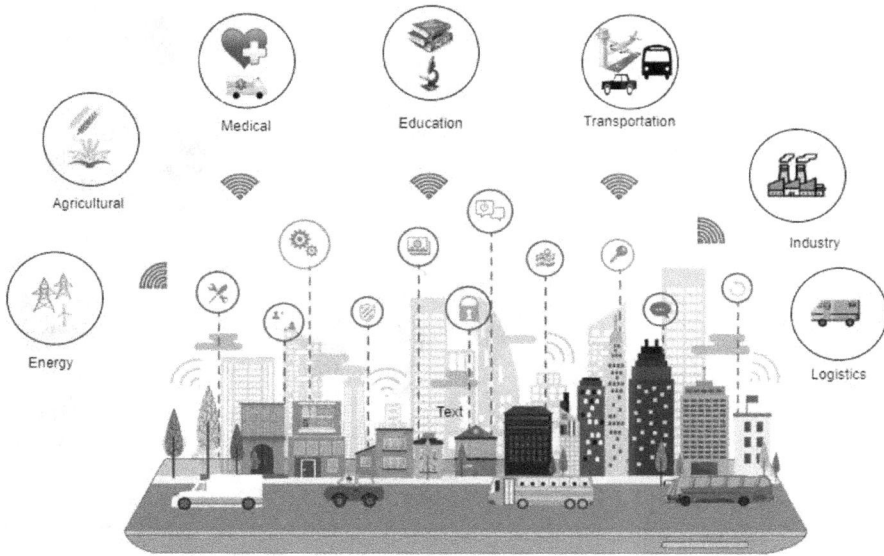

FIGURE 7.5   Smart cities.

the advanced techniques used in the agricultural field. Smart farming applications can help monitor the soil for moisture level and nutrients deficiency.

Farmers can keep track of the crops and are alerted to follow a timetable for watering, fertilizing, and harvesting the crop at the appropriate time. Arable and Semios are smart crop-management devices (Vongsingthong et al. 2014). Smart irrigation devices like GreenIQ, and fertilization alert systems enable growth monitoring for maximized yield. allMETEO, Smart Elements, and Pycno are IoT powered devices that predict weather conditions. Farmapp and Growlink are devices that facilitate greenhouse automation. Even the livestock can be monitored with IoT's sensing tools. SCR by Allflex and Cowlar are smart devices used to monitor the health of pets.

7.4.5   SMART SUPPLY CHAIN

IoT has transformed the traditional supply chain system into a smart system, more efficient in terms of time, money, and energy. Figure 7.8 shows the smart supply chain with this innovative technology. The system uses GPS technology and assists consumers with the tracking of their shipments easily and cost-effectively. The system consumes less manpower and optimizes the output (Wan et al. 2016).

7.4.6   SMART RETAIL

The modern retail system wholly draws upon the IoT framework and is a never seen before experience for both the consumers and the retailers. IoT has enabled us to shop anything and from anywhere just by a click on our smartphones. Where retailers can

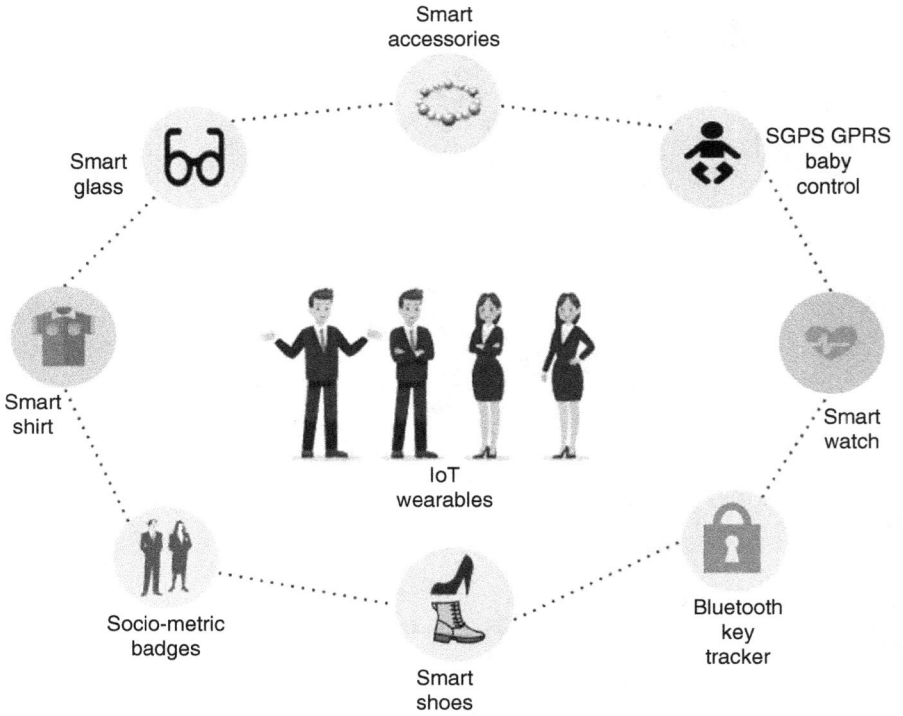

FIGURE 7.6 Wearables.

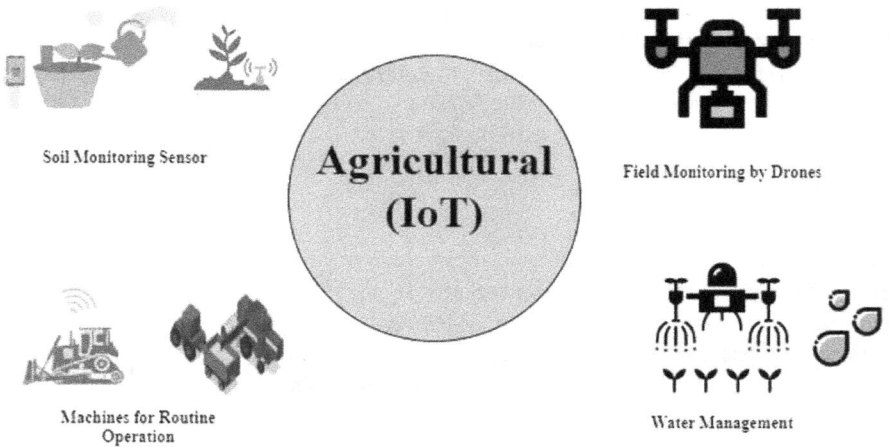

FIGURE 7.7 Agriculture and IoT.

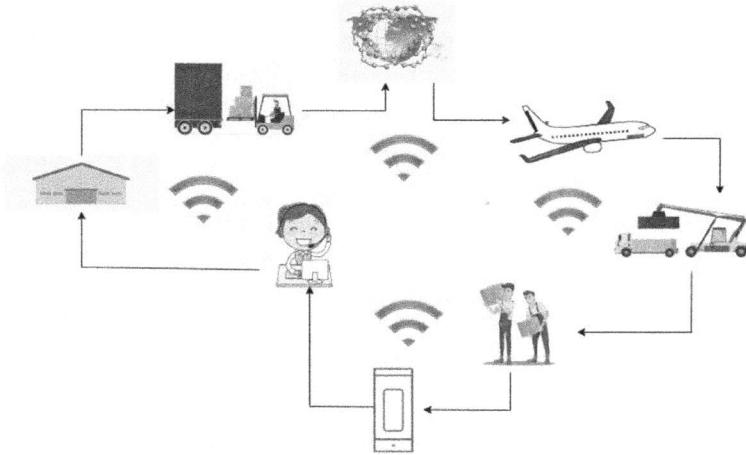

FIGURE 7.8 Smart supply chain.

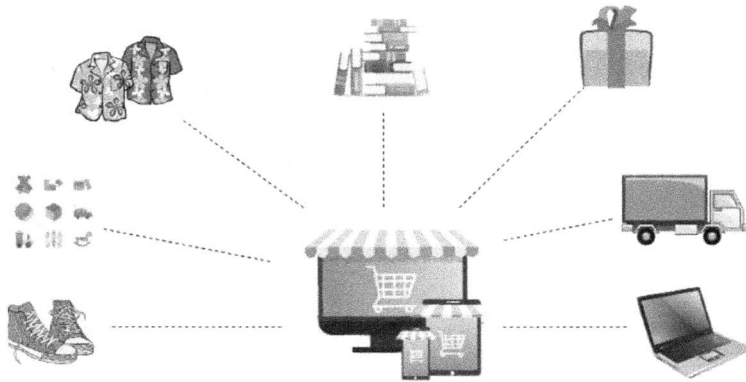

FIGURE 7.9 Smart retail.

directly approach a huge number of customers, showcase and sell their products, the consumers can also have the opportunity to buy a product of their preference from a wealth of available stores. One can pay or receive online and even keep track of his orders. Figure 7.9 shows the retailing system.

7.4.7 Transportation

The modern transportation system has been greatly enhanced by IoT technology. Smart cars of today are equipped with smart tools and sensors that empower drivers to control and monitor their driving (Xia et al. 2012).

The in-car computer assists the driver with auto climate control, navigation, parking, speed and fuel alerts, traffic congestion, road information, and much more. Even taxis use IoT applications to provide smart pick-up and drop facilities to their customers.

7.4.8 SMART GRID

IoT has conceptualized smart and reliable power supply in the form of smart grids. The system not only regulates energy efficiently but also collects a lot of information about the consumers. Figure 7.10 shows the smart grid system.

Recording data like energy consumption statistics, consumer behavior, detecting power outrages makes it an intelligent system to enhance energy utility services. It maintains smooth and uninterrupted power distribution by keeping a check on power failure, overloading due to meter tampering, short-circuiting, etc. IoT-enabled devices like WebNMS and Sense, etc. facilitate smart energy consumption monitoring and optimize household energy conservation.

7.4.9 HEALTHCARE

IoT technology with wearables collects data about an individual's health and fitness. Smart watches, fitness bands, and stress detectors are some of the IoT tools that continuously monitor our physical and mental health (Agarwal and Lau, 2010). The connected devices analyze and process data to provide health alerts and strategies to combat undesirable health conditions (Kumar et al. 2019 a).

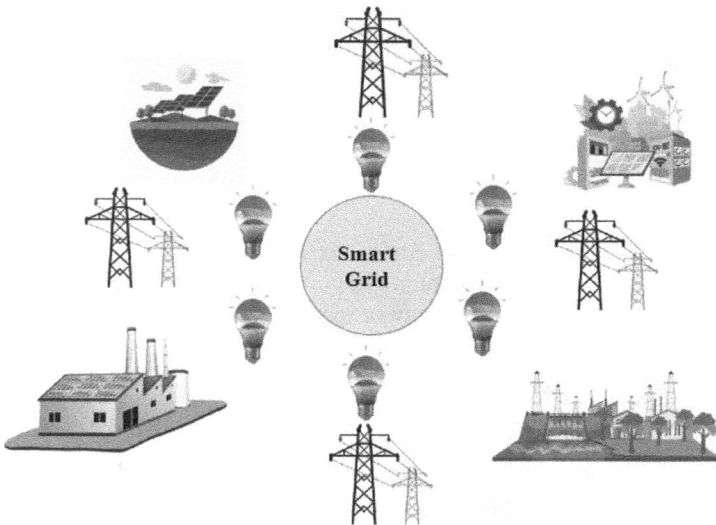

FIGURE 7.10   Smart grid

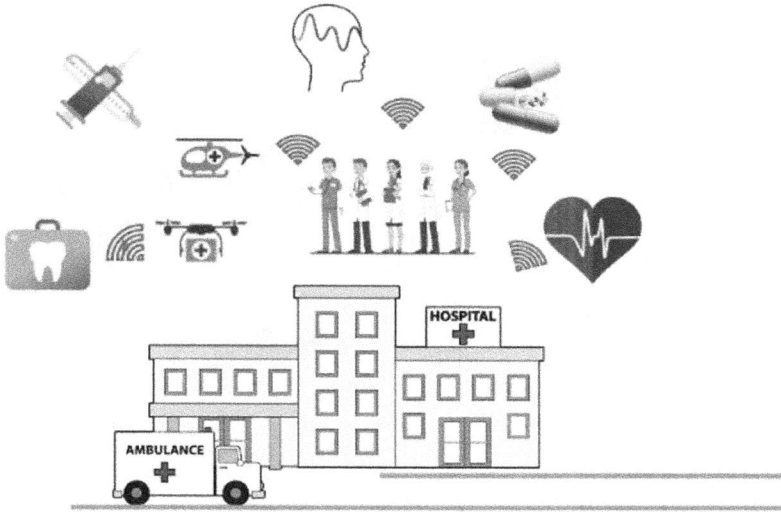

FIGURE 7.11 Healthcare.

QardioCore, a wearable ECG monitor, monitors blood pressure and cholesterol in patients (Kumar et al. 2019 b). Zanthion sends an alert to a patient's family members if he or she falls off the bed, or becomes motionless for a long time. Up by Jawbone tracks a person's fitness including weight, diet, and sleep disorders. NHS test beds keep track of a patient's health and recovery on a routine basis. Propeller's Breezhaler device helps in monitoring and treating asthma patients. UroSense, a transmitter-fitted catheter, helps to monitor and manage patients with urine problems, diabetes, or prostate cancer. There are many such IoT devices used to support healthcare services. Figure 7.11 shows the healthcare system scenario. These healthcare devices smartly convey information about one's critical condition to concerned people or emergency medical services departments so that timely assistance can be provided to patients. IoT has enhanced the medical diagnostics with sensing tools.

7.4.10 INDUSTRY

IoT has enhanced industries with smart engineering, powered by sensors, software, and intelligent machines. The smart machines can now communicate data more accurately and efficiently compared to humans. IoT enabled machines use sensors and tracking devices to yield desirable outputs. This in turn helps industries to achieve enhanced, accurate, and transparent productions. Damage detection sensors help improve product quality. SAGE developed a cloud-based SCADA interface 'STRATUS' that can be used in various industry practices like site-monitoring and manhole monitoring, remote power generation monitoring and control. In addition to that, SAGE proposed transport radio network performance monitoring dashboard,

FIGURE 7.12   IoT industry.

and recycling sorting and data system for data probity. Other IoT-driven applications in manufacturing industry include predictive repairing, remote production control, asset tracking, logistics management, digital twins, etc. Figure 7.12 shows IoT based advanced Industry system.

7.5   IOT APPLICATIONS FOR SOCIAL SAFETY AND SECURITY

7.5.1   URBAN DATA PRIVACY

In smart cities, most public services have been digitized. People access them through smartphones, computers, or other devices via diverse networks (Sharma et al., 2019). These transmissions are prone to cyberattacks. It is therefore imperative to secure urban data. IBM provides IN3 (instrumented, interconnected, intelligent), a cyber security paradigm that tracks privacy issues based on how information is used within IN3. A smartphone application tracks any cyber breach and conveys a panic notification to the police. A mobile computing solution uses geolocation data, video streaming, and note-taking, and clouds multiple systems onto a single user interface that can be accessed via smartphones and Intel-powered computers. Historical data can be used to create an intelligent database that can be accessed to visualize established relationships among people and places. Figure 7.13 shows the applications of IoT for social safety and security.

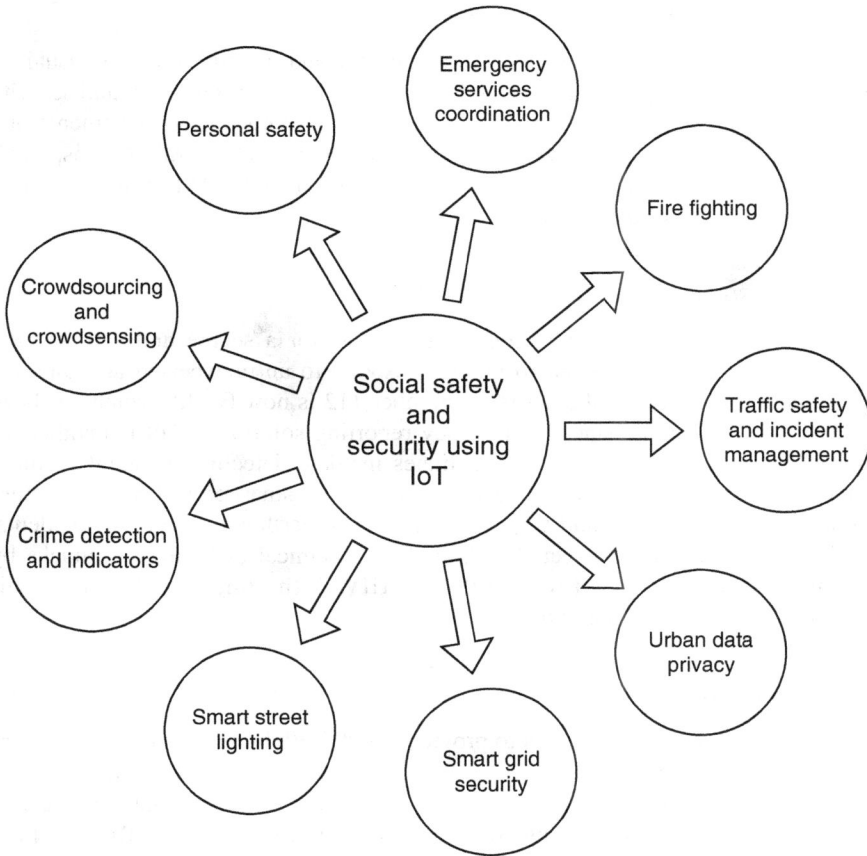

FIGURE 7.13   IoT applications for social safety and security.

7.5.2   Personal Safety

Personal safety is a prime condition for a secure society. Personal safety refers to a condition of being safe from intentional criminal offenses and stays protected from or not exposed to them. The IP surveillance system uses smart cameras plugged into Ethernet LAN for creating a video surveillance network. VERETOS cloud is an evidence management solution that can store large video data. It is controlled by the Microsoft Azure government cloud and help police to capture crime scenes or doubtful activities on the go either with a wearable body camera or an in-car camera that can identify the criminal's car's license plate. The CISCO Smart+ Connected City Safety and Security solution can facilitate location monitoring, incident detection, and management, administration, and analytics to curb the criminal menaces.

7.5.3 Crime Detection and Indicators

An intelligent video surveillance system detects and identifies any unusual or alarming activity in a particular area. It is an advanced integrated safety and security system that can foresee crime by detecting and analyzing object movements and their trajectories. Photonics, thermal imaging, facial and behavioral systems, satellite monitoring, biometric systems, retinal scans are some other IoT applications that have revolutionized security systems.

7.5.4 Emergency Services Coordination

A unified response plan to emergencies is a must for a secure and safe society. Location detection is important to quickly respond to an unexpected and undesirable situation. The universal emergency number 112 is now E-112, which has been updated as a location-enhanced emergency reporting solution. eCall is another in-vehicle emergency call system that facilitates incident detection, incident verification, incident response, and clearance. CITYSCAPE is a smart protection architecture solution to save buildings and their occupants from earthquake and fire incidents. Another IoT application protects buildings from chemical or biological attacks by shutting down air intake and reverse flow of HIVAC (heating, ventilation, and air conditioning) systems of the building.

7.5.5 Firefighting

IoT has created many applications to provide smart firefighting solutions. Roadmap, Frisco, and Firecast are the firefighting systems that detect a fire site and equip fire personnel with information like a roadmap, hazardous material, hydrant location, etc. ALIVE (advanced learning in integrated visual environments) and FIREPLAN are IoT enabled systems that are effective tools to train fire personnel in administrative, reporting, and alerting tasks.

7.5.6 Crowdsourcing and Crowdsensing

Smart devices, sensors, and smartphones can be networked on the web to sense the crowd and share information about it. This application is quite helpful in making decisions such as evacuation, etc. in case of any emergency. This is a participatory framework where every individual and even those who are illiterate can also share information with their smartphone.

7.5.7 Smart Grid Security

Electricity supply grids are now upgraded to be smart grids. Smart grids supply electricity to consumers using digital technology. They, however, require an effective control and communication system to rule out cyber-attacks. Most smart grids use SCADA systems to ensure the secure and effective functioning of smart grids.

7.5.8 SMART STREET LIGHTING

Street lighting has become smart and energy-efficient with the IoT framework. Instead of turning on at night, they will now turn on sensing an approaching vehicle or a pedestrian. Additionally, they can adjust brightness as per the requirement that makes them cut energy usage. LED lights with IoT technology can provide environmental alerts and emergency notifications.

7.5.9 COUNTER-TERRORISM

The USA's cell-all project has installed chemical agent detectors in phones and has linked them to security networks. This enables the security agencies to detect any terror threat to urban populations and rapidly respond to the emergency.

7.5.10 TRAFFIC SAFETY AND INCIDENT MANAGEMENT

Traffic management and safety have been enhanced with IoT technology. VANET (Vehicular Adhoc Networks) is a position security system that detects the position and announced coordinates of neighboring vehicles with onboard radar. Automated speed enforcement and speed adaptation, safety belt interlock system, alcohol sensors, and programmed smart cards to regulate vehicular performance are some IoT solutions for safe traffic. Dynamic circulation lane allocation system with the help of multi-lane sensors helps manage the traffic congestion by allocating lanes to the vehicles as per requirement. The SRM system collects information about the vehicular position, traffic rule violation, accidents, etc., and sends appropriate information to drivers and police for caution or decision making.

A crossing road monitoring system uses two cameras. A long-range camera detects an illegal vehicle that violates the distance between the vehicle and the lane lines. The close-range camera captures its license plate. Qualcomm Vehicle-to-Pedestrian technology makes use of smartphones and in-car alert technology to make car-drivers and pedestrians aware of each other's position through short-range communications.

7.6 CASE STUDY (INTRUSION DETECTION AND TACKLING SYSTEM FOR AGRICULTURAL FIELDS)

Extending our discussion on the IoT applications for security, this section presents a case study that deals with the development and deployment of an IoT device to ward off stray animals from entering crop-ready fields.

7.6.1 ANIMAL INTRUSION AND HANDLING

With the rise in stray animals, their intrusion into harvest-ready fields has become a big problem. Handling this menace is both unethical and costly. Farmers generally thrash stray animals to save their crops from being ruined. Another method they employ is protecting the fields using barbed wire or another possible fencing. This is a costly solution. The fencing cost comes out to be nearly 25,000 per acre along with wastage of considerable time and labor. This is not a foolproof solution as even after the fencing, farmers need guards to look after the system. Moreover, warding off birds remains a problem.

7.6.2 DEVELOPMENT OF IDTS

The Intrusion Detection and Tackling System, to deter animals from entering fields, is a foolproof and cost-effective IoT-based device. The security application does not even harm the animals. The device uses Raspberry pi, a camera, and speakers. The device continuously monitors fields using real-time motion detection and image processing via deep learning techniques. After classifying the animal, it inhibits their entry by producing sounds at a specific frequency that is not bearable by the targeted animals, and thus prevents them from entering the fields. In addition to this, the system also sends text SMS to farmers to alert them about intrusions with the help of the GSM module.

7.6.3 DEPLOYMENT AND DISCUSSION

The system was deployed in real time and was found to be successful in resolving the problem of animal intrusion. The farmers admired the application and accepted it readily. The application, they commented, was a revolutionary, cost-effective, and ethical one since it did not harm animals and could even fend off birds.

7.7 PRIVACY AND SECURITY ISSUES WITH IOT

Many security and privacy issues plague IoT technologies.

7.7.1 INSUFFICIENT TESTING AND UPDATING

With a massive increase of IoT devices over 20 billion, the issues with their testing and updating have also increased. In the race of producing devices to meet the rising demand, manufacturers often ignore the security of the devices. They either provide firmware updates for a short period or do not provide security updates at all, which makes their devices insecure and prone to hacking in due course of time. Moreover, they use outdated hardware and software and unsupported Linux kernels that make their products vulnerable to security risks. Proper testing and regular updates can save both the manufacturers and consumers from these security threats.

7.7.2 DEFAULT PASSWORDS AND CREDENTIALS

Selling and purchasing devices with default passwords is a major security issue with IoT devices. Mirai botnet, the most disruptive DDos attack, can best exemplify this issue. Despite instructions from the government, the manufacturers do not pay heed to this issue. Devices with factory default credentials and login details become prone to a cyber breach and can be hacked and infected easily.

7.7.3 Malware and Ransomware

With the rise in IoT devices, there came malware and eavesdropping ransomware to misappropriate them. Unlike encryption-based hacking, the hybridized malware and ransomware attacks are more fatal as they can paralyze devices and steal data simultaneously.

An IP camera, for example, can be used to steal your vital information from different locations. Your security devices can be hacked and the information can be stolen using an infected web address. You can then be asked for a ransom to return data or unlock your security system.

7.7.4 Data Security and Privacy Concerns (Mobile, Web, Cloud)

In today's digital world billions of devices constantly exchange, process, and store data. Companies use this IoT networked data to leverage their business. Companies generally share or even sell public data and violate our rights of data privacy and security. Dedicated compliance and privacy rules therefore must be enforced upon the companies so that our identity is kept intact. One way is to anonymize the data before sharing it. Cached or insignificant data must be dumped securely. Data storage must follow legal regulations.

7.7.5 Small IoT Incursions That Bypass Detection

Besides big attacks, IoT is more feared for small incursions that may bypass detection Instead of stealing bigger chunks of data, the hackers may breach sites to cause micro leaks of information.

7.7.6 Data Handling and Security

Handling huge data across billions of IoT networked devices is a big challenge. The IoT administrators are using the help of artificial intelligence and automation to deal with this issue. Great caution and precision are, however, required to employ autonomous solutions for data handling. Even a single small error in an algorithm or a code can be fatal enough to pull down the whole network. Another big issue is keeping IoT data safe from cyber-attacks and thefts. Firm regulations to rein companies involved in IoT businesses are very much required and defaulters need to be penalized.

7.7.7 Home Seizures

The smart home is a popular IoT invention that uses many IoT application including access to the house. This makes your home vulnerable to unauthorized access too.

Any criminal outfit can breach your IP address and can have your passwords to intrude your property.

7.7.8 REMOTE VEHICLE ACCESS

Smart cars are going to become a reality in view of connected IoT devices. But, their IoT association makes them prone to hijacking. An accomplished hacker might use remote access to overpower your car and might use it for criminal activities.

7.7.9 UNRELIABLE COMMUNICATIONS

The biggest IoT challenge is when IoT devices are seized to send unencrypted messages. All companies must ensure encryption of their devices as well as cloud services. They can achieve this by using transport encryption and TLS like standards. Alternatively, they can use different networks to isolate different devices. Using private communication can also ensure security and confidentiality.

7.8 CONCLUSION

IoT comes into play when innovation meets science. IoT applications have transformed traditional human practices into revolutionary buildouts. It has lent the modern world a not seen phenomenon where inanimate objects can assume the brain and partake in human endeavors. It has provided intelligent solutions to various fields like production, transportation, communication, retail, energy, agriculture, healthcare, security, and many more. IoT has leveraged progress by enhancing human activities through time, money, and labor-saving solutions. The technology, like any other invention, is also not free of flaws. Despite rendering a wealth of benefits, the technology has its downsides too. Since it is a framework of a huge number of interconnected devices, it is quite prone to risks. The IoT web uses and shares large data so it has developed privacy and security issues. There is a need, therefore, to use this technology wisely. Solutions needed to be worked out to make it a more secure and reliable source for the progress and development of the human race.

REFERENCES

Agarwal, S., & Lau, C. T. (2017). Remote health monitoring using mobile phones and Web services. *Telemedicine and e-Health*, *16*(5), 603–607.

Asghari, P., Rahmani, A. M., & Javadi, H. H. S. (2019). Internet of Things applications: A systematic review. *Computer Networks*, 148, 241–261.

Bhatt, Y., & Bhatt, C. (2017). Internet of things in healthcare. In *Internet of things and big data technologies for next generation HealthCare* (pp. 13–33). Springer, Cham.

Brito, L. M. (2017). Introduction to the Internet of Things. *Internet of Things* (pp. 3–32). Chapman and Hall/CRC.

García, C. G., Meana-Llorián, D., & Lovelle, J. M. C. (2017). A review about smart objects, sensors, and actuators. *International Journal of Interactive Multimedia & Artificial Intelligence*, 4(3), 7–10.

Kumar, N., Panda, S., Pradhan, P., & Kaushal, R. (2019). IoT based hybrid system for patient monitoring and medication. *EAI Endorsed Transactions on Pervasive Health and Technology*, 5(19), 1–7.

Kumar, N., Panda, S. N., Pradhan, P., & Kaushal, R. (2019). IoT based E-critical care unit for patients in-transit. *Indian Journal of Public Health Research & Development*, 10(3), 46–50.

Mohanty, S. P., Choppali, U., & Kougianos, E. (2016). Everything you wanted to know about smart cities: The internet of things is the backbone. *IEEE Consumer Electronics Magazine*, 5(3), 60–70.

Sharma, L., & Garg, P. K. (Eds.). (2019). *From visual surveillance to internet of things: technology and applications*. CRC Press.

Vongsingthong, S., & Smanchat, S. (2014). Internet of things: a review of applications and technologies. *Suranaree Journal of Science and Technology*, 21(4), 359–374.

Wan, J., Tang, S., Shu, Z., Li, D., Wang, S., Imran, M., & Vasilakos, A. V. (2016). Software-defined industrial internet of things in the context of industry 4.0. *IEEE Sensors Journal*, 16(20), 7373–7380.

Xia, F., Yang, L. T., Wang, L., & Vinel, A. (2012). Internet of Things. *International journal of communication systems*, 25(9), 1101.

8 Artificial Intelligence
Encouraging Students in Higher Education to Seize Its Potential

*Rajul Kumar, Karan Sehgal, and
Ankit Lal Meena*

CONTENTS

DOI: 10.1201/9781003153405-8

8.1 INTRODUCTION

There is a contemporary rise in the substantial requirement of higher education in the knowledge-based environment and society. Knowledge of aspects of IT and other related technologies is extremely important for growth and development in the society. Equally important is the establishment of new ideas and innovations in the related fields (Valluru, 2009; Haveman, 2006; JICA, 2004). Now the movement towards mass education and lifetime learning with greater diversity among subjects, students, and institutions can be seen as a global trend and there is an essential need for progress of the healthy civil community, lending government and commercial district representatives, improving aspects of life with income and availability of ample choices (Auter, 2017; Shuaibi, 2014; Frey, 1988).

Higher education covers entire graduation, research and subsequent-secondary training provided by recognized academic institutions. There is a broad diversity of courses and streams enclosing history, geography, political science, economics, sociology, e-commerce, marketing-management, finance-accounting, medicine and healthcare, and engineering sciences (David, 2018). In particular, engineering has several streams, for example, electrical, computer, civil, mechanical, environmental, biomedical, chemical, aerospace, paint and leather, metallurgy, navel-architecture, textile, petroleum, agriculture, and food technologies (Blackwell, 2001; Kettunen, 2006). However, it has been realized over a considerable period that the enrollment of students in higher studies has reduced drastically (Lynöe, 2003; Tomlinson, 2008), and the possible reasons behind this reduced participation may be the apathy and misconception of students about higher education courses that these fields are restricted towards mathematical formulas and graphs. But apart from all this they need to know that the streams and courses in higher education can relate to human intelligence and fuzziness, nature-influenced metaheuristic algorithms, ecological optimization, prey-hunting and mating behavior of predator animals, roosting traits of birds, mine blast, chemical reaction inspired optimizations, etc. Artificial intelligence using machine and deep learning of computer machinery results in the inefficient conduct of psychological and logical tasks of an individual, solely obtaining the optimized solution, capable of taking selections, actions, thinking and planning quickly (Lupo, 2005; Beauchamp, 1999; Fairlie, 2002; Brown, 1991).

The imminent intention of AI is to optimize all social practices and give a better solution to constraints than human logic can provide. Nowadays, AI is currently used in automatic vehicles and robotics, medical surgery, dismantling bombs, virtual teaching, and many more for reducing human efforts in the inevitable trend (ElMaraghy, 1987; Holzinger et al., 2007).

8.2 ETHOS FOR HIGHER EDUCATION

It is well known that the domain of higher education is not restricted to a precise range, but it shows a wide variety of diversification among subjects, branches, and steams (David, 2018). Broadly, it can be classified into humanities, commerce and business administration, medical and engineering science, as given in Figure 8.1.

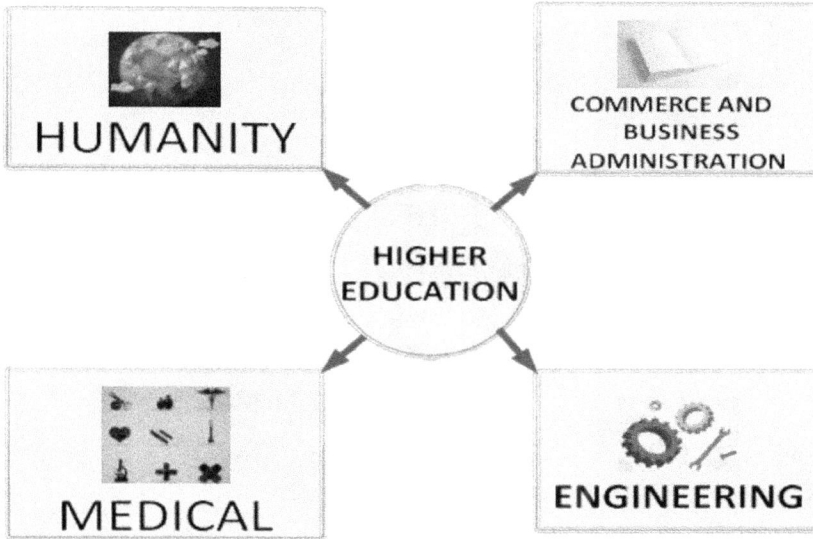

FIGURE 8.1   Gantt chart for a software development project.

All four branches of higher education shown in Figure 8.1 are furthermore classified into their respective fields such as humanities and social science into literature, art, music, philosophy, religion, language, geography, political science, economics, and sociology. Commerce and business administration consist of trade, banking, insurance, transportation, marketing, HR, supply chain and total quality management whereas medical sciences (Albrecht, 1992; de Lima, 2006) consists of cardiology, medicine, hematology, endocrinology, dentistry, and gynecology, etc. And in the last engineering subsists diversified streams shown in Figure 8.2.

All these distinct branches and streams of various disciplines integratively contribute towards the development of economic, social, and self-development.

In spite of this broad diversification, the reasons behind less participation in higher education may be the lack of interest in subjects and branches, and the belief that these branches are restricted to mathematical formulas, calculations, and mugging of theories and their drawbacks. Another important reason may be the long period for accomplishment of the enrolled courses, less respect in society for academicians as well as researchers, and unawareness of some exciting and real-world inspired subjects like artificial intelligence (AI).

AI in the present era is improving and gaining popularity with its new modernization for solving real-world problems in an optimized way. It allows us to replace fixed routine and multiple human tasks with AI coded computer machinery and robotics through continuous knowledge skills, i.e., machine and deep learning. These learning methods are inspired by human creativeness; nature inspires algorithms, robust ecological colonization of unwanted plants, etc. Some of these AI algorithms and techniques are discussed in the next section.

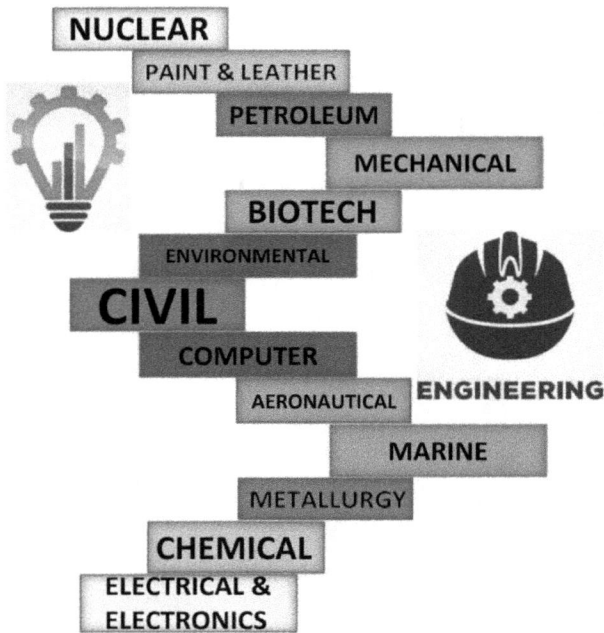

FIGURE 8.2   Streams of engineering.

8.3   SOME BIO-INSPIRED AND HUMAN FUZZY TECHNIQUES IN AI

AI emulates human creative thinking, nature-inspired hunting, and surviving behaviors of animals and predators for learning to perform the given task precisely. Some examples of these algorithms and techniques are explained briefly.

8.3.1   FUZZY LOGIC TYPES 1 AND 2

Fuzzy logic is based on human fuzzy logic that the probability of an event going to happen is not always crisp values. Still, it can also lie between the prescribed range, i.e., it can be ambiguous and fuzzy (Valluru et al., 2011). Fuzzy logic Type-1 deals with real-world problems that are more uncertain, complex, and more non-linear. The process between the prescribed limit of the universe of discourse can be shown in the form of triangular, trapezoidal, Gaussian, bell, etc., shaved membership functions, as shown in Figure 8.3.

Crisp values are converted into fuzzy sets by a fuzzifier. The inference engine simulates fuzzy rules controlled by the rule base in the Mamdani or Takagi-Sugeno approach, and after processing these fuzzy values again converted into human- understandable form. The advanced and more capable fuzzy logic form is Type-2 fuzzy, which contains a footprint of uncertainty in-universe of the discourse of membership functions (Cazarez-Castro, 2012) shown in Figure 8.4.

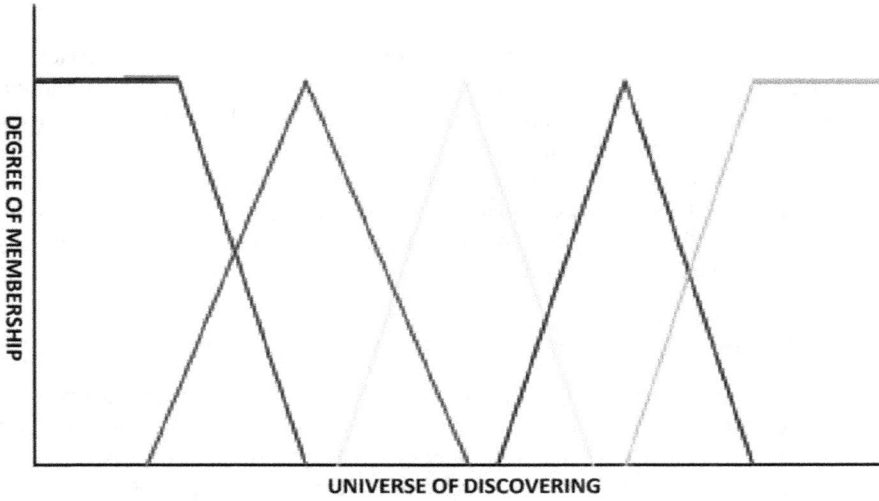

FIGURE 8.3 Membership function of Fuzzy Type 1.

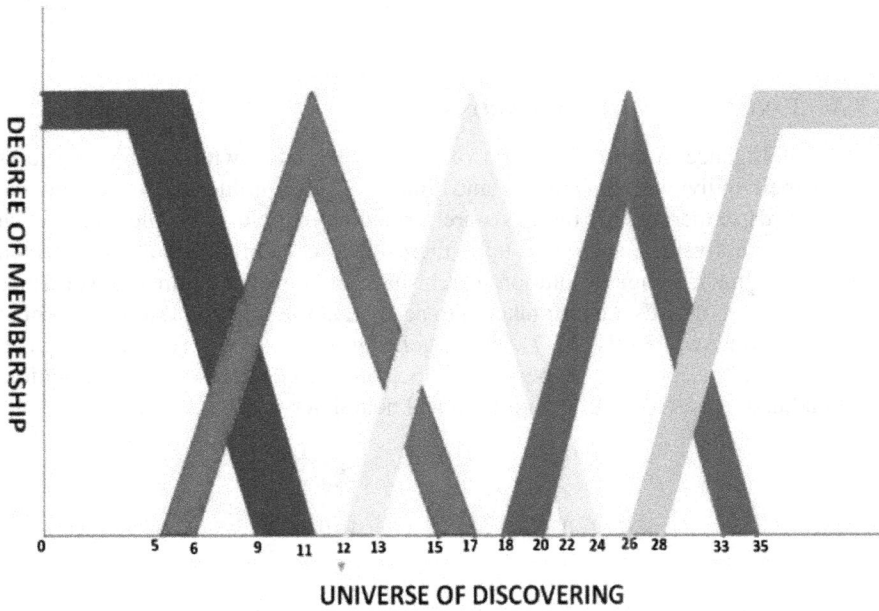

FIGURE 8.4 Membership function of Fuzzy Type 2.

8.3.2 Genetic Algorithm

This is a metaheuristic bio-inspired optimization technique, which is based on Charles Darwin's theory of innate progression. The process starts with the initialization of community formed through genes encircled by chromosomes, then the choice of an individual based upon fitness function followed by random crossover and mutation among genes shown in Figure 8.4. The algorithm is terminated if the offspring produced by the parent generation gives the fittest solution to our real-world benchmarked problems or if the population of the community gets completely converged (Xin-She, 2010; Valluru, 2011).

8.3.3 Artificial Neural Network

ANN is fabricated to identify complex patterns, similarly, as the human brain operates. It comprises initial layers formed from estimation units called neurons. These neurons are initially multiplied by weights, and later on, they acquire required input patterns and get mapped to the output layer, as explained pictorially in Figure 8.5. The input pattern may or may not contain the potentially required desired output. However, the hidden layer tunes the input weights by iterative methodology based on fitness function until the error gets diminished to a required level (Falco et al., 1998). It is a very efficient and agile scheme, capable of learning intricate patterns and adapt to new system environments quickly.

8.3.4 Particle Swarm Optimization

Swarm intelligence comprises groups of birds synergizing with complex systems through their individual capabilities and fitness. PSO is influenced by communal traits of birds gathering. All the birds are termed as particles with their individual computed fitness values, and based upon their fitness capabilities; they try to search the global optima in assigned solution search space, as shown in Figure 8.6. Velocities and acceleration of birds are also taken into account at the time of iterative searching of solution (Mavrovouniotis, 2017). This algorithm seems to be very elementary and takes less time for computation because it does not include crossover and mutation, weight adjustment as in genetic algorithm and neural network.

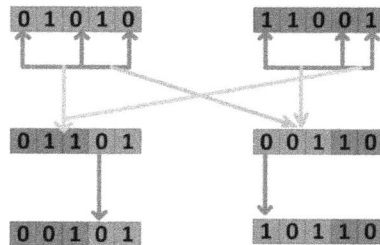

FIGURE 8.5   Crossover and mutation in GA.

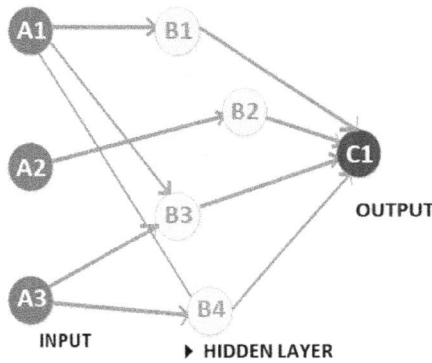

FIGURE 8.6   Neurological layers in ANN.

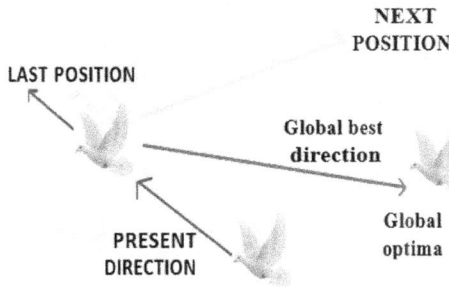

FIGURE 8.7   Swarms seeking the global optima in PSO.

8.3.5   INVASIVE WEED OPTIMIZATION

IWO is an ecological metamorphic optimization algorithm based upon the robust acclimating behavior of invasive weeds in cultivation (Xing, 2014). Weeds are undesirable shrubs or saplings found in our solution search space or cultivation area, and their invasive traits pose a critical hazard to the plantation. They are very adaptive and flexible to their surrounding and environmental changes; they seize the cultivation search area by permanent colonization and are improved through interacting with their peer weeds based upon their capabilities and linear fitness enhancement of whole community fitness. They reproduce by sex cells through fertilization of pollen eggs and can choose R or K force selections for surviving. An iterative process of seed dispersion takes place with the elimination of weeds with the lowest fitness based upon standard deviation, mean, and variance.

8.3.6   GRAY WOLF OPTIMIZATION

GWO is one of the swarm intelligence technique based on the near-optimum hunt ability of canis lupus gray wolfs. Gray wolfs hunt for prey in groups of 5–11 and are

FIGURE 8.8 Spreading of sex seeds in outer orbits.

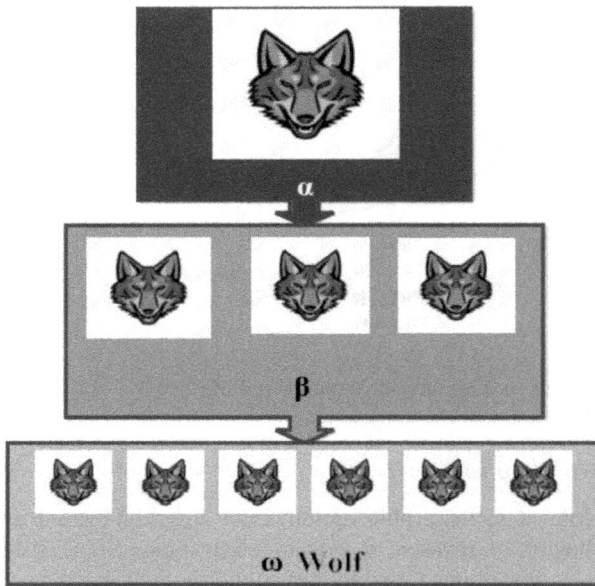

FIGURE 8.9 Hierarchy of canis lupus grey wolfs.

considered as top predators. They decode their hierarchy as alpha, beta, and omega wolfs based upon their hunt capability of the optimum solution, dominant nature, and potential to enhance cooperation, federation, and acquiescence in respective squads. The alpha wolf is considered as the head of the squad and makes the majority of decisions like a solution searching regime and trailing, etc. Beta wolfs assist and follow the orders of alpha leaders in solution hunting and guide omega wolfs, which occupy the last ranking in the hierarchy as shown in Figure 8.8. It is a very reliable and potent nature influenced technique capable of having near-optimum search efficacy (Mirjalili, 2014).

8.3.7 Lion Optimization Algorithm

LOA is a population-based meta-heuristic algorithm, which is influenced by the social behavior of the group "Lion prides." The process begins with initialization in which the total population of lions and lioness form a pride based on their fitness value. Further for hunting, all-female lions are sorted according to their fitness value and grouped into three groups (chasers, wingers, and cheaters) to catch prey. Then an intergroup interaction is introduced by eliminating the weak male lion according to their fitness value. Mating and migration are also considered while forming an optimized pride (Yazdani, 2016). This algorithm is adequate in solving a complex and non-linear optimization problem.

8.3.8 Mine Blast Optimization

The mine blasting algorithm is the metaheuristic algorithm based on the concept of mine detonations. When a mine is detonated, its shell pieces get spread in many directions with different velocities. These shells lead to more explosions of other mines when nearby mines get activated on collision with the shells of the previously exploded mines. This technique starts with the initialization in which the first mine gets detonated, and an optimal point is formed there, and its shell leads to explosions of other mines. The ratio of the number of mines exploding to the number of shells is quite large, this leads to the explosion of other mine as well (which may/may not have high detonation power). The overall number of casualties thus gets increased. These casualties are considered as fitness values for approaching many iterations (Siskind, 1980). This optimization is efficient for solving complicated bounded optimization and engineering problems.

8.3.9 Whale Optimization Algorithm

Whale optimization algorithm is a nature-influenced meta-heuristic optimization technique, which is based on their social-demeanor like hunting, mating, etc. Whales' hunting comprises different methods such as seeking prey, flanking the prey and air-globules. Generally, whales perform these methods in two different manners either by diving down from the surface about few meters and then producing an air-globule in a helical shape around the prey and then diving up towards the surface or by beating the surface of the water with strikes to surround the prey (Mirjalili, 2016; Prakash, 2018). This technique is very feasible for treating highly non-linear dynamical systems.

These algorithms are highly capable of optimization and solving other real-world tasks. Some of these applications of AI are discussed in the next section.

8.4 APPLICATION OF ARTIFICIAL INTELLIGENCE IN REAL WORLD PROBLEMS

AI simulates the creative and astute computing capabilities of human behavior for solving complex real-world tasks. It is used in industries, engineering problems,

$\mathcal{P}$ $\mathcal{N}$

$\mathcal{R}$ $\mathcal{O}$

$\mathcal{I}$ $\mathcal{M}$

$\mathcal{D}$ $\mathcal{A}$

$\mathcal{E}$ $\mathcal{D}$

FIGURE 8.10   Classification in LOA.

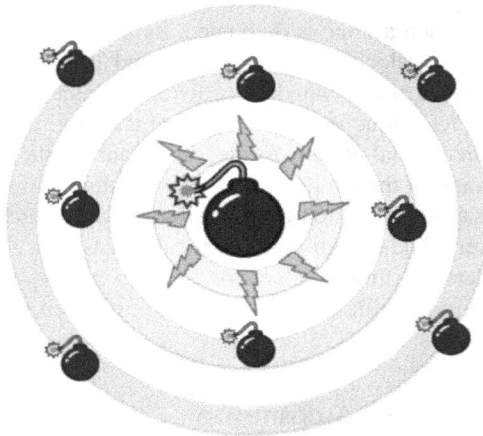

FIGURE 8.11   Dispersal of mine shells.

oriented to business tasks, booking of hotels, and many more. Some of the major applications of AI are explained theoretically as well as pictorially below.

8.4.1   E-COMMERCE MARKETING

Recently, online marketing and advertisement have gained popularity because of advanced stochastic AI algorithms solely fabricated for marketing and brand advertisement of industries. Nowadays, it is straightforward to search an online product by knowing its properties and also, we get the list of the products with similar ratings and applications, which is possible because of the pattern recognition capability of AI techniques. In Bao-he and Yu (2012; Balachandran, 2015), Fuzzy logic is used to elaborate on the subjective preferences of humans by their initial single choices and negotiation. The genetic algorithm is used by Chen and Shahabi (2002) to reduce the gap between the physical properties of data with the user's perception by utilizing

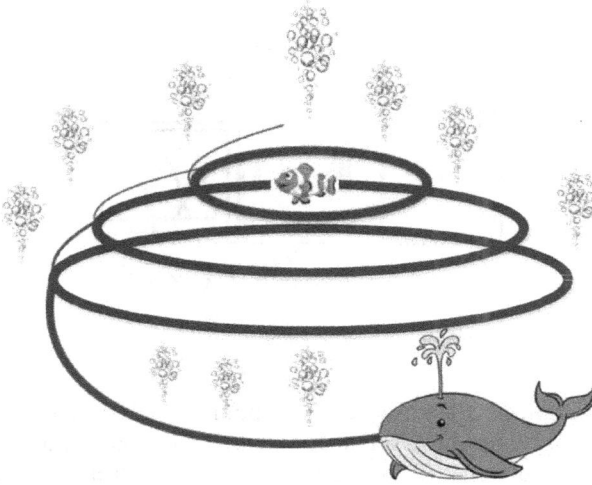

FIGURE 8.12 Helical trajectory in WOA.

FIGURE 8.13 Elements of e-commerce and advertisement.

relevant feedback from the customer. Also, Park (2000) uses neural network data mining-based strategy for clustering similar customers into groups for target and direct marketing. For an efficient task, scheduling LOA is used by Almezeini (2017) by sharing resources with internet users through service providers.

8.4.2 Control Engineering

AI plays a significant role in control engineering practices for the optimization of controller parameters. After optimization through the bio-inspired evolutionary algorithms the controller performs better in terms of rise time, maximum overshoot, and steady state error. Controller performances may be validated by the time taken to stabilize after the given disturbances for any benchmarked systems like a cart inverted pendulum, quad-copter, uncrewed aerial, and autonomous underwater vehicles. Various controllers are equipped with control problems, for example, PID, FOPID,

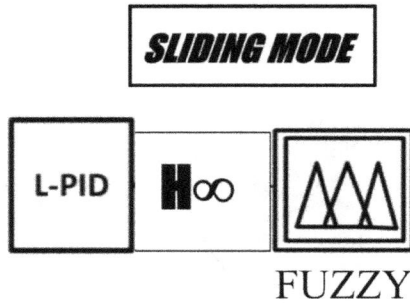

FIGURE 8.14   Different control schemes in control engineering.

SMC, H∞, etc. Sharma (2019) use LOA for optimizing FO-PI controller parameters for load frequency control of two interconnected multi-source power system areas. Also, Patel (2019) applies WOA for tuning Fuzzy-PI controllers for the LFC issue in thermal, hydro-wind interconnected system. GWO is applied by Şen (2018) for optimal tuning of the PID controller for single footstep tracking in the desired path for a quadruped robot. Mandava (2018) used a modified IWO based PID controller to control the motors of the biped robot by walking on a flat surface.

8.4.3 Surgery and Operations in Medical Science

With the inferior learning and pattern reorganization capabilities of AI techniques, bots and computer machinery are trained to perform operations and surgeries in dental, cardiovascular, pulmonary, and esthetic domains. Earlier robotics are used for giving local anesthesia and simple wounding tasks. Still, now AI and its precise computing techniques have changed the trend of medical science from human resources to pattern learning capability of robots. Beligiannis et al. (2006) use a GA based intelligent driving system for the diagnosis of male sexual dysfunction function, which is presented with detailed and accurate results. The genetic algorithm also applies contemporary theories for achieving successful treatment (Hackenberger, 2019) with a maximum 20% probability of harming other organs. Also, dark arrow lines of fuzzy membership function are used for medical diagnosis of drugs required for a patient under treatment (Sundaresan, 2018).

8.4.4 AI in Agriculture

AI enhances the agriculture productivity and lowers the non-environment friendly impacts. Pattern recognition techniques in AI help farmers to predict soil humidity, solar broadcast, speed, and direction of wind for superior outcomes in agriculture production. Also, they identify the requirement and over nourishment of particular nutrients and minerals in crops. AI bots examine the unwanted invasive weeds and rectify them by the use of appropriate separation methods by the desired production. However, it is well known that the implementation of AI trained bots does not eradicate the farmer's job, rather it helps and improves the quality of production. In Wang

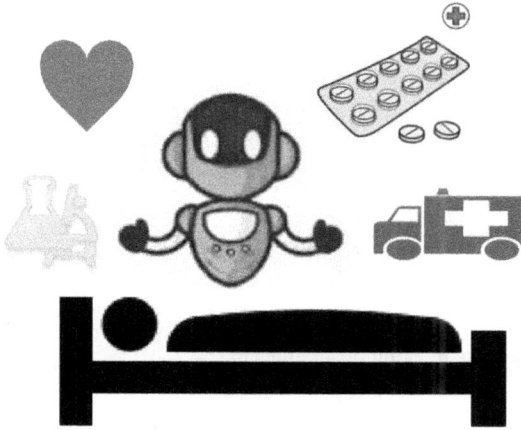

FIGURE 8.15 AI in medical science.

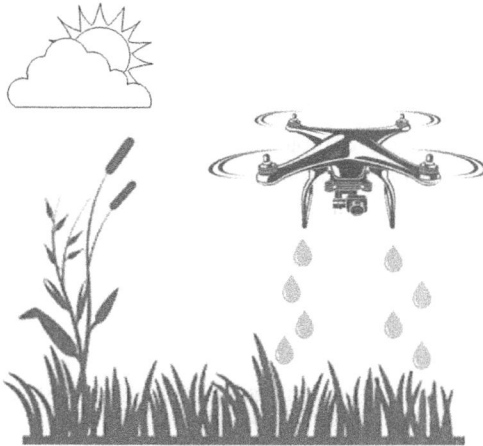

FIGURE 8.16 Automated drone irrigation in agriculture.

(2012) for water-deficient regions, PSO is used for optimizing and adjusting a water saving-plant structured for having a maximum grain yield and higher water productivity. Image detection using fuzzy logic and neural network (Murmu and Biswas, 2015) is done for crop mapping and classification so that appropriate irrigation can be done according to the crop identified.

8.4.5 In Aviation and Space Organizations

Space agencies from all over the world use AI for optimization of their machine parameters and also for trajectory tracking of satellite defined paths over no man's land. Even in air aviation, it is used in drill simulators and support panels, the

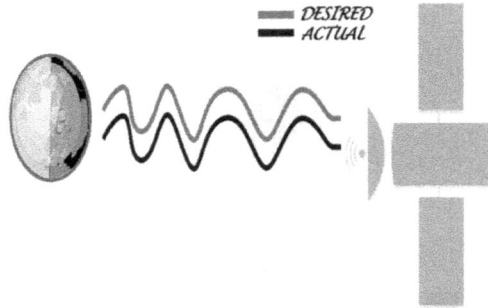

FIGURE 8.17   Trajectory tracking of satellites in space.

simulated air war, and detecting wind direction, humidity, and temperature during hovering of aerial vehicles by past pattern recognition. AI-enabled autopilot in aircraft and helicopters is also a recent trend because of some unwanted causalities faced because of naïve acts of human pilots like flying above and below operational ranges, a distraction from concentration, some misunderstandings between co-pilots, family and work stress. Such pilot errors can be rectified to some extent with the partial help of automated system enabled sensors and trackers. The neural network is used to harness unsteady aerodynamic effects, parallel processing of 1000s of sensors for detection, and approaching complex non-linear problems in aeronautics (Faller, 1996). Also, space trajectory optimization is achieved using hidden genes in GA for optimal trajectory tracking (Darani, 2018) of one celestial orbit into others.

8.5 CONCLUSION

It can be seen worldwide that after secondary school education, the enrollment and interest of students in pursuance higher education has reduced drastically. And some of the possible reasons behind the reduced participation may be the lack of interest and creativity or bare implementation of mathematical formulas, equations, and theories which they have learned in secondary education for solving real-world problems or for helping humanity. Towards the enhancement of participation and interest of students in higher studies, with unique perception, AI technology is presented with its application in real-world problems. Some AI techniques that relate to human thinking and natural phenomena for solving complex real-world problems are presented theoretically as well as pictorially. We hope our scant efforts develop some interest and motivation among students for enhancing higher education pursuance.

REFERENCES

Albrecht, D., & Ziderman, A. (1992). Funding mechanisms for higher education: Financing for stability, efficiency, and responsiveness. http://files.eric.ed.gov/fulltext/ED344518.pdf.
Almezeini, N., & Hafez, A. (2017). Task scheduling in cloud computing using lion optimization algorithm. *International Journal of Advanced Computer Science and Applications.* 8(11), 77–83. DOI: 10.14569/IJACSA.2017.081110.

Auter, Z. (2017). Half of college students say their major leads to a good job. www.gallup.com/ education/231731/half-college-students-say-major-leads-good- job.aspx.

Balachandran, B. M., & Mohammadian, M. (2015). Development of a fuzzy-based multi-agent system for e-commerce settings. *Procedia Computer Science*, 60(1), 593–602. https:// doi.org/10.1016/j.procs.2015.08.186.

Bao-he, Z., & Yu, S. (2012). Multi-agent negotiation with fuzzy logic in e- commerce. *Advanced Technology in Teaching*. 163, 551–561. https://doi.org/10.1007/978-3-642-29458-7_80.

Beauchamp, G. (1999). The evolution of communal roosting in birds: origin and secondary losses. *Behavioral Ecology*. 10(1). 675–687. https://doi.org/10.1093/beheco/10.6.675.

Beligiannis, G., Hatzilygeroudis, I., & Koutsojannis, C. (2006). A GA driven intelligent system for medical diagnosis. In: *International Conference on Knowledge-Based and Intelligent Information and Engineering Systems*, pp. 968–975. https://doi.org/10.1007/ 11892960_116.

Blackwell, A., Bowes, L., Harvey, L., Hesketh, A. J., & Knight, P. T. (2001). Transforming work experience in higher education. *British Educational Research Journal*. 27(3), 269–285. https://doi: 1080/01411920120048304.

Brown, F. A. (1991). An ecological optimization criterion for finite-time heat engines. *Journal of Applied Physics*. 69(11), 7465–7469. https://doi.org/10.1063/1.347562.

Cazarez-Castro, N. R., Aguilar, L. T., & Castillo, O. (2012). Designing Type-1 and Type-2 fuzzy logic controllers via fuzzy Lyapunov synthesis for nonsmooth mechanical systems. *Engineering Applications of Artificial Intelligence*. 25(5). 971–979. https:// doi.org/10.1016/j.engappai.2012.03.003.

Chen, Y. S., & Shahabi, C. (2002). Improving user profiles for e-commerce by genetic algorithms. In: Segovia J., Szczepaniak P.S., Niedzwiedzinski M. (eds) *E-Commerce and Intelligent Methods, Studies in Fuzziness and Soft Computing, vol 105*. Physica, Heidelberg. 215–229. https://doi.org/10.1007/978-3-7908-1779-9_13

Darani, S. A., & Abdelkhalik, O. (2018). Space trajectory optimization using hidden genesgenetic algorithms. *Journal of Spacecraft and Rockets*, 55(3), 764–774. https:// doi.org/10.2514/1.A33994.

David, S., & Ashley, B. (2018). *The Integration of the Humanities and Arts with Sciences, Engineering, and Medicine in Higher Education: Branches From The Same Tree*. Washington DC: National Academies Press.

de Lima, A. A., Bettati, M. I., & Baratta, S., et al. (2006). Learning strategies used by cardiology residents: Assessment of learning styles and their correlations. *Education for Health*. 19(3), 289–297. https://doi.org/10.1080/13576280600937788.

ElMaraghy, H. A. (1987). Artificial intelligence and robotic assembly. *Engineering with Computers*. 2. 147–155. https://doi.org/10.1007/BF01201262.

Fairlie. G., & Bergeron, D. (2002). Numerical simulation of mine blast loading on structures. In: 17th International Symposium on the Military Aspects of Blast and Shock June 10-14, 2002 Las Vegas, Nevada, Washington, DC: US Dept. of Defense, 17–22.

Falco, D., Cioppa, A. D., & Natale, P. (1998). *Soft Computing in Engineering Design and Manufacturing: Artificial Neural Networks Optimization by means of Evolutionary Algorithms*. London: Springer. https://doi.org/10.1007/978-1-4471-0427-8_1.

Faller, W. E., & Schreck, S. J. (1996). Neural networks: Applications and opportunities in aeronautics. *Progress in Aerospace Sciences*. 32(5), 433–456. https://doi.org/10.1016/ 0376-0421(95)00011-9.

Frey, Donald E. (1988). *Educational research and policy*. Edited by T. Husen and M. Kogan. Oxford, Pergamon Press. *Economics of Education Review,* Elsevier, vol. 7(2), 261–261, April.

Hackenberger, B. K. (2019). Genetics without genes: Application of genetic algorithms in medicine. *Croatian Medical Journal.* 60(2), 177–180. https://dx.doi.org/10.3325%2Fcmj.2019.60.177.

Haveman, R., & Smeeding, T. (2006). The role of higher education in social mobility. *Future of Children.* 16(2), 125–150. https://doi.org/10.1353/foc.2006.0015.

Holzinger, A., Biemann, C., Pattichis, C. S., & Kell, D. B. (2017). What do we need to build explainable AI systems for the medical domain?. 1–28. https://arxiv.org/abs/1706.07979.

Kettunen, J. (2006). Strategies for the cooperation of educational institutions and companies in mechanical engineering. *International Journal of Educational Management.* 20(1), 19–28. http://dx.doi.org/10.1108/09513540610639567.

Lupo, K. D., & Schmitt, D. N. (2005). Small prey hunting technology Small prey hunting technology and zooarchaeological measures of taxonomic diversity and abundance: Ethnoarchaeological evidence from Central African forest foragers. *Journal of Anthropological Archaeology.* 24(4), 335–353. DOI: 10.1016/j.jaa.2005.02.002.

Lynöe, N., Höyer, K., & Mjörndal T. O. (2003) Research Ethics Committees – who is to represent the community? *Lakartidningen.* 100, 1987–1988.

Mandava, R. K., & Vundavilli, P. R. (2018). Implementation of modified chaotic invasive weed optimization algorithm for optimizing the PID controller of the biped robot. *Sadhana – Academy Proceedings in Engineering Sciences* . 43: 1–18.

Mavrovouniotis, M., Li, C., & Yang, S. (2017). A survey of swarm intelligence for dynamic optimization: algorithms and applications. *Swarm and Evolutionary Computation.* 33, 1–17. https://doi.org/10.1016/j.swevo.2016.12.005.

Mirjalili, S. & Lewis, A. (2016). The whale optimization algorithm. *Advances in Engineering Software.* 95, 51–67. https://doi.org/10.1016/j.advengsoft.2016.01.008.

Mirjalili, S., Mirjalili, S. M., & Lewis, A. (2014). A grey wolf optimizer. *Advances in Engineering Software.* 69, 46–61. https://doi.org/10.1016/j.advengsoft.2013.12.007.

Murmu, S., & Biswas, S. (2015). Application of fuzzy logic and neural network in crop classification: A review. *Aquatic Procedia*, 4, 1203–1210. https://doi.org/10.1016/j.aqpro.2015.02.153.

Park, S. (2000). Neural networks and customer grouping in e-commerce: A framework using fuzzy ART. In: *Academia/Industry Working Conference on Research Challenges: Next Generation Enterprises: Virtual Organizations and Mobile/Pervasive Technologies*, 331–336. https://doi.org/10.1109/AIWORC.2000.843312.

Patel, N. C., & Debnath, M. K. (2019). Whale optimization algorithm tuned fuzzy integrated PI controller for LFC problem in thermal-hydro-wind interconnected system. In: Mishra, S., Sood, Y., Tomar, A. (eds) *Applications of Computing, Automation and Wireless Systems in Electrical Engineering. Lecture Notes in Electrical Engineering.* Springer, Singapore. 553, 67–77. doi:10.1007/978-981-13-6772-4.

Prakash, D. B., & Lakshminarayana, C. (2018). Multiple DG placements in radial distribution system for multi objectives using whale optimization algorithm. *Alexandria Engineering Journal.* 57(4), 2797–2806 https://doi.org/10.1016/j.aej.2017.11.003.

Şen, M. A., & Kalyoncu, M. (2018). Optimal tuning of PID controller using grey wolf optimizer algorithm for quadruped robot. *Balkan Journal of Electrical and Computer Engineering.* 6(1), 29–35. DOI: 10.17694/bajece.401992.

Sharma, D., & Yadav, N. K. (2019). Lion algorithm with Levy update: Load frequency controlling scheme for two-area interconnected multi-source power system. *Transactions of the Institute of Measurement and Control.* 41(14), 4084–4099. https://doi.org/10.1177%2F0142331219848033.

Shuaibi, A. Al. (2014). The importance of education. *Qatar Chronicle.* www.qatarchronicle.com/forum/47660/the-impo.

Siskind, D. E., Stagg, M. S., & Kopp, J. W. (1980). *Structure Response and Damage Produced By Ground Vibration From Surface Mine Blasting*, Epub ahead of print 1980. DOI: 10.1016/0148-9062(81)91353-x.

Sundaresan, T., Sheeja, G., & Govindarajan, A. (2018). Different treatment stages in medical diagnosis using fuzzy membership matrix. In: *Journal of Physics: Conference Series*. Epub ahead of print 2018. DOI: 10.1088/1742-6596/1000/1/012094.

Tomlinson, M. (2008). The degree is not enough: Students' perceptions of the role of higher education credentials for graduate work and employability. *British Journal of Sociology of Education*. 29(1). 49–61. https://doi.org/10.1080/01425690701737457.

Valluru, S., & Narayana, M. S. (2009). Ethical values in professional education-need of the hour. *The Indian Journal of Technical Education*. 32(1), 85–89.

Valluru, S. K., & Rao, T.N. (2011). *Introduction to Neural networks, Fuzzy Logic, and Genetic Algorithms*. 2nd ed. Mumbai: Jaico Publishing House.

Valluru, S. K., Singh, M., & Kumar, N. (2011). National Electrical Engineering Conference: The Inventive Use of Genetic Algorithms in Neuro Fuzzy Applications, 1–6.

Wang, Y., Wu, P., & Zhao, X. (2012). Water-saving crop planning using multiple objective chaos particle swarm optimization for sustainable agricultural and soil resources development. *Clean-Soil, Air, Water*. 40(12), 1376–1384. https://doi.org/10.1002/clen.201100310.

Xing, B., & Gao, W. J. (2014). *Innovative Computational Intelligence: A Rough Guide to 134 Clever Algorithms: Invasive Weed Optimization Algorithm*. Switzerland: Springer International Publishing. DOI: 10.1007/978-3-319-03404-1.

Xin-She, Y. (2010). *Engineering Optimization an Introduction with Metaheuristic Application*. Hoboken, NJ: John Wiley & Sons. DOI: 10.1002/9780470640425.

Yazdani, M., & Jolai, F. (2016). Lion Optimization Algorithm (LOA): A nature-inspired metaheuristic algorithm. *Journal of Computational Design and Engineering*. 3(1), 24–36. https://doi.org/10.1016/j.jcde.2015.06.003.

9 Thermal Performance of Natural Insulation Materials for Energy Efficient Buildings

R.K. Pal, Parveen Goyal, and Shankar Sehgal

CONTENTS

9.1 INTRODUCTION

A major part of the total energy consumption and carbon dioxide emission production is by the building sector. Carbon dioxide emissions of more than 40 billion tons will be produced and around 33% of world energy expenditure will be consumed by buildings by 2030 (Ofori et al. 2000; Wang et al. 2018). The building sector consumes the highest amount of energy in the world (Pérez-Lombard et al. 2008). Energy efficient buildings can be used to reduce the energy consumption. Energy efficient design uses natural methods such as insulation on walls and roofs to reduce energy consumption in buildings (Al-Badi and Al-Saasi 2020). Natural materials are sustainable and

DOI: 10.1201/9781003153405-9

of great importance for green and energy efficient buildings (Yu et al. 2018; Jiang et al. 2018). Sustainable materials and methods can be used for construction and insulation of buildings. They can play a major role in conserving the energy in the world (Liang et al. 2007). Production of building materials and construction of buildings causes 40% of the overall pollution (Ofori et al. 2000). In comparison to structures made from concrete or bricks, buildings constructed from wood cause lower carbon dioxide emissions during their life cycle (Upton et al. 2008). Man-made buildings in a city can increase the temperature by 2°C by creating a heat island and this can be controlled by constructing energy efficient buildings using sustainable and recyclable materials (Besir and Cuce 2018). It is very much clear that contemporary building design is not sustainable for a long time (John et al. 2005). Energy consumption for indoor thermal comfort in buildings is rising continuously and energy savings can be obtained by providing insulation in the buildings thereby reducing heat loss or gain for the buildings (Liu et al. 2017). Thermal insulation materials and wall materials are the major constituents of buildings (Zuo and Zhao 2014). The research has been carried mostly on green buildings and the literature availability is rare on thermal insulation materials though it is a key constituent of green buildings (Wang et al. 2018). A major role can be played by high level of insulation combined with thermal mass in maintaining comfortable indoor temperature at lower costs (John et al. 2005). Insulation of buildings is becoming a major energy saving method because of advantages like easy to use, lower cost, good efficiency and high viability (Meng et al. 2018; Xing et al. 2019). The heat loss or gain from buildings can be reduced greatly by employing insulation layer in cold or hot climates (Besir and Cuce 2018). The materials with lower thermal conductivity act as better insulators and a thinner layer needs to be provided for the same indoor air temperature (Hurtado et al. 2016). The energy efficiency of a building can be improved by insulating the building (Erkmen et al. 2020; Keynakli 2012). Insulating the roof and walls of buildings thermally can reduce the energy consumption by 30% and 20%, respectively (Balaras et al. 2000). Thermally insulated buildings can provide more comfortable indoor conditions to live in (Wang and Yan 2004). Stone wool, glass fiber, polyurethane foam and expanded polystyrene are traditional insulations used though these maintain high thermal comfort and have large embodied energy (Hurtado et al. 2016). Utilization of the insulation materials in buildings has an impact on the environmental (Heeren and Hellweg 2019). The insulation materials available in the market are mainly produced from synthetic materials (Orlik-Kożdoń 2017; Zach et al. 2012). Insulation materials used in buildings are generally made from petrochemicals and their production process causes a lot of environmental pollution (Wang et al. 2018). The petrochemical based insulations also cause lot of environmental pollution during the life cycle (Mahmood 2020). Nearly half of the present global greenhouse gases are produced due to natural resource extraction (Wiprächtiger et al. 2020). Further thermal insulations like polystyrene, polyurethane and polyethylene produced from petrochemicals are not of a renewable nature and thermal insulations like rock wool, perlite and vermiculite made from rocks also create a lot of pollution and consume a great deal of energy during their production (Liu et al. 2017). Recovery and reuse of these materials are not easy and have many problems though some of these materials like extruded polystyrene board, foamed polystyrene board, glass wool board and rock wool board, etc. exhibit a good performance (Wang et al. 2018). The chemical nature of some organic materials used for buildings can affect indoor air quality as some materials emit volatile organic compounds and chemical bonds can be broken by ultraviolet radiation of sun in an organic material, which can change composition and influence the properties in a big way (John et al. 2005). On the other hand, naturally occurring building materials exhibit a better prospect than these materials and are recyclable and environment friendly (Wang et al. 2018). The present energy and environmental crisis have resulted in renewed interest in sustainable materials utilization for building applications (Asdrubali et al. 2017). The awareness of high environmental impact of synthetic materials has led to the development of materials that can replace these synthetic materials (Girijappa et al. 2019). Natural bio-based material has a good potential to be used as insulation (Binici et al. 2020). An increase in the use of natural materials for production of insulation has been observed in the recent past (Erkmen et al. 2020). This will reduce the environment effect due to use of petrochemical based insulation materials (Lithner et al. 2011). Bio-insulations based on biomasses are also gaining interest for insulating buildings lately (Binici et al. 2014; Martínez-García et al. 2020; Pinto et al. 2011; Valverde et al. 2013; Wei et al. 2015). These natural materials have advantages of easy availability, abundance and lower cost over the synthetic materials (Arpitha et al. 2017; Madhu et al. 2019). These materials have lower weight, lower density and lower cost in comparison to synthetic insulation materials (Abu-Jdayil et al. 2019). Natural bio-based insulation materials are renewable, sustainable and are biodegradable at the end of their life span as compared to mineral and petroleum-based insulation materials (Latif 2020). Materials based on wood fiber, wood shavings and wood chips can also be used as bio-insulations for buildings (Binici and Aksogan 2016; Cosereanu et al. 2010; Latif et al. 2015; Liu et al. 2017). Natural insulations can be used to provide thermally comfortable, healthier and more sustainable indoor environment (Volf et al. 2015). Insulation boards prepared from cotton stalk, bagasse and coconut husk, corncob and natural bark can also be used for insulating buildings (Knapic et al. 2016; Paiva et al. 2012; Palomo et al. 1999; Pasztory et al. 2017; Zhou et al. 2010). Insulation prepared from date palm wood can be utilized as well for insulating buildings (Agoudjil et al. 2011). A lot of research has been carried on wood as a building insulation material (Liu et al. 2017). Wood is a renewable, reusable, recyclable material that can be utilized with high sustainability rates. Wood has thermal insulation properties due to low thermal conductivity and can be used for insulating buildings and energy savings up to 40% are obtainable by employing wood as insulation due to its low thermal conductivity (Asdrubali et al. 2017). The thermal insulations employed in the buildings can cover its cost through savings in energy expenditure in terms of lesser air conditioning usage (Li et al. 2016).

Keeping the available literature in view the present study is focused on assessment of the suitability of the various natural materials like wood, wood fiber, wood waste and coconut husk etc. for insulating the buildings from sustainable, environmental, thermal performance perspectives, reduce the cooling load and conserve energy. The main parameters considered are the thermal conductivity and thermal diffusivity as performance parameters, and these are either taken from literature or computed using the available literature and compared for various natural materials between themselves and with synthetic and mineral based materials. The sustainable and environmental aspects of these natural materials will also be viewed from the available literature.

9.2 INSULATION FOR THERMAL COMFORT, THERMAL PERFORMANCE OF INSULATION AND SUSTAINABLE INSULATION MATERIALS

The indoor air temperature inside a building becomes very high in the summer and may fall to a very low level in the winter. In these conditions the thermal comfort can be provided inside a building using insulation materials. The performance of these insulation materials is measured in terms of thermal performance parameters like thermal conductivity, thermal resistance, thermal transmittance and thermal diffusivity. These insulation materials have an impact on the environment, have embodied energy consumed during production process and may be sustainable or not from these points of views. In this section the use of insulation for thermal comfort, thermal performance of insulations and sustainable materials are elaborated.

9.2.1 INSULATION FOR THERMAL COMFORT

The estimates foretell that the average earth surface temperature may rise by around 6°C by the end of 2100 due to carbon dioxide emissions (Aditya et al. 2017). This rise in earth surface temperature will increase the indoor air temperature of buildings. This high indoor air temperature of buildings will make the indoor environment very uncomfortable to live in for occupants. In order to make it comfortable for living a lot of energy will be needed for space cooling using air conditioners in hot summers. This will also generate lot of environmental pollution in addition to the increased energy expenditure. Another method is to provide insulation materials to stop the heat flow to the inside of the buildings. Thermal insulations can also be utilized in buildings in cold climate regions. Thermal insulation is used to check heat loss from buildings and provide thermal comfort inside a building in cold climates (Hurtado et al. 2016). Indoor air temperature can be maintained independent of ambient temperature fluctuations using superior insulation materials (Kawasaki and Kawai 2006). This will help to keep the indoor air temperature within the comfortable limits without any additional expenditure of energy. The desired thermal resistance can be obtained using adequate thickness of the insulating material being used (Kawasaki and Kawai 2006). Thickness of insulation provided should be optimum thickness as the savings decrease if the thickness of insulation is increased beyond this limit (Aditya et al. 2017). Therefore thermal insulation can be used for providing the desired thermal comfort and it is also environment friendly (Aditya et al. 2017).

9.2.2 THERMAL PERFORMANCE

It is an established fact that the thermal comfort inside a building can be increased using the thermal insulation. The parameters in terms of which the thermal performance of an insulation material is usually measured are thermal conductivity, thermal transmittance offered by the insulation materials in steady state and in terms of thermal diffusivity in unsteady state heat flow through a material (Asdrubali et al. 2015). The thermal performance of a component of building envelope is measured in terms of its resistance to heat flow or overall heat transfer coefficient (Kosny et al. 2014). Thermal conductivity is an important parameter to work out the thermal resistance of an insulation material (Wu et al. 2020). The insulation material for buildings must have thermal conductivity lesser than 0.065 W/mK (Florea and Manea 2019). The performance of a wall from a thermal point of view is considered good if it can improve thermal comfort inside a building consuming a minimum amount of energy (Barrios et al. 2012; Kosny et al. 2014). The thermal conductivity of an insulation should be low and thermal resistance must be high in steady state heat flow conditions. Thermal diffusivity of an insulation material should be low for dynamic conditions of heat flow. Thermal conductivity (W/mK) of a homogeneous material is defined as steady state heat flow per second, through 1 m^2 area, 1 m thickness and for a temperature difference of 1 K. Thermal transmittance (W/m^2K) of a material is defined as the heat flow in steady state through a 1 m^2 area by a temperature difference of 1 K. Thermal diffusivity (m^2/s) is defined as the ratio of thermal conductivity to the heat capacity of a material. An additional parameter that is of prime importance and must be considered is solar absorptivity of the insulation material when used on the exterior of a building. The solar absorptivity of various wood species varies from 0.308 to 0.526 with an average value of 0.396 (Kang et al. 2011).

9.2.3 SUSTAINABLE INSULATION MATERIALS

Sustainable building materials that are environment friendly and have low embodied energy should be utilized for insulation in the buildings. Construction of buildings consume a major share of the total resource consumption all over the world. Insulation and wall materials are the main constituents of building materials. Studies reveal that construction of buildings use up around 24% of the raw materials utilized the world over (Bribian et al. 2011). Extraction of raw materials has a serious effect on the health of the people in the vicinity of the extraction site and a lot of primary energy is consumed during the extraction process, which is counted as the preliminary embodied energy of the raw material (Ding 2014). In the past, while choosing a building material generally the environmental properties of the material were not taken into consideration (Guggemos and Horvath 2005). Recently, interest in the research on sustainable building material is increasing to make building construction sustainable (Ding 2014). Renewable natural materials that improve human health, comfort and are environment friendly during their life cycle are termed as sustainable materials (Ding 2014). The environmental impact of the building material during its lifetime and problems like excessive resource utilization are evaluated using the life cycle assessment method (Ahn et al. 2010). The environmental effects mainly taken into account for a construction product in life cycle assessment are climate change, eutrophication, acidification, photochemical ozone depletion, stratospheric ozone depletion and other effects measured are water consumption, renewable and non-renewable primary energy, resource depletion, etc. (Cao 2017). The influence of lack of environmental qualities for the life cycle can also be generated using life cycle analysis (Puetmann and Wilson 2005). Sustainable building materials should have low environment impact in addition to the performance parameters (Ip and Miller 2012). The study of embodied energy is becoming more significant nowadays (Dixit et al. 2012). Materials used for insulation purposes in buildings should have low embodied energy. Materials with high embodied energy consume a lot of energy during production and give rise to greenhouse gases (Bribian et al. 2011). Utilizing the lower embodied energy natural materials can make the final product sustainable (John et al. 2005). Sustainable materials should be used for insulation in buildings so that the environmental impact is low.

9.3 NATURAL INSULATION MATERIALS, THEIR THERMAL PROPERTIES, ENVIRONMENTAL IMPACT AND SUSTAINABLE ASPECTS

The various natural materials that can be used for insulation of buildings are bagasse, coconut husk, wood, wood-based materials like wood fiber, wood crust, recycled cellulose, oriented strand boards, plywood and hardboard. The various natural insulation materials, their thermal properties and environmental and sustainable aspects of these materials used for insulation are presented in the coming sections.

9.3.1 NATURAL INSULATION MATERIALS

Many natural based materials are used for construction and thermal insulation of buildings. Thermal insulation materials like wood fiber, recycled cellulose and wood crust based on wood are used for insulating purpose other than solid wood (Krišták et al. 2014; Mitterpach et al. 2019). Insulating materials like wood waste, wood bark and coconut fibers are also tried for insulation purpose due to their lesser impact on the environment (Lakrafli et al. 2013; Niro et al. 2016; Pasztory and Ronyecz 2013; Tudor et al. 2018). Wood will play a major role as a building material for construction as it has characteristics like high strength, light weight, high thermal resistance and warmth keeping (Kawasaki and Kawai 2006). Wood can be used to construct building with excellent thermal insulation properties along with lighter buildings constructions (Corduban et al. 2017). Wood-based structures have the benefits like low cost, easy availability of material, fast construction rate, simple processing and lower pollution level caused to the environment (Mitterpach et al. 2019). Solid wood can be used for structural purposes or for making a roof, floors in building construction, other insulating materials can increase the thermal resistance of a wall (Asdrubali et al. 2017). A roof and floors made from solid wood can act as insulation. This is because the insulating properties of wood due to lower thermal conductivity of the solid wood makes it a good insulation material (Kosny et al. 2014). In wood the heat transfer is influenced by parameters like geometry, moisture content and porosity, etc. (Igaz et al. 2017). The parameters like porosity and geometry of a material are generally fixed after their preparation. The moisture content of the material varies with ambient humidity and weather conditions. The rate of multi-functional building construction from wood is on the rise (Müller et al. 2016). Life cycle analysis of buildings made from wood as construction materials shows 25% lesser pollution and around 18% lesser energy expenditure than conventional buildings (Heeran et al. 2015). Tree bark has properties like low thermal conductivity, low density, high fungi resistance and fire resistance. Also, the thermal diffusivity of bark is very low and as a result bark is suitable as an insulating material (Kain et al. 2013a). Insulation panels made from black locust tree bark of the trees have a thermal conductivity of 0.06 W/mK and can be used for insulating buildings and the tree bark is available abundantly and has no other use rather than simply dumping or burning (Pásztory et al. 2017). The strength of the boards prepared of bark is low and only composite fiberboards can be prepared due to the inadequate strength of the bark boards (Pásztory et al. 2017). On the other hand, the insulation properties of the bark boards are very good (Kain et al. 2012; Kain et al. 2013b; Kain et al. 2014). Insulation boards prepared from binder less bagasse and coconut husk have thermal conductivity close to the cellulose and mineral wool insulation and bagasse boards prepared with a density of 350 kg/m^3 can be utilized as an insulation material in buildings (Panyakaew and Fotios 2011). Reconstructed panels like plywood, particle board, hardboard, fiberboard and oriented strand board can be used for insulation in buildings (Asdrubali et al. 2017). The thermal properties of natural wood-based materials are comparable with conventional materials like date palm and oil palm fiber, etc. (Asdrubali et al. 2015). Oriented strand boards can also be used as thermal insulation for insulating the buildings. The oriented strand boards are materials prepared from wood strands joined by adhesives at a certain pressure and different temperatures. The thermal conductivity and thermal diffusivity of the oriented strand boards do not vary appreciably with moisture content and these boards are also used in the construction of wooden houses due to their good mechanical properties (Igaz et al. 2017). The most significant disadvantage of concrete-wood composite is the environmental pollution caused by them but they have some better functional properties than wood (Mitterpach et al. 2019).

9.3.2 SUSTAINABILITY OF NATURAL INSULATION MATERIALS

The sustainability concept in building materials has led to development of insulation materials from natural and recyclable sources (Asdrubali et al. 2015). Natural materials like wood, straws, coconut, wood fiber, cellulose and wood-based materials, etc. are sustainable materials that have very less impact on the environment and have low embodied energy. The embodied energy for the natural materials can be recovered within a very short span of time (Rojas et al. 2019). A very good progress has been seen in the field bio-insulators in the recent past; a lot of research has been done on materials like wood, straws and coconut (Liu et al. 2017). Insulation materials like mineral wool can be replaced by natural insulation materials prepared from wood fiber in particular as it has comparable thermal conductivity and slightly thicker layer will have the same thermal resistance and can provide thermal comfort inside a building (Volf et al. 2015). Wood fiber is a natural and recyclable material, and has properties similar to the synthetic materials and is already commercialized with scope for further increase in diffusion (Asdrubali et al. 2015). Interest in the organic insulation materials is increasing as these materials are renewable, recyclable and environment friendly (Kymäläinen and Sjöberg 2008). Organic insulation materials like wood fiber, wood wool and cellulose etc. are derived from renewable natural vegetation while inorganic materials like mineral wool, aerated concrete and perlite, etc. are made from non-renewable materials (Aditya et al. 2017). Organic insulation materials are sustainable and recyclable and reusable. The thermal insulation offered by inorganic compounds is higher and they are cheaper and are more fire resistant (Aditya et al. 2017). Cellulose fiber is a renewable, recyclable and environment friendly material derived from natural materials. Cellulose fiber insulation needs the lowest amount of energy for manufacture as compared to other insulation materials like glass wool and rock wool and the primary energy use can be reduced by 6–7% when cellulose fiber insulation is used instead of rock wool (Tettey et al. 2014). Wood can also be used as an insulation material in the buildings as it has good insulation properties. Lesser carbon dioxide emissions and lower embodied energy are reported in the case of wood and it also a sustainable material (Falk 2009). Natural wood resources can be used as a sustainable construction material (Kosny et al. 2014). Wood is a renewable and sustainable material provided an equilibrium is maintained between utilized and cultivated wood. It is reusable for furniture making or for combustion, which reduces the influence on the environment and the embodied energy of wood-based materials is lesser than concrete and steel (Asdrubali et al. 2017). Utilizing wood as building construction material is much better option to decrease the carbon emission significantly but an increase in wood use should be complemented with growth in forest areas for sustainability (Buchanan and Levine 1999). It is to be noted that the life expectancy of wooden houses is much less than concrete houses and it may give rise to more pollution in case the wood is to be imported from a distance (Guardigli 2014). The carbon dioxide absorbed by trees during growth remains stored in the solid wood until dumping and a major part of the total emission is during paper and panel manufacture (Asdrubali et al. 2017). Certain wood-based materials like mineralized wood fiber have high embodied energy comparable with synthetic materials like expanded polystyrene and glass wool (Asdrubali 2009). The building constructed from steel reinforced concrete has a higher effect on the environment as compared to wood-based constructions (Gerilla et al. 2007). The heat flow to the building made from concrete can be significantly reduced using coconut fiber as insulation materials and a considerably lower indoor temperature can be maintained (Rodríguez et al. 2011). A net carbon dioxide emission drop can be achieved by increasing the proportion of wood-based natural construction materials and replacing the concrete (Gustavsson et al. 2006). High water absorption and fire resistance are the major disadvantages of the natural-based thermal insulation materials but the degradation of natural-based thermal insulations can be decreased by application hydrophobic agents to the natural insulation (Zach et al. 2013).

9.3.3 ENVIRONMENTAL IMPACT OF NATURAL INSULATION MATERIALS

Construction of buildings impact the environment during the utilization of raw materials, renovation, maintenance and throughout the life cycle of the buildings (Balaras et al. 2005). Building construction gives rise to greenhouse gases and impacts the environment badly in addition to exhaustion of natural resources (Ding 2014). The impact of the insulation material on the environment is a major parameter for selection of an insulation at present. Insulation with less effect on the environment should be selected as it can provide thermal comfort and at the same time also be friendly to the environment. As stated earlier, the insulation materials generally used are made from petrochemicals like polystyrene, polyurethane and polyethylene. The production of these materials emits lot pollution into the environment. Thermal insulations like rock wool, perlite and vermiculite also produce a lot of pollution during production. Naturally occurring insulation materials are recyclable and these insulation materials can provide thermal comfort. Insulation materials based on natural resources are environment friendly, have a lower impact on health and have energy efficient production (Ahmed et al. 2019; Ali et al. 2020; Florea and Manea 2019; Liu et al. 2019). Life cycle assessment shows that wooden house construction materials have lesser harmful effect on the environment as compared to the brick house construction materials (Mitterpach and Štefko 2016). The carbon footprint in the environment is double in the case of brick house construction materials as compared to wooden house construction materials (Mitterpach and Štefko 2016). Cellulose fiber insulation can reduce carbon dioxide emission by 6–8% when cellulose fiber insulation is used instead of rock wool (Tettey et al. 2014). Therefore the environmental effect can be greatly reduced by using the wood-based materials for insulating the buildings in addition to getting the thermal comfort. Studies have shown that in some particular structures the environmental effect can be reduced by around 60% by changing the materials of construction (Eštoková and Porhinčák 2015).

9.3.4 THERMAL PERFORMANCE OF NATURAL INSULATION MATERIALS

The thermal performance of insulation materials is measured in terms of thermal conductivity of the material in static heat transfer and in terms of thermal diffusivity in dynamic heat transfer. Thermal conductivity tells us how much heat will be transferred through the insulation material and thermal diffusivity tells how fast the heat will be conducted through an insulation material. Thermal conductivity range and densities for various insulation materials has been presented in Table 9.1. Presently, insulation materials based on inorganic materials like expanded polystyrene, extruded polystyrene and polyurethane foam are utilized in buildings. The environmental effect of these materials during production process and during the life cycle period of buildings are usually high and in order to reduce the environment effect and improving the life cycle performance of buildings bio-based materials must be used as an insulation (Cetiner and Shea 2018). The thermal conductivities of natural materials like wood fiber, coconut husk, kenaf insulation board and wood waste are very close to that of the extruded polystyrene, rock wool and expanded polystyrene. Wood fiber and wood waste insulation materials with different densities have thermal conductivity in the range of 0.040–0044 W/mK and 0.048–0.05 W/mK respectively. The thermal conductivity of natural materials like oriented strand board, hardboard, particle board and solid wood is somewhat higher than rock wool, extruded and expanded polystyrene insulations. The thermal conductivity of oven dried solid birch wood at 21°C in longitudinal and transverse directions vary from 0.291–0.323 and 0.177–0.214 W/mK respectively (Suleiman et al. 1999).

The mean thermal conductivity of various insulation materials in sorted form is shown in Table 9.2 and Figure 9.1. The mean thermal conductivity of wood fiber is almost the same as that of materials like rock wool, expanded and extruded

TABLE 9.1
Thermal conductivity and density of various insulation materials

S. No.	Material of insulation	Thermal conductivity (W/m-K)	Density (kg/m³)	Source
1	Extruded polystyrene (XPS)	0.032–0.037	32–40	(Asdrubali et al. 2015)
2	Expanded polystyrene (EPS)	0.031–0.038	15–35	(Asdrubali et al. 2015)
3	Rock wool	0.033–0.040	40–200	(Asdrubali et al. 2015)
4	Binder less coconut husk insulation board	0.046–0.068	250–350	(Panyakaew and Fotios 2011)
5	Kenaf insulation board	0.051–0.058	150–200	(Xu et al. 2004)
6	Wood waste	0.048–0.055	117–167	(Cetiner and Shea 2018)
7	Solid wood	0.177–0.323	443–680	(Suleiman et al. 1999)
8	Wood fiber	0.040–0.044	146–152	(Vololonirnia et al. 2014)
9	Oriented strand board (OSB)	0.094–0.1	562–602	(Vololonirnia et al. 2014)
10	Plywood	0.12–0.15	540–700	(Asdrubali et al. 2017)
11	Hardboard	0.08–0.29	600–1000	(Asdrubali et al. 2017)
12	Particleboard	0.098–0.17	750–1000	(Asdrubali et al. 2017)

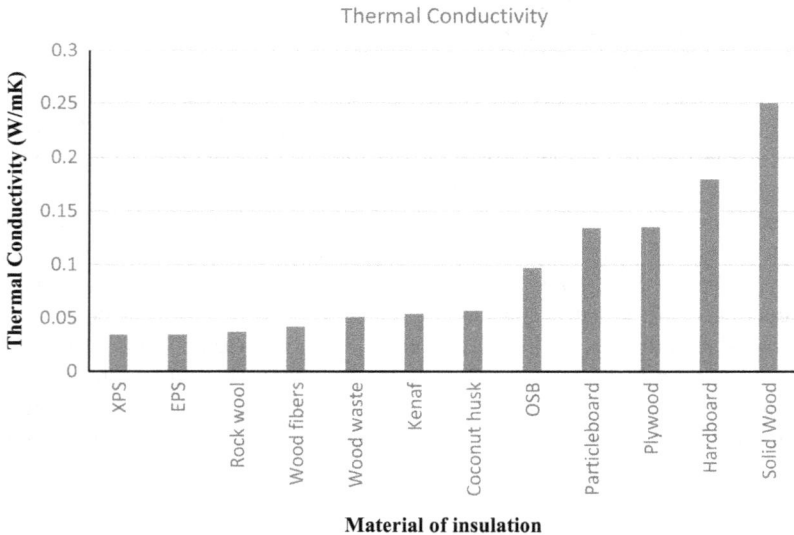

FIGURE 9.1   Mean thermal conductivity of various insulation materials.

TABLE 9.2
Mean thermal conductivity of various insulation materials

S. No.	Material of insulation	Thermal conductivity (W/m-K)
1	Extruded polystyrene foam (XPS)	0.0345
2	Expanded polystyrene (EPS)	0.0345
3	Rock wool	0.037
4	Wood fiber	0.042
5	Wood waste	0.051
6	Kenaf insulation board	0.054
7	Binder less coconut husk insulation board	0.057
8	Oriented strand board (OSB)	0.097
9	Particleboard	0.134
10	Plywood	0.135
11	Hardboard	0.18
12	Solid wood	0.25

polystyrene. The mean thermal conductivities of insulation materials like wood waste, kenaf insulation board and coconut husk insulation board are comparable with rock wool, expanded and extruded insulations. Though inorganic based insulation materials may have slightly lesser thermal conductivities than wood waste insulation materials the latter has the advantage of having a lower cost (Cetiner and Shea 2018). The mean thermal conductivity of oriented strand board, particleboard, plywood, hardboard and solid wood is somewhat higher than the materials like rock wool, expanded and extruded polystyrene. Therefore these natural materials can be used as insulation materials in buildings in place of synthetic materials like rock wool, extruded polystyrene and expanded polystyrene as the thermal conductivity of wood-based materials like wood fiber, wood waste, kenaf insulation board and coconut husk insulation board is either same or comparable.

The mean thermal diffusivity for different insulation materials is presented in Table 9.3 and Figure 9.2. The mean thermal diffusivity of natural materials is much less compared to synthetic materials like rock wool, expanded and extruded poly-styrene. The mean thermal diffusivity is much lower for wood waste, oriented strand board, particle board and wood fiber compared to other natural materials, rock wool and extruded polystyrene and expanded polystyrene. The thermal diffusivity of materials like particleboard, hardboard, plywood, solid wood and kenaf insula-tion board is somewhat higher than other natural materials but is less than synthetic materials like rock wool, expanded and extruded polystyrene. The thermal diffusivity of oven dried solid birch wood at 21°C in longitudinal and transverse directions vary from 0.496–0.513 and 0.418–0.537 mm^2/K respectively (Suleiman et al. 1999). The thermal diffusivity of the synthetic materials is higher than all natural materials in the list. Hence the natural materials like coconut husk, wood waste, oriented strand board and wood fibers, etc. are more suitable for use as insulation in buildings than the syn-thetic materials like rock wool, expanded and extruded polystyrene. The radiation heat transfer is negligible in the case of wood compared to conduction and convection

TABLE 9.3
Mean thermal diffusivity of various insulation materials

S. No.	Material of insulation	Thermal diffusivity ($\times 10^{-6}$ m²/s)
1	Binder less coconut husk insulation board	0.073
2	Wood waste	0.086
3	Oriented strand board (OSB)	0.108
4	Wood fiber	0.111
5	Particleboard	0.118
6	Hardboard	0.135
7	Plywood	0.166
8	Solid wood	0.186
9	Kenaf insulation board	0.187
10	Rock wool	0.343
11	Extruded polystyrene foam (XPS)	0.618
12	Expanded polystyrene (EPS)	1.104

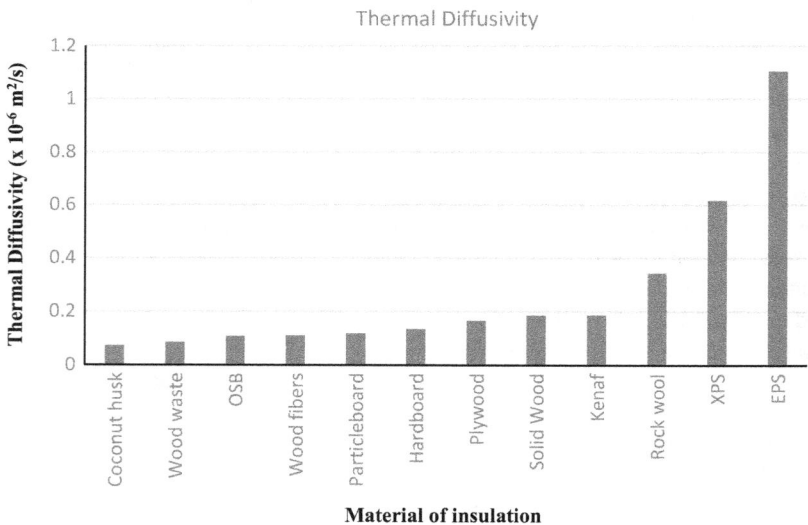

FIGURE 9.2   Mean thermal diffusivity of various insulation materials.

heat transfer (Kaemmerlen et al. 2010). Therefore wood-based materials can also be used as an exterior insulation material and it can reduce the absorption of heat from the solar radiation and transfer of the heat into the indoor of a building. Hence thermal comfort inside buildings can be improved by using natural insulation to a great extent as the thermal conductivity of many of these is comparable or only somewhat higher than the synthetic materials and thermal diffusivity is lower than synthetic materials. This comparable thermal conductivity and lower thermal diffusivity will reduce the heat flow to the inside of a building and thereby improve the thermal comfort by maintaining a lower temperature of the indoor air.

9.3.5   REDUCTION IN COOLING LOAD

Cooling load is the amount of heat required to be removed from the buildings in order to maintain the thermally comfortable indoor environment. It is clear that man-made buildings in a city can raise the temperature of the city and increase the cooling load of a building. This can be controlled by energy efficient buildings using sustainable materials. The heat interaction from buildings can be reduced significantly by using insulation layer and comfortable indoor temperature conditions can be maintained inside buildings. The heat transfer in buildings is of dynamic nature and is dependent on thermal diffusivity of the insulation materials. Natural materials have lower thermal diffusivity than synthetic and mineral-based materials as discussed in the previous section. Therefore the natural materials can perform much better under the dynamic heat interaction conditions like insulating the buildings. This better performance of the natural insulation materials will lead to a reduction in the cooling load and more thermal comfort inside buildings.

9.3.6   REDUCTION IN ENERGY CONSUMPTION

It is obvious from the literature that a great deal of energy is consumed in the building sector and energy efficient buildings can reduce the energy consumption in the buildings. The energy efficient natural methods like insulation on walls and roofs can reduce energy consumption in buildings. Insulating the roofs and walls of buildings can reduce the energy consumption by 30% and 20% respectively. Natural materials are sustainable and of great importance for green and energy efficient buildings. Synthetic and mineral-based thermal insulations materials like polystyrene, polyurethane and rock wool create a lot of pollution and consume a great deal of energy during their production. The natural and sustainable materials and methods can play a major role in conserving the energy in the world. The lesser cooling load by utilization of natural insulation materials under dynamic heat transfer conditions like building insulation can reduce the energy expenditure on space cooling significantly in the buildings.

9.4   CONCLUSION

Natural materials are renewable, sustainable, have less environmental impact and have lower embodied energy. The materials like rock wool, expanded and extruded polystyrene are of non-renewable nature, have a larger environmental impact and high embodied energy. The natural insulation materials have less effect on the environment compared to materials like rock wool, expanded and extruded polystyrene. The mean thermal conductivity of natural materials like wood fiber is almost the same and that of insulation materials like wood waste, kenaf insulation board and coconut husk insulation board is comparable with that of materials like rock wool, expanded and extruded polystyrene. The other natural derived materials like oriented strand board, particleboard, plywood, hardboard and solid wood have a mean thermal conductivity slightly higher than materials like rock wool, expanded and extruded polystyrene. Natural materials have a very low value of mean thermal diffusivity in comparison to materials like rock wool, expanded and extruded polystyrene. Natural materials are a better option from renewability, sustainability, environmental impact and embodied energy perspectives. Natural insulation materials like wood fiber, wood waste, kenaf insulation board and coconut husk have comparable thermal performance to that of synthetic insulation materials under static conditions of heat transfer and other natural materials have slightly lower thermal performance than synthetic materials under static heat transfer conditions. All the natural materials have better thermal performance under dynamic heat transfer conditions compared to synthetic materials due to lower value of mean thermal diffusivity. The mean thermal conductivity of natural materials like kenaf and coconut husk is higher than natural derived materials like wood fiber, wood waste and the mean thermal conductivity is lower than other natural derived materials like oriented strand board, particleboard, plywood and hardboard. The mean thermal conductivity of natural solid wood is higher than all the natural derived materials and other natural materials. The natural material coconut husk has similar mean thermal diffusivity to the natural derived materials like wood waste, oriented strand board and wood fiber and have a significantly lower mean thermal diffusivity in comparison to other natural materials like particleboard, hardboard, plywood. All the natural derived materials have significantly lower mean thermal diffusivity than natural materials like solid wood and kenaf. The natural materials like wood, natural derived wood-based materials are reusable while wood fiber is a reusable and recyclable material. The natural derived materials like wood fiber, wood waste, oriented strand board have better performance under static heat flow conditions compared to natural materials like kenaf and coconut husk and other natural derived materials like particleboard, plywood and hardboard have lower performance than these natural materials. Solid wood has lowest thermal performance amongst all the materials considered in the chapter under static conditions. The thermal performance of natural derived materials like wood fiber, wood waste, oriented strand board is comparable to coconut husk and better than other natural materials under dynamic conditions. Coconut husk has better thermal performance than natural derived materials like particleboard, hardboard, plywood. Wood fiber and wood waste are better insulation materials compared to other insulation materials considering all the parameters. These materials have similar thermal performance compared to materials like rock wool, expanded, extruded polystyrene and better thermal performance than other materials like kenaf, coconut husk, oriented strand board, particleboard, plywood, hardboard under static heat transfer conditions. Wood fiber and wood waste have similar thermal performance to that of coconut husk and are better than all other materials under dynamic conditions. Wood fiber is also a recyclable material and can be utilized again after recycling. Hence for enhancing thermal comfort in buildings which comes under dynamic heat transfer conditions, natural derived insulation materials like wood fiber, wood waste and coconut husk can be utilized. Hence using the natural insulation materials under the dynamic heat interaction conditions like insulating the buildings can lead to better thermal performance and reduction in the cooling load and more thermal comfort inside energy efficient buildings. The smaller cooling load by utilization of natural insulation materials under dynamic heat transfer conditions like building insulation can reduce the energy expenditure on space cooling significantly. Therefore the natural and sustainable materials can play a major role in conserving the energy in energy efficient buildings and green pollution free cities. Hence it can be generalized that natural insulation materials can be utilized to reduce the cooling load, conserve energy in the buildings and save the environment from excessive pollution. It can be concluded that the natural insulation materials utilization can improve thermal comfort inside energy efficient buildings and make the buildings and the cities energy efficient and green.

REFERENCES

Abu-Jdayil, B., Mourad, A. H., Hittini, W., Hassan, M., Hameedi, S. 2019. Traditional, state-of-the-art and renewable thermal building insulation materials: An overview. *Construction and Building Materials*. 214, pp. 709–735.

Aditya, L., Mahlia, T.M.I, Rismanchi, B., Ng, H.M., Hasan, M.H., Metselaar, H.S.C., Muraza, O, Aditiya, H.B. 2017. A review on insulation materials for energy conservation in buildings. *Renewable and Sustainable Energy Reviews*. 73, pp. 1352–1365.

Agoudjil, B., Benchabane, A., Boudenne, A., Ibos, L., Fois, M. 2011. Renewable materials to reduce building heat loss: Characterization of date palm wood. *Energy and Buildings*. 43, pp. 491–497.

Ahmed, A., Qayoum, A., Mir, F.Q. 2019. Investigation of the thermal behavior of the natural insulation materials for low temperature regions. *Journal of Building Engineering*. 26, 100849.

Ahn, C., Lee, S.H., Pena-Mora, F., Abourizk, S. 2010. Toward environmentally sustainable construction processes: the US and Canada's perspective on energy consumption and GHG/GAP emission. *Sustainability*. 2, pp. 354–370.

Al-Badi, A.H., Al-Saadi, S.N. 2020. Toward energy-efficient buildings in Oman. *International Journal of Sustainable Energy*. 39, 5, pp. 412–433.

Ali, M., Alabdulkarem, A., Nuhait, A., Al-Salem, K., Iannace, G., Almuzaiqer, R., Al-turki, A., Al-Ajlan, F., Al-Mosabi, Y., Al-Sulaimi, A. 2020. Thermal and acoustic characteristics of novel thermal insulating materials made of Eucalyptus Globulus leaves and wheat straw fibers. *Journal of Building Engineering*. 32, 101452.

Arpitha, G.R., Sanjay, M.R., Senthamaraikannan, P., Barile, C., Yogesha, B. 2017. Hybridization effect of sisal/glass/epoxy/filler based woven fabric reinforced composites. *Experimental Techniques*. 41, pp. 577–584.

Asdrubali, F. 2009. The role of Life Cycle Assessment (LCA) in the design of sustainable buildings: Thermal and sound insulating materials. *Euronoise 2009*, Edinburgh, Scotland.

Asdrubali, F., D'Alessandro, F., Schiavoni, S. 2015. A review of unconventional sustainable building insulation materials. *Sustainable Materials and Technologies*. 4, pp. 1–17.

Asdrubali, F., Ferracuti, B., Lombardi, L., Guattari, C., Evangelisti, L., Grazieschi, G. 2017. A review of structural, thermo-physical, acoustical, and environmental properties of wooden materials for building applications. *Building and Environment*. 114, pp. 307–332.

Balaras, C.A., Droutsa, K., Dascalaki, E., Kontoyiannidis, S. 2005. Heating energy consumption and resulting environmental impact of European apartment buildings. *Energy and Buildings*. 37, 5, pp. 429–442.

Balaras, C.A., Droutsa, K., Argiriou, A.A., Asimakopoulos, D.N. 2000. Potential for energy conservation in apartment buildings. *Energy and Buildings*. 31, 2, pp. 143–154.

Barrios, G., Huelsz, J., Rojas, J., Ochoa, M., Marinic, I. 2012. Envelope wall/roof thermal performance parameters for non air-conditioned buildings. *Energy and Buildings*. 50, pp. 120–127.

Besir, A. B., Cuce, E. 2018. Green roofs and facades: A comprehensive review. *Renewable and Sustainable Energy Reviews*. 82, pp. 915–939.

Binici, H., Aksogan, O., Dıncer, A., Luga, E., Eken, M., Isikaltun, O. 2020. The possibility of vermiculite, sunflower stalk and wheat stalk using for thermal insulation material production. *Thermal Science and Engineering Progress*. 18, 100567.

Binici, H., Aksogan, O. 2016. Eco-friendly insulation material production with waste olive seeds, ground PVC and wood chips. *Journal of Building Engineering*. 5, pp. 260–266.

Binici, H., Eken, M., Dolaz, M., Aksogan, O., Kara, M. 2014. An environmentally friendly thermal insulation material from sunflower stalk, textile waste and stubble fibres. *Construction and Building Materials*. 51, pp. 24–33.

Bribian, I.Z., Capilla, A.V., Uson, A.A. 2011. Life cycle assessment of building materials: comparative analysis of energy and environmental impacts and evaluation of the eco-efficiency improvement potential. *Building and Environment*. 46, pp. 1133–1140.

Buchanan, A.H., Levine, S.B. 1999. Wood-based building materials and atmospheric carbon emissions. *Environmental Science and Policy*. 2, 6, pp. 427–437.

Cao, C. 2017. *21 – Sustainability and life assessment of high strength natural fibre composites in construction*. Woodhead Publishing. 529–544.

Cetiner, I., Shea, A.D. 2018. Wood waste as an alternative thermal insulation for buildings. *Energy and Buildings*. 168, pp. 374–384.

Corduban, C.G., Dumitrascu, A.I., Hapurne, T., Nica, R.M., Gheban, C., 2017. Innovative wooden platform framing structure for a near-zero energy house. *In International Multidisciplinary Scientific Geoconference Surveying Geology and Mining Ecology Management*, SGEM. 17, 62, 223–230.

Cosereanu, C., Lazarescu, C., Curtu, I., Lica, D., Sova, D, Brenci, L.M. Research on new structures to replace polystyrene used for thermal insulation of buildings. *Materiale Plastice*. 47, 3, pp. 341–345.

Ding, G.K.C. 2014. 3 – Life Cycle Assessment (LCA) of Sustainable Building Materials: An Overview. *Eco-efficient construction and building materials*, Woodhead Publishing. 38–62.

Dixit, M.K., Fernández-Solís, J.L., Lavy, S., Culp, C.H. 2012. Need for an embodied energy measurement protocol for buildings: A review paper. *Renewable and Sustainable Energy Reviews*. 16, pp. 3730–3743.

Erkmen, J., Yavuz, H. I., Kavci, E., Sari, M. 2020. A new environmentally friendly insulating material designed from natural materials. *Construction and Building Materials*. 255, 119357.

Eštoková, A., Porhinčák, M. 2015. Environmental analysis of two building material alternatives in structures with the aim of sustainable construction. *Clean Technologies and Environmental Policy*. 17, 1, pp. 75–83.

Falk, B. 2009. Wood as a sustainable building material. *Forest Products Journal*. 59, 9, pp. 6–12.

Florea, I., Manea, D. L. 2019. Analysis of thermal insulation building materials based on natural fibers. *Procedia Manufacturing*. 32, pp. 230–235.

Gerilla, G.P., Teknomo, K., Hokao, K. 2007. An environmental assessment of wood and steel reinforced concrete housing construction. *Building and Environment*. 42, 7, pp. 2778–2784.

Girijappa, Y.G.T., Rangappa, S.M., Parameswaranpillai, J., Siengchin, S. 2019. Natural fibers as sustainable and renewable resource for development of eco-friendly composites: A comprehensive review. *Frontiers in Materials*. 6, 226.

Guardigli, L. 2014. 17 – Comparing the environmental impact of reinforced concrete and wooden structures. *Eco-Efficient Construction and Building Materials*, Woodhead Publishing. 407–433.

Guggemos, A.A., Horvath, A. 2005. Comparison of environmental effects of steel- and concrete-framed buildings. *Journal of Infrastructure Systems*. 11, 2, pp. 93–101.

Gustavsson, L., Pingoud, K., Sathre, R. 2006. Carbon dioxide balance of wood substitution: comparing concrete- and wood-framed buildings. *Mitigation and Adaptation Strategies for Global Change*. 11, pp. 667–691.

Heeren, N., Hellweg, S. 2019. Tracking construction material over space and time: prospective and geo-referenced modeling of building stocks and construction material flows. *Journal of Industrial Ecology*. 23, 1, pp. 253–267.

Heeren, N., Mutel, C.L., Steubing, B., Ostermeyer, Y., Wallbaum, H., Hellweg, S. 2015. Environmental Impact of Buildings – What Matters? *Environmental Science and Technology*. 49, 16, pp. 9832–9841.

Hurtado, P. L., Rouilly, A., Vandenbossche, V., Raynaud, C. 2016. A review on the properties of cellulose fibre insulation. *Building and Environment*. 96, pp. 170–177.

Igaz, R., Krišták, Ľ., Ružiak, I., Gajtanska, M., Kučerka, M. 2017. Thermophysical Properties of OSB Boards *versus* Equilibrium Moisture Content. *BioResources*. 12, 4, pp. 8106–8118.

Ip, K., Miller, A. 2012. Life cycle greenhouse gas emissions of hemp-lime wall constructions in the UK, Resources. *Conservation and Recycling*. 69, pp. 1–9.

Jiang, F., Li, T., Li, Y., Zhang, Y., Gong, A., Dai J., Hitz E., Luo W., Hu L. 2018. Wood-based nanotechnologies toward sustainability. *Advanced Materials*. 30, 1, 1703453.

John, G., Clements-Croome, D., Jeronimidis, G. 2005. Sustainable building solutions: a review of lessons from the natural world. *Building and Environment*. 40, pp. 319–328.

Kaemmerlen, A., Asllanaj, F., Sallée, H., Baillis, D., Jeandel, G. 2010. Transient modeling of combined conduction and radiation in wood fibre insulation and comparison with experimental data. *International Journal of Thermal Science*. 49, pp. 2169–2176.

Kain, G., Barbu, M. C., Hinterreiter, S., Richter, K., Petutschnigg, A. 2013a. Using Bark as a Heat Insulation Material. *BioResources*. 8, 3, pp. 3718–3731.

Kain, G., Heinzmann, B., Barbu, M.C., Petutschnigg, A. 2013b. Softwood bark for modern composites. *ProLigno*. 9, 4, pp. 460–468.

Kain, G., Barbu, M.C., Teischinger, A., Musso, M., Petutschnigg, A. 2012. Substantial Bark Use as Insulation Material. *Forest Products Journal*. 62, 6, pp. 480–487.

Kain, G., Güttler, V., Barbu, M.C., Petutschnigg, A., Richter, K., Tondi, G. 2014. Density related properties of bark insulation boards bonded with tannin hexamine resin. *European Journal of Wood and Wood Products*. 72, pp. 417–424.

Kang, W., Lee, Y.H., Kang, C.W., Chung, W.Y., Xu, H.L., Matsumura, J. 2011. Using the inverse method to estimate the solar absorptivity and emissivity of wood exposed to the outdoor environment. *Journal – Faculty of Agriculture Kyushu University*. 56, 1, pp. 139–148.

Kawasaki, T., Kawai, S. 2006. Thermal insulation properties of wood-based sandwich panel for use as structural insulated walls and floors. *Journal of Wood Science*. 52, pp. 75–83.

Keynakli, O. 2012. A review of the economical and optimum thermal insulation thickness for building applications. *Renewable and Sustainable Energy Reviews*. 16, pp. 415–425.

Knapic, S., Oliveira, V., Machado, J.S., Pereira, H. 2016. Cork as a building material: A review. *European Journal of Wood and Wood Products*. 74, pp. 775–791.

Kosny, J., Asiz, A., Smith, I., Shrestha, S., Fallahi, A. 2014. A review of high R-value wood framed and composite wood wall technologies using advanced insulation techniques. *Energy and Buildings.* 72, pp. 441–456.

Krišták, Ľ., Igaz, R., Brozman, D., Réh, R., Šiagiová, P., Stebila, J., Očkajová, A. 2014. Life cycle assessment of timber formwork: A case study. *Advanced Materials Research.* 1001, pp. 155–161.

Kymäläinen, H.R., Sjöberg, A.M. 2008. Flax and hemp fibres as raw materials for thermal insulations. *Build and Environment.* 43, 7, pp. 1261–1269.

Lakrafli, H., Tahiri, S., Albizane, A., Bouhria, M., El Otmani, M.E. 2013. Experimental study of thermal conductivity of leather and carpentry wastes. *Construction and Building Materials.* 48, pp. 566–574.

Latif, E. 2020. A review of low energy thermal insulation materials for building applications. *Proceedings of International Conference*, Bangkok, Thailand, February 2020, 55–59.

Latif, E., Ciupala, M.A., Tucker, S., Wijeyesekera, D.C., Newport, D.J. 2015. Hygrothermal performance of wood-hemp insulation in timber frame wall panels with and without a vapor barrier. *Build and Environment.* 92, pp. 122–134.

Li, T., Zhu, M., Yang, Z., Song, J., Dai, J., Yao, Y., Luo, W., Pastel, G., Yang, B., Hu, L. 2016. Wood composite as an energy efficient building material: Guided sunlight transmittance and effective thermal insulation. *Advanced Energy Materials.* 6, 22, 1601122 (pp. 1–7).

Liang, J., Li, B., Wu, Y., Yao, R. 2007. An investigation of the existing situation and trends in building energy efficiency management in China. *Energy and Buildings.* 39, 10, pp. 1098–1106.

Lithner, D., Larsson, Å., Dave, G. 2011. Environmental and health hazard ranking and assessment of plastic polymers based on chemical composition. *Science of Total Environment.* 409, pp. 3309–3324.

Liu, L. F., Li, H. Q., Lazzaretto, A., Manente, G., Tong, C. Y., Liu, Q. B., Li, N. P. 2017. The development history and prospects of biomass-based insulation materials for buildings. *Renewable and Sustainable Energy Reviews.* 69, pp. 912–932.

Liu, L., Zou, S., Li, H., Deng, L., Bai, C., Zhang, X., Wang, S., Li, N. 2019. Experimental physical properties of an eco-friendly bio-insulation material based on wheat straw for buildings. *Energy and Buildings.* 201, pp. 19–36.

Madhu, P., Sanjay, M.R., Senthamaraikannan, P., Pradeep, S., Saravanakumar, S.S., Yogesha, B. 2019. A review on synthesis and characterization of commercially available natural fibers: Part II. *Journal of Natural Fibers.* 16, pp. 25–36.

Mahmood, R.A. 2020. Assessment of a sustainable building using eco-friendly insulation materials. *International Journal of Advances in Mechanical and Civil Engineering.* 7, 2, pp. 16–20.

Martínez-García, C., González-Fonteboa, B., Carro-López, D., Pérez-Ordóñez, J.L. 2020. Mussel shells: A canning industry by-product converted into a bio-based insulation material. *Journal of Cleaner Production.* 269, 122343.

Meng, X., Luo, T., Gao, Y., Zhang, L., Huang, X., Hou, C., Shen, Q., Long, E. 2018. Comparative analysis on thermal performance of different wall insulation forms under the air-conditioning intermittent operation in summer. *Applied Thermal Engineering.* 130, pp. 429–438.

Mitterpach, J., Igaz, R., Štefko, J. 2019. Environmental evaluation of alternative wood-based external wall assembly. *Acta Facultatis Xylologiae Zvolen.* 61, 2, pp. 133–149.

Mitterpach, J., Štefko, J. 2016. An environmental impact of a wooden and brick house by the LCA method. *Key Engineering Materials.* 688, pp. 204–209.

Müller, A., Kolb, H., Brunner, M. 2016. Modern timber houses, *Lignocellulosis Fibers and Wood Handbook: Renewable Materials for Today's Environment*, Scrivener Publishing. 595–610. ISBN 978-111877372-7

Niro, J.F.D.V.M., Kyriazopoulos, M., Bianchi, S., Mayer, I., Eusebio, D.A., Arboleda, J.R., Lanuzo, M.M., Pichelin, F. 2016. Development of medium- and low-density fibreboards made of coconut husk and bound with tannin-based adhesives. *International Wood Products Journal*. 7, 4, pp. 208–214.

Ofori, G., Briffett, C., Gang, G., Ranasinghe, M. 2000. Impact of ISO 14000 on construction enterprises in Singapore. *Construction Management and Economics*. 18, pp. 935–947.

Orlik-Kożdoń, B. 2017. Assessment of the application efficiency of recycling materials in thermal insulations. *Construction and Building Materials*. 156, 23, pp. 476–485.

Paiva, A., Pereira, S., Sá, A., Cruz, D., Varum, H., Pinto, J. 2012. A contribution to the thermal insulation performance characterization of corn cob particleboards. *Energy and Buildings*. 45, pp. 274–279.

Palomo, A., Grutzeck, M.W., Blanco, M.T. 1999. Alkali-activated fly ashes—A cement for the future. *Cement and Concrete Research*. 29, pp. 1323–1329.

Panyakaew, S., Fotios, S. 2011. New thermal insulation boards made from coconut husk and bagasse. *Energy and Buildings*. 43, pp. 1732–1739.

Pásztory, Z., Mohácsiné, I. R., Börcsök, Z. 2017. Investigation of thermal insulation panels made of black locust tree bark. *Construction and Building Materials*. 147, pp. 733–735.

Pasztory, Z., Mohacsine, I.R., Borcsok, Z. 2017. Investigation of thermal insulation panels made of black locust tree bark. *Construction and Building Materials*. 147, pp. 733–735.

Pasztory, Z., Ronyecz, I. 2013. The thermal insulation capacity of tree bark. *Acta Silv. Lign. Hung*. 9, 1, pp. 111–117.

Pérez-Lombard, L., Ortiz, J., Pout, C. 2008. A review on buildings energy consumption information. *Energy and Buildings*. 40, 3, pp. 394–398.

Pinto, J., Paiva, A., Varum, H., Costa, A., Cruz, D., Pereira, S., Fernandes, L., Tavares, P., Agarwal, J. 2011. Corn's cob as a potential ecological thermal insulation material. *Energy and Buildings*. 43, 8, pp. 1985–1990.

Puettmann, M.E., Wilson, J.B. 2005. Life-cycle analysis of wood products: cradle-to-grate LCI of residential wood building materials. *Wood and Fibre Science*. 37, pp. 18–29.

Rodríguez, N.J., Yáñez-Limón, M., Gutiérrez-Miceli, F.A., Matadamas-Ortiz, T.P., Lagunez-Rivera, L., Feijoo, J.A.V. 2011. Assessment of coconut fibre insulation characteristics and its use to modulate temperatures in concrete slabs with the aid of a finite element methodology. *Energy and Buildings*. 43, 6, pp. 1264–1272.

Rojas, C., Cea, M., Iriarte, A., Valdés, G., Navia, R., Cárdenas-R, J.P. 2019. Thermal insulation materials based on agricultural residual wheat straw and corn husk biomass, for application in sustainable buildings. *Sustainable Materials and Technologies*. 17, e00102.

Suleiman, B.M., Larfeldt, J., Leckner, B., Gustavsson, M. 1999. Thermal conductivity and diffusivity of wood. *Wood Science and Technology*. 33, pp. 465–473.

Tettey, U.Y.A., Dodoo, A., Gustavsson, L. 2014. Effects of different insulation materials on primary energy and CO_2 emission of a multi-storey residential building. *Energy and Buildings*. 82, pp. 369–377.

Tudor, E.M., Barbu, M.C., Petutschnigg, A., Reh, R. 2018. Added value for wood bark as a coating layer for flooring tiles. *Journal of Cleaner Production*. 170, pp. 1354–1360.

Upton, B., Miner, R., Spinney, M., Heath, L.S. 2008. The greenhouse gas and energy impacts of using wood instead of alternatives in residential construction in the United States. *Biomass and Bioenergy*. 32, pp. 1–10.

Valverde, I.C., Castilla, L.H., Nuñez, D.F., Rodriguez-Senín, E., Ferreira, R.M. 2013. Development of new insulation panels based on textile recycled fibers. *Waste and Biomass Valorization*. 4, pp. 139–146.

Volf, M., Diviš, J., Havlík, F. 2015. Thermal, moisture and biological behaviour of natural insulating materials. *Energy Procedia*. 78, pp. 1599–1604.

Vololonirina, O., Coutand, M., Perrin, B. 2014. Characterization of hygrothermal properties of wood-based products – Impact of moisture content and temperature. *Construction and Building Materials*. 63, pp. 223–233.

Wang, H., Chiang, P., Cai, Y., Li, C., Wang, X., Chen, T., Wei, S., Huang, Q. 2018. Application of wall and insulation materials on green building: A review. *Sustainability*. 10, 3331.

Wang, J.C., Yan, P. Y. 2004. Influence of external thermal insulation compound system on the indoor temperature and humidity. *Low Temperature Architecture Technology*. 2, pp. 80–81.

Wei, K.C., Lv, C.L., Chen, M.Z., Zhou, X.Y., Dai, Z.Y., Shen, D. 2015. Development and performance evaluation of a new thermal insulation material from rice straw using high frequency hot-pressing. *Energy and Buildings*. 87, pp. 116–122.

Wiprächtiger, M., Haupt, M., Heeren, N., Waser, E., Hellweg, S. 2020. A framework for sustainable and circular system design: Development and application on thermal insulation materials. *Resources, Conservation and Recycling*. 154, 104631.

Wu, S., Yan, T., Kuai, Z., Pan, W. 2020. Thermal conductivity enhancement on phase change materials for thermal energy storage: A review. *Energy Storage Materials*. 25, pp. 251–295.

Xing, Q., Hao, X., Lin, Y., Tan, H., Yang, K. 2019. Experimental investigation on the thermal performance of a vertical greening system with green roof in wet and cold climates during winter. *Energy and Buildings*. 183, pp. 105–117.

Xu, J.Y., Sugawara, R., Widyorini, R., Han, G.P., Kawai, S. 2004. Manufacture and properties of low-denstiy binderless particleboard from kenaf core. *Journal of Wood Science*. 50, pp. 62–67.

Yu, Z.L., Yang, N., Zhou, L.C., Ma, Z.Y., Zhu, Y.B., Lu, Y.Y., Qin, B., Xing, W.Y., Ma, T., Li, S.C., Gao, H.L., Wu, H.A., Yu, S.H. 2018. Bioinspired polymeric woods. *Science Advances*. 4, 8, 7223.

Zach, J., Hroudová, J., Brožovský, J., Krejza, Z., Gailius, A. 2013. Development of thermal insulating materials on natural base for thermal insulation systems. *Procedia Engineering*. 57, pp. 1288–1294.

Zach, J., Korjenic, A., Petránek, V., Hroudová, J., Bednar, T. 2012. Performance evaluation and research of alternative thermal insulations based on sheep wool. *Energy and Buildings*. 49, pp. 246–253.

Zhou, X.Y., Zheng, F., Li, H.G., Lu, C.L. 2010. An environment-friendly thermal insulation material from cotton stalk fibers. *Energy and Buildings*. 42, pp. 1070–1074.

Zuo, J., Zhao, Z.Y. 2014. Green building research – current status and future agenda: A review. *Renewable and Sustainable Energy Reviews*. 30, pp. 271–281.

10 Maximum Power Point Tracking Techniques for PV Framework under Partial Shaded Conditions
Towards Green Energy for Smart Cities

*Mohammad Junaid Khan and
Anurag Choudhary*

CONTENTS

10.1 INTRODUCTION

The photo-voltaic (PV) framework has become one of the most successful resources of renewable energy in recent years, due to technological advances. The PV device creates nonlinear properties. The major challenge in the PV system is therefore the technique of maximum power point tracking (MPPT) under standardized and increasing weather states. Partial shading conditions (PSC) is triggering multiple peaks in a P-V curve. This chapter presents a comparative study of two popular MPPT algorithms, namely Incremental Conductance (INC) and Perturb & Observe (P&O), under different operating conditions for application towards green energy for smart cities. The complete PV architecture is conceived using modeling/Simulink with

DOI: 10.1201/9781003153405-10

a boost converter and MPPT algorithms. Under standard experiments and various meteorological conditions, both the suggested MPPT solutions have been tested to prove their efficiency by looking at their maximum power monitoring across the maximum power point.

PV solar power is among the most promising renewable energy sources that generate power from the sun (Khan 2019). It is environmentally friendly, safe to use and needs very little maintenance. Nonetheless, there are several disadvantages to the PV system, such as affecting PV output under PSCs. MPPT techniques, therefore, have to be applied to maximize output capacity under various weather conditions. MPPT is an algorithm used to ensure that the framework of the PV system performs with better efficiency at all times (Boukenoui 2015). On the MPP path, that duty-cycle has been either enhanced or shortened. When different cells are exposed to different levels of irradiance, PSC is referred to, including examples such as building shadow, PV cell dust, partially cloudy, trees, etc. (Saravanan, 2016), Nonetheless, some MPPT algorithms can allow the PV system to operate under varying operating conditions at high performance, as most MPPT algorithms find the maximum power that results in extracted power loss. To get more power out of a PV module, this study uses P&O and INC algorithms as the MPPT controllers. The PV model is designed to operate to a greater efficiency (Kumar et al. 2017a). The power output and electrical efficiency of the PV module are linearly dependent on irradiation and temperature. Alternatively, they compare the two MPPT techniques to look at their efficacy, complexity, settling time around the MPP and convergence.

In the past, different approaches have been employed to ensure high-performance operation of the PV system. One of the most revered and used approaches in the past few years was the Fractional Open Circuit Voltage (FOCV). This approach employs found that the PV array is directly proportional to open circuit voltage (Voc) at MPP (Khan 2020). That depends on the source, manufacturing and weather conditions of the solar cells (Palariya 2016). By switching off the power converter the voltage level at MPP is calculated by calculating the Voc. In recent years, several researchers have implemented the short circuit current (Isc) system. This methodology operates along the same lines as the Voc method. This employs the linear relation theory between PV array current at maximum power point (Impp) and Isc (Rai et al. 2016). Under PSCs, these two methods are not essential. Both methods have low convergence speed and experience temporary power loss. But less accurate under PSC and fail to work at full power due to the features of the P-V output and real power peaks cannot be discerned.

Another solution is to transfer control of the ripple correlation process. This technique uses the ripples created by converter switching to control maximum power in current and voltage. To obtain full power the relation between ripple current or voltage waveform and power has been utilized. This reasoning is laid out by Gupta (2014). This technique depends on the P-V curve with the lack of local maxima and the input voltage is a non-zero sum of ripples at all times. The downside of this technique is that due to the local maxima, incorrect MPP tracking will occur. A couple of studies have been published using artificial intelligence-based techniques (Khan 2019; Khan 2020), in a robust way to overcome the aforementioned disadvantages. Regardless of the ability of these algorithms to function on a nonlinear characteristic, more computation is required (Farayola 2018; Kumar et al. 2017b). Although becoming the most popular method, it does have some drawbacks, such as ambiguous structure, difficult to implement, and requiring detailed and costly computation to be implemented (Farayola et al. 2018; Bansal et al. 2021; Khan 2017). The commonly used PV systems usually have efficiency varying from 6 to 16 percent and the difference in their output depends on the equipment used (Jaen 2008; Roshan 2013; Farayola et al. 2018). In order to get high conversion efficiency, there are primarily two approaches. The first is focused on both mechanical and electrical (electro-mechanical) instruments, also recognized as sun trackers, which focus on supporting the best location during the operating period for the solar PV module.

As well, the second is entirely based on electronic systems that help to adjust the electrical characteristics of the solar PV module's performance, enabling the PV module to work at the optimum operating stage. A fundamental solar cell V-I characteristic is nonlinear is seen in Figure 10.1.

Based on various environmental factors, such as shading, wind velocity, and solar insolation angle, PV cells generate varied amounts of power in partially shaded conditions (Masoum 2002). In order to get the most power out of PV systems, MPPT techniques are equipped with proper controllers. For running PV modules at maximum power, MPPT methods come in a variety of shapes and sizes. Nevertheless, in instantly changing weather conditions, the output of the specific technique relies on its tracking capacity. The techniques for classical MPPT (Ahmad 2019; De Brito et al. 2013) include incremental conductance (InC), fractional short and open circuit voltage, P&O, hill climbing (HC), adaptive reference and constant voltage, and MPPT-based ripple correlation control (RCC). Due to their lower complexity in the algorithm, these methods are quickly applied. However, these implementations have intense perturbations around the MPP, resulting in power loss. In comparison, the influence of PSCs resulting in the inability to trace the genuine MPP is ignored above all by these classical strategies.

This chapter is structured as follows: The P&O MPPT technique is described in Section 10.2. The INC MPPT algorithm is presented in Section 10.3. The simulation

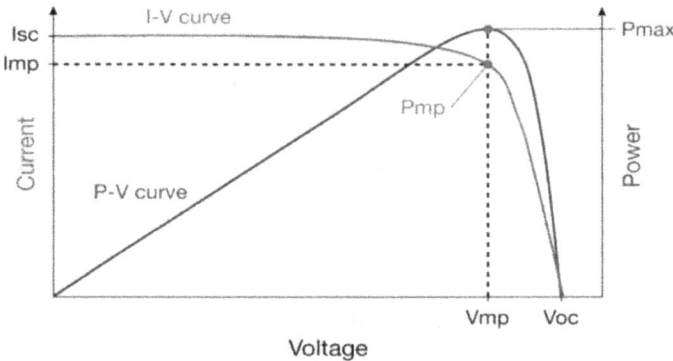

FIGURE 10.1 The V-I curve of solar cells is nonlinear.

model is defined in Section 10.4. Section 10.5 displays the results and discussions. This chapter concludes with Section 10.6.

10.2 PERTURB & OBSERVE MPPT TECHNIQUE

P&O is the simplest and most common technique that many researchers have been studying. It operates by voltage sensing and PV system current. It checks the voltage, current and calculates the power (Khan 2020). This contrasts new power with previous energy stored in a memory. The current voltage would be equivalent to the previous one if power was more than the actual power output. Because the difference is higher than zero, the voltage is changed in the optimistic direction to achieve MPP. This fluctuation the voltage in the downward direction unless the power difference is below zero. Until the achieved higher-efficiency PV system operates (Abdelsalam et al. 2011), the procedure is continued. As illustrated in Figure 10.2, the flowchart demonstrates the developed P&O technique.

The advantages of this approach are its less complexity, being easy to develop and being more reliable. However, under steady-state activity, it has a downside of oscillating around MPP, which results in power loss.

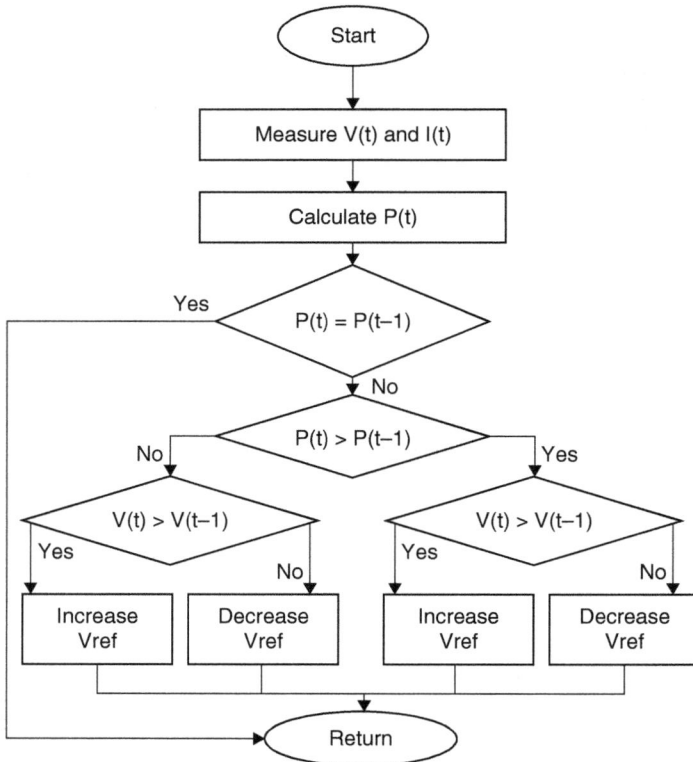

FIGURE 10.2 P&O MPPT flowchart.

10.3 INCREMENTAL CONDUCTANCE MPPT TECHNIQUES

INC is a widely used MPPT technique, which compares the contrast of instant conductance, i.e. I/V to the INC, or dI/dV. It's also based on the formulation of the PV module's power to the voltage, resulting in a zero balance (Gil-Antonio 2016). The MPP is attained from Figure 10.2 when the INC is equivalent to the immediate conductance but in the opposite polarity. To hit the MPP, the service period is prolonged or shortened by a minor factor (Gupta and Saxena 2016). The advantages of the INC MPPT algorithm are that it is simple to put into practice, with a higher speed and performance of tracking (Karami 2017).

10.4 SIMULATION MODEL OF PROPOSED SYSTEM

The implemented simulation is performed to justify and evaluate the effectiveness of the two approaches, the speed and precision of both the MPPT algorithm under PSCs, such as P&O and INC as illustrated in Figure 10.3. The MATLAB / SIMULINK program was used for modelling and simulation of the MPPT algorithm discussed above.

The specification of the Zhejiang Kingdome Solar Energy Technique KD-P230 PV panel and boost DC-DC converter are discussed in Table 10.1 and Table 10.2 respectively. Figure 10.4 shows the simulation model of the proposed work and Figure 10.5 shows the subsystem of the P&O MPPT technique and the subsystem of INC based MPPT technique as shown in Figure 10.6.

A boost converter is being used to match the impedance of the source and load. The duty-cycle is being used in the circuit to control the power. The irradiance and temperature are the corresponding inputs of the PV system. At the switching frequency of 11 kHz, the Pulse Width Modulation (PWM) generator was used and the PWM generator supplies a triangle waveform as discussed by Bounechba (2014). The simulation time of all the cases is one second.

10.5 RESULTS AND DISCUSSION

The simulated results from the MATLAB/S IMULINK model are referenced using both P&O and INC MPPT techniques. The results are the output power and voltage of the proposed system using different MPPT techniques. The thermal signature of all the cases is constant at 27°C, but solar irradiance is varied as shown in Figure 10.7. Figure 10.8 shows the output power characteristics of a proposed system using different MPPT techniques and output voltage characteristics of a proposed system using different MPPT techniques as shown in Figure 10.9.

Figure 10.8 and Figure 10.9 demonstrate the schematic effects of each of the MPPT algorithms derived from the MATLAB simulation. The advisable algorithm over P&O is the INC MPPT strategy. On all simulation outcomes, the INC algorithm outperformed P&O in terms of consolidation speed and monitoring a real MPP under PSC. P&O has a quicker settle period around MPP. To verify the MPPT tracking performance of the proposed work, a simulation was developed, as illustrated in Figure 10.4.

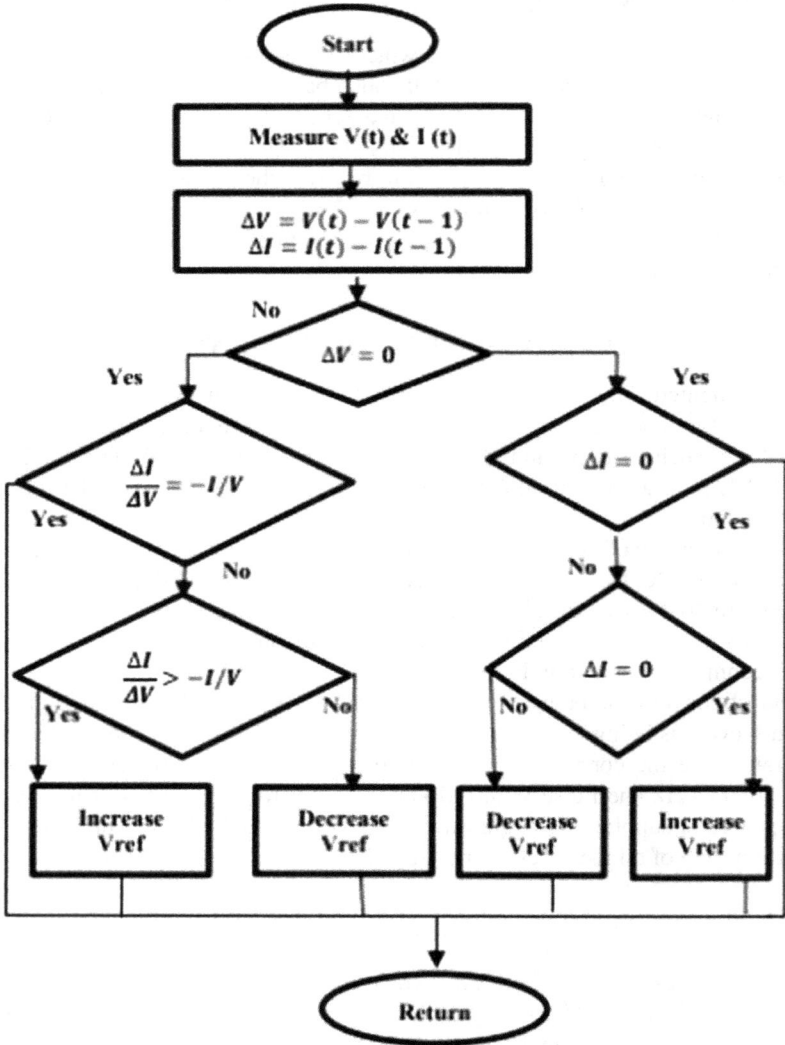

FIGURE 10.3 Flowchart of INC MPPT technique.

As the solar PV source, an I-V curve method was used to program various PSC patterns. To give the output of PV source to a resistive load, a DC-DC boost converter powered using a PWM signal is used. The boost converter's specifications are described in Table 10.2. A MATLAB/SIMULINK platform was used to simulate and implement the MPPT techniques for evaluation. Voltage and current sensors on the virtual platform were used to test the boost converter's input/output voltage and current. A PC software program written in MATLAB/ SIMULATION was used to perform and monitor all assessment experiments. Figures 10.5 and 10.6 show the

TABLE 10.1
Specification of Zhejiang Kingdome Solar Energy Technique KD-P230

S. No.	Parameters	Values
1.	Maximum power	226.78
2.	No. of cells per module	60
3.	Open circuit voltage	36.3
4.	Short circuit current	8.3
5.	The voltage at maximum power point	29
6.	Current at maximum power point	7.82
7.	Temperature coefficient of open circuit voltage	-0.33
8.	Temperature coefficient of short circuit	0.05
9.	Irradiations	Varies
10.	Temperature in °C	27
11.	Parallel string of modules	01
12.	Series string of modules	03

TABLE 10.2
Specification of DC-DC converter

S.No.	Parameters	Values
1.	Inductor	1mH
2.	Capacitor	40 Micro Farad
3.	Resistive load	15 Ohm
4.	Input resistance	0.1 Ohm
5.	No. of IGBT	01
6.	No. of diode	01
7.	Switching frequency	11kH
8.	Duty cycle	Varies

MPPT controller for a PV module with partial shading. The DC-DC power converter had a slew rate of around 0.112 V/ms. Furthermore, the highest voltage change needed to switch the operating points for the two experimental algorithms was the maximum. As an output result, the overall transition time needed to achieve steady state is 0.01 seconds. As a result, for each voltage change there was a 0.02 ms delay time. The execution process for one iteration of the experimental algorithms is depicted in Figure 10.2 (flowchart). The algorithms measure the individual operating voltages at the start of each iteration. The PWM signal's duty-cycle is then modified for each person so that the PV array is at fixed voltage. Because of the differences in conversion speeds, we summed the net computation times of the algorithms as the execution times for comparison. The time constants of the converter would have a large impact on the overall MPPT time.

Although the traditional PO can precisely extract the MPP under PSCs, the complicated calculation makes implementation difficult. Though the measurement may be simplified, precision decreases as processing time decreases. Furthermore,

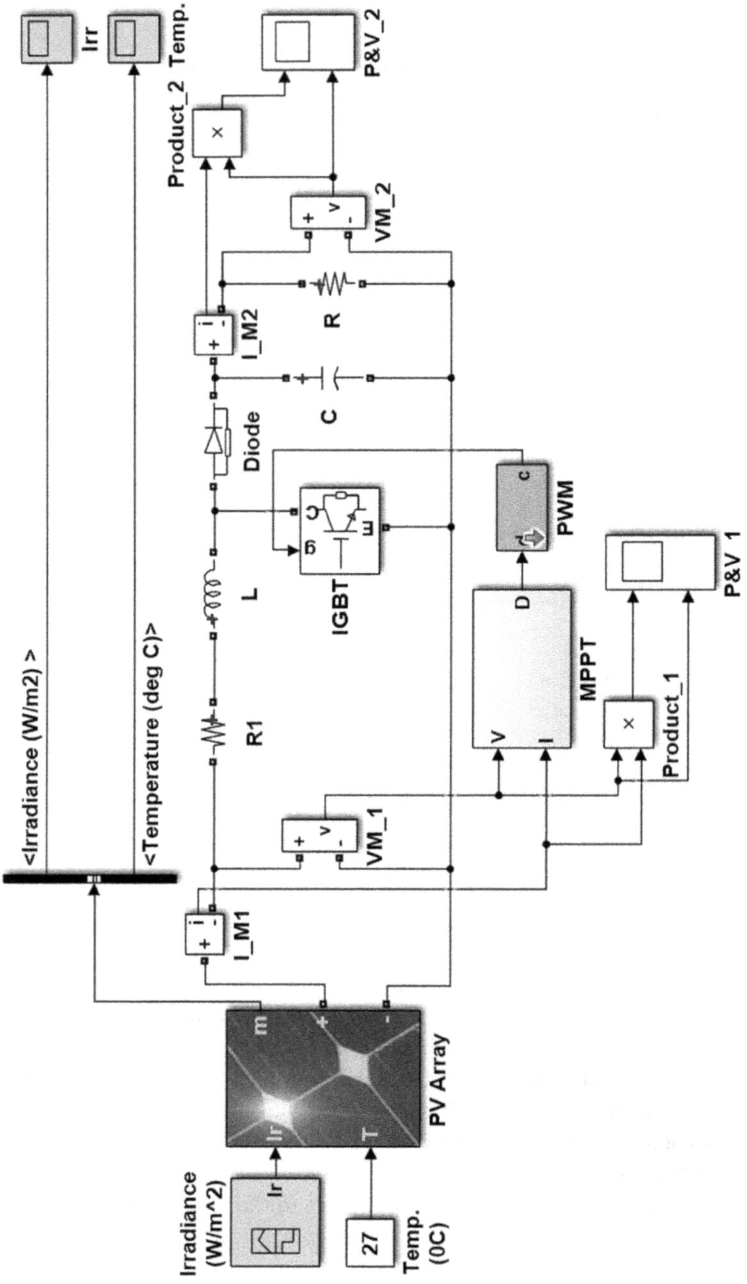

FIGURE 10.4   Simulation model of proposed system.

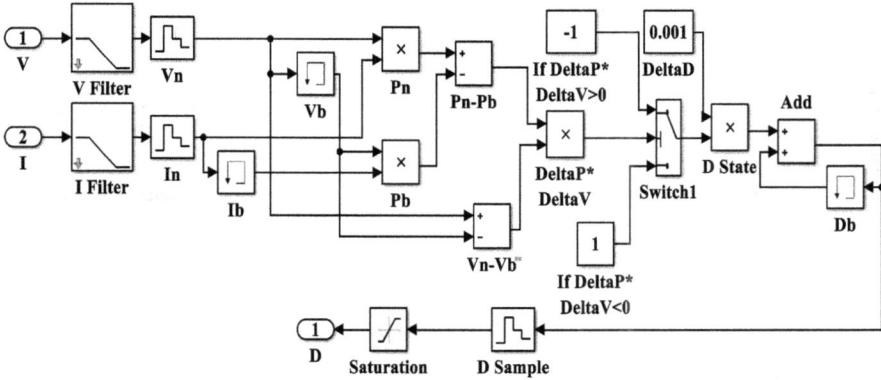

FIGURE 10.5 Subsystem of P&O MPPT technique.

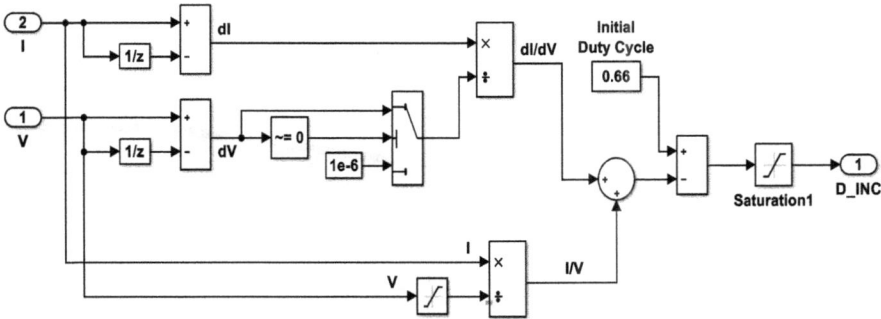

FIGURE 10.6 Subsystem of INC MPPT technique.

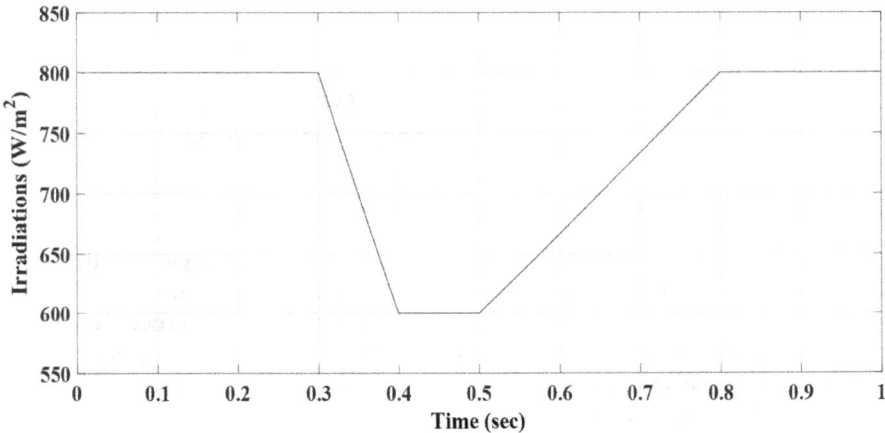

FIGURE 10.7 Irradiations pattern for PV module.

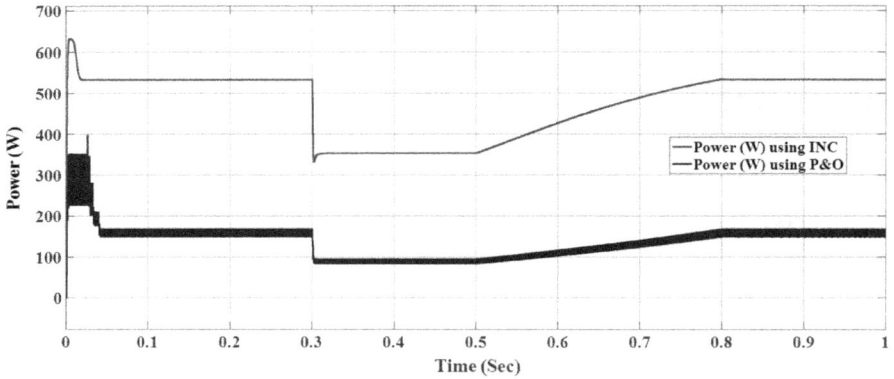

FIGURE 10.8 Output power characteristics of proposed system using different MPPT techniques.

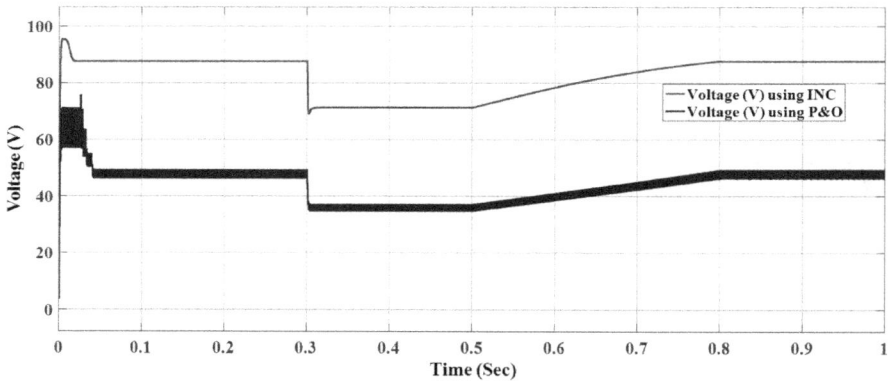

FIGURE 10.9 Output voltage characteristics of proposed system using different MPPT techniques.

simpler calculation processes can result in search results being stuck at MPP. To address this issue, this chapter proposes an INC-based MPPT algorithm that outperforms traditional PO-based MPPT techniques in terms of response time.

10.6 CONCLUSIONS

It is quite clear that two MPPT algorithms will harvest maximum power under PSCs, following the simulation being performed. Both techniques are not PV based and can both locate the real MPP with variable insolation and the INC process converges towards MPP has better then P&O. Furthermore, INC offers the best remedy against P&O for PSC. The INC approach enhances the performance of the system being proposed and also reduces the tracking process. P&O is simple, easy to implement and fast response around MPP compared to INC algorithm.

REFERENCES

Abdelsalam, A.K., Massoud, A.M., Ahmed, S. and Enjeti, P.N., 2011. High-performance adaptive perturb and observe MPPT technique for photovoltaic-based microgrids. *IEEE Transactions on power electronics*, 26(4), pp.1010–1021.

Ahmad, R., Murtaza, A.F. and Sher, H.A., 2019. Power tracking techniques for efficient operation of photovoltaic array in solar applications–A review. *Renewable and Sustainable Energy Reviews*, *101*, pp.82–102.

Bansal, S., Mago, J., Gupta, D. and Jain, V., 2021. Parametric optimization and analysis of cavitation erosion behavior of Ni-based+ 10WC microwave processed composite clad using Taguchi approach. *Surface Topography: Metrology and Properties*, 9(1), p.015011.

Bounechba, H., Bouzid, A., Nabti, K. and Benalla, H., 2014. Comparison of perturb & observe and fuzzy logic in maximum power point tracker for PV systems. *Energy Procedia*, *50*, pp.677–684.

Boukenoui, R., Bradai, R., Mellit, A., Ghanes, M. and Salhi, H., 2015, November. Comparative analysis of P&O, modified hill climbing-FLC, and adaptive P&O-FLC MPPTs for microgrid standalone PV system. In *2015 International Conference on Renewable Energy Research and Applications (ICRERA)* (pp. 1095–1099). IEEE.

De Brito, M.A.G., Galotto, L., Sampaio, L.P., e Melo, G.D.A. and Canesin, C.A., 2012. Evaluation of the main MPPT techniques for photovoltaic applications. *IEEE Transactions on Industrial Electronics*, 60(3), pp.1156–1167.

Farayola, A.M., Hasan, A.N. and Ali, A., 2018. Efficient photovoltaic MPPT system using coarse Gaussian support vector machine and artificial neural network techniques. *International Journal of Innovative Computing Information and Control (IJICIC)*, *14*(1), pp.323–329.

Gil-Antonio, L., Saldivar-Marquez, M.B. and Portillo-Rodriguez, O., 2016, June. Maximum power point tracking techniques in photovoltaic systems: A brief review. In *2016 13th International Conference on Power Electronics (CIEP)* (pp. 317–322). IEEE.

Gupta, A., Chanana, S. and Thakur, T., 2014. Power quality improvement of solar photovoltaic transformer-less grid-connected system with maximum power point tracking control. *International Journal of Sustainable Energy*, *33*(4), pp.921–936.

Gupta, A.K. and Saxena, R., 2016, February. Review on widely-used MPPT techniques for PV applications. In *2016 International Conference on Innovation and Challenges in Cyber Security (ICICCS-INBUSH)* (pp. 270–273). IEEE.

Jaen, C., Moyano, C., Santacruz, X., Pou, J. and Arias, A., 2008, September. Overview of maximum power point tracking control techniques used in photovoltaic systems. In *2008 15th IEEE International Conference on Electronics, Circuits and Systems* (pp. 1099–1102). IEEE.

Karami, N., Moubayed, N. and Outbib, R., 2017. General review and classification of different MPPT Techniques. *Renewable and Sustainable Energy Reviews*, *68*, pp.1–18.

Khan, M., 2019. Artificial Intelligence Based Maximum Power Point Tracking Controller for Fuel Cell System [J]. *European journal of electrical engineering*, 21(3), pp.297–302.

Khan, M.J., 2020. Review of recent trends in optimization techniques for hybrid renewable energy system. *Archives of Computational Methods in Engineering*, pp.1–11.

Khan, M.J. and Mathew, L., 2020. Comparative study of optimization techniques for renewable energy system. *Archives of Computational Methods in Engineering*, 27(2), pp.351–360.

Khan, M.J. and Mathew, L., 2017, October. Artificial intelligence based maximum power point tracking algorithm for photo-voltaic system under variable environmental conditions. In *2017 Recent Developments in Control, Automation & Power Engineering (RDCAPE)* (pp. 114–119). IEEE.

Kumar, R., Choudhary, A., Koundal, G. and Yadav, A.S.A., 2017a. Modelling/simulation of MPPT techniques for photovoltaic systems using Matlab. *International Journal*, 7(4), 178–187.

Kumar, R., Choudhary, A. and Shimi, S.L., 2017b, August. Smooth Starter for DC Shunt Motor using Buck-Boost Power Converter. In *2017 International Conference on Innovations in Control, Communication and Information Systems (ICICCI)* (pp. 1–8). IEEE.

Masoum, M.A., Dehbonei, H. and Fuchs, E.F., 2002. Theoretical and experimental analyses of photovoltaic systems with voltage and current-based maximum power-point tracking. *IEEE Transactions on Energy Conversion*, 17(4), pp.514–522.

Palariya, A.A.K., Choudhary, A. and Yadav, A., 2016. Modelling, control and simulation of MPPT for wind energy conversion using Matlab/Simulink. *Eng. J. Appl. Scopes*, 1(2), pp.9–13.

Rai, A., Awasthi, B., Singh, S. and Dwivedi, C.K., 2016. A Review of maximum power point tracking techniques for photovoltaic system. *Int. J. Eng. Res*, 5, pp.539–545.

Roshan, R., Yadav, Y., Umashankar, S., Vijayakumar, D. and Kothari, D.P., 2013, April. Modeling and simulation of Incremental conductance MPPT algorithm based solar photo voltaic system using CUK converter. In *2013 International Conference on Energy Efficient Technologies for Sustainability* (pp. 584–589). IEEE.

Saravanan, S. and Babu, N.R., 2016. Maximum power point tracking algorithms for photovoltaic system–A review. *Renewable and Sustainable Energy Reviews*, 57, pp.192–204.

11 Forecasting Energy Consumption Using Deep Echo State Networks Optimized with Genetic Algorithm

*V. Dinesh Reddy, Kurinchi Nilavan,
G.R. Gangadharan, and Ugo Fiore*

CONTENTS

11.1 INTRODUCTION

The residential and commercial buildings sector represents the majority of energy consumption at 28% and it was about 21 quadrillion British thermal units in 2019 (according to the U.S. Energy Information Administration). It is clear that in view of the technological developments, residential energy consumption is expected to grow in future (Guneralp et al. 2017). With the proliferation of cloud computing, Internet of Things technologies, and social media applications, data centers play an important role in business computing (Reddy et al. 2018; Reddy et al. 2020). The global greenhouse gas emissions in the Information and Communication Technology sector is around 0.86 metric gigatons of carbon emissions and accounts for 2% of global CO_2 emissions (Malmodin et al. 2018).

DOI: 10.1201/9781003153405-11

Residential energy consumption is largely an undefined energy sink and a complex issue (Amber et al. 2018). It is highly related to house size, geometry, occupant behavior, energy systems within these homes, and macro-economic factors such as fuel prices and inflation. In today's rapidly growing digital economy, the demand by consumers for more data and high-bandwidth content is fueling the upsurge in business of data centers. Due to economic, social and environmental issues, the high energy consumption of data centers is attracting ever more attention. We need to develop proactive approaches to identify and stop the growth of these silent threats. This will abate the energy expenditures and reduce inefficiencies of data centers.

Machine learning involves building models or finding trends in previous examples of system behavior, with as little expert intervention as possible. In the absence of application/system level measures in a data center, we can use various methods of machine learning to model the missing information (Lee et al. 2014; Hazelwood et al. 2018). There is an opportunity to explore advanced machine learning techniques for energy usage modeling in both industrial and residential buildings. The ability to manage high-dimensional data, agility and high predictive accuracy is needed for the relevant model. Deep learning is an emerging area of machine learning, that learns from multiple levels of representations (Lecun et al. 2015; Schmidhuber 2015). Deep learning methods detect the representations required with multiple levels of abstraction, with a deeper stack of coupling layers.

Deep learning makes significant strides in high-dimensional data by discovering complex structures. It has shown highly promising results for the solution of complex problems in many domains such as signal processing, natural language understanding, etc. This chapter describes an improved deep echo state network optimized with genetic algorithm (GA-DeepESN) for forecasting energy consumption.

11.2 RELATED WORK

Gassar et al. (2019) have developed data-driven models for forecasting electricity and gas usage in the residential buildings of London with the multi-layer artificial neural network, random decision forests, gradient boosting machine, and multiple regression algorithms. The research indicates the characteristics of the house, number of households, and household revenue were by far the most significant predictors of energy and gas consumption. Kialashaki et al. (2013) proposed two energy-demand prediction models for the residential sector using multiple regression and artificial neural networks. The conditional demand analysis (CDA) was used in (Aydinalp-Koksal and Ugursal 2008) to estimate the national level of residential energy usage. They reported that the impact of socio-economic variables is best calculated using the neural network model rather than the CDA model.

Wong et al. (2010) proposed a neural network model for predicting daily electricity use of an office building considering the weather and building characteristics. With one hidden layer, the model has shown a root mean square error varying from 3 to 5.6%. Ekici et al. (2009) proposed an approach using artificial neural networks for analyzing the energy demand of buildings based on various factors such as insulation thickness, orientation, and transparency ratio. Further, they experimented with a finite difference approach for brick walls with and without insulation of one-dimensional transient conduction. Yu et al. (2010) proposed a model for predicting and classifying energy usage intensity (EUI) levels of buildings using the decision tree method. The results demonstrate that this model can identify significant factors influencing the building EUI and classify EUI levels accurately (up to 93%). However, authors have not provided insights on modeling the energy demand of building.

Liu et al. (2019) discussed ANN and support vector machines (SVM) models in detail and presented their structure and accuracy comparison. Further, they proposed a hybrid machine learning method to estimate the building energy consumption. Jahani et al. (2020) used the genetic algorithm based numerical moment matching approach for estimating the electricity usage of a large dataset of single-family homes. Zhao et al. (2010) proposed a parallel implementation of SVM method with Gaussian kernel to predict the energy consumption of multiple buildings. Nishi et al. (2017) proposed a supervised machine learning approach to cluster data with minute granularity based on power variation trend models in a modern data center. Versick et al. (2013) described a novel approach to estimating the energy usage of computer systems in reliance on software-generated workloads in various execution environments. They developed a power measurement model for virtual machines that are self-adaptable to the underlying hardware to produce accurate results.

Based on the literature review, we observe that statistical models and forward engineering models require fixed assumptions and algorithmic approximations. The neural network with backpropagation and its variations may not always find an optimal solution. Regression models are attempting to forecast outcomes depending on a number of independent variables. However, if it has incorrect independent variables, the model would have less prediction accuracy. Further, logistic regression demands that each data point be independent of some other data point. If the observations are related to each other, the model will appear to overweigh the importance of those observations.

11.3 DEEP ECHO STATE NETWORK OPTIMIZED WITH GENETIC ALGORITHM

In this section, we describe our proposed deep learning approach: a deep echo state network optimized with genetic algorithm (GA – DeepESN) for forecasting energy consumption. The DeepESN combines the nonlinear time series modeling ability of ESN and the efficient learning ability of deep learning (Gallicchio et al. 2017; Hu et al. 2020). A DeepESN consists of an input layer, some stacked reservoir layers, and an output layer. The hyperparameters of DeepESN mainly include the number of input unit k, the spectral radius ρ, and the connectivity rate α for each reservoir layer as similar to basic ESN. In this chapter, for the purpose of obtaining optimized hyperparameters, we have used the genetic algorithm (GA).

GA can mimic the process of evaluation with the objective to minimize or maximize an objective function (Holland 1984) that can be used to find optimal solutions for the time series models. A neural network approach integrating GA with the echo state network is described (Zhang et al. 2018). It has been studied that GA performs better in getting optimized solutions for time series models (Bouktif et al. 2018), so it is expected to obtain better performance with GA to optimize DeepESN. Generally, a deep echo state network is multiple echo state networks stacked sequentially or in parallel. In this chapter, we consider a sequential deep echo state network. We evaluate the trained model with evaluation data and determine the RMSE value and check for fitness criteria. If the RMSE score can be improved, then the new value for random variables is selected and the process is repeated until the fitness criteria is satisfied. By this process, we get an optimized window size and number of reservoirs for the model. Finally, we train the DeepESN model with the optimized hyper-parameters and predict the energy consumption for the test datasets. The procedure for our proposed GA-DeepESN model is described as follows (see Figure 11.1):

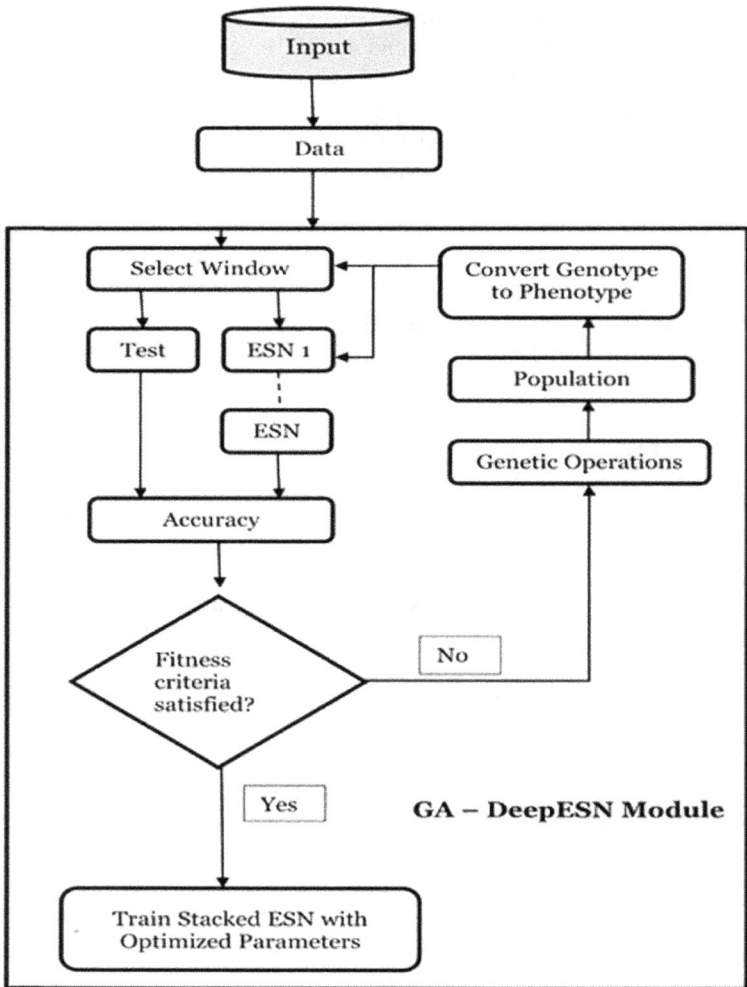

FIGURE 11.1   Flow chart for deep echo state network optimized with genetic algorithm.

Step 1. Initialize the random variable in the genetic algorithm using any distribution which is the number of reservoir (*n_res*) in the ESN and window size (*s*) of the input data.

Step 2. Prepare the input data for the windows size (*s*).

Step 3. Train the DeepESN, a multi-layer sequentially stacked ESN with *n_res* number of reservoirs.

Step 4. Evaluate the model using the root mean squared error (*RMSE*).

Step 5. Check the fitness function of the genetic algorithm. If it does not clear, then mutate and select the random variable. Repeat steps 2 to 5 until it reaches the desired fitness value.

Step 6. With the optimized parameters *n_res* and window size (*s*), train and test ESN.

11.4 DESCRIPTION OF DATASETS AND PREPROCESSING

We use the following datasets for our experiments.

Individual household power consumption (IHPC) dataset We used the dataset on measurements of electric power consumption in a household with a one-minute sampling rate over a period of almost 4 years from December 2006 to November 2010, stored on the UC Irvine Machine Learning Repository (Georges et al. 2012). This dataset contains 2,075,259 measurements. The data contains four electrical quantities and three sub-metering values.

Data center power consumption (DCPC) dataset We used adequate data on measurements of a data center power consumption with a one-day sampling rate over a period of almost 5 years from January 2013 to October 2017 collected from a real data center. This dataset has power consumption of chillers, lighting, air handling units (AHU), UPS, rack and loss incurred during the transformations.

To handle the missing data, we used the mean substitution method where the mean is calculated for each variable over all examples in the dataset. The mean of the variable in the dataset is used whenever that variable is missing. Some inputs to neural network might not have a 'naturally defined' range of values. So, feeding the raw value to the neural network will not work very well. Normalizing the input data avoids the chances of getting stuck in local optima and the training becomes faster. For normalization of the input data, we consider the Min-Max normalizer, Binning normalizer, and Gaussian normalizer (Shalabi et al. 2006).

We have implemented the proposed GA-DeepESN in Python running on an AMD Thread Ripper, 4 × 16GB 3200 MHz DDR4 RAM, and 4 TB SATA HDD system with two cores.

11.5 EXPERIMENTAL ANALYSIS OF DATA CENTER POWER CONSUMPTION DATASET

We train our models using a data center power consumption of 4 years (2013–2016) and we forecast the energy consumption of chiller for the year 2017. The performance

TABLE 11.1
Performance comparison of various prediction models

Prediction model	Root mean squared error	Coefficient of determination
Neural networks	7.641626	0.813066
Linear regression	10.329192	0.658453
Boosted decision trees	7.458507	0.821917
DPSGD	4.874361	0.923941
GA-DeepESN	2.322854	0.993756

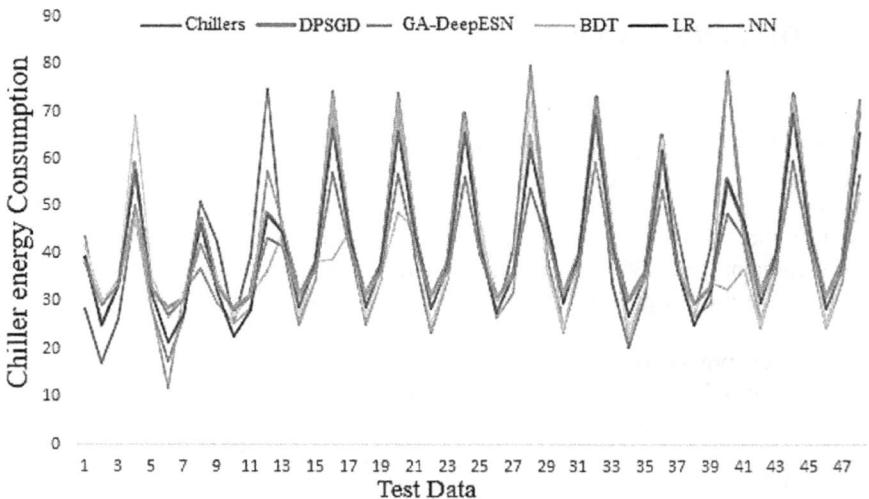

FIGURE 11.2　Experimental data vs predictions for test data.

of our proposed GA-DeepESN model is compared with a boosted decision tree (BDT) regression, neural network, linear regression, and deep learning with parallel stochastic gradient descent (DPSGD) approaches. From Table 11.1, we can see that our proposed model has low RMSE and high COD. This prediction accuracy is promising and comparable with the reported results.

Figure 11.2 and Figure 11.3 present the prediction results of the said models for the test data and the training data respectively. These figures present the predicted versus the observed chiller energy consumption with varying loads. In Figure 11.2 and Figure 11.3, we see that the predicted values of the GA-DeepESN approach has a similar pattern with the observed chiller energy consumption. Further, the proposed model has shown high accuracy in high and low power consumption conditions. It is obvious that the GA-DeepESN algorithm offers a gain in the RMSE analysis relative to the canonical ESN.

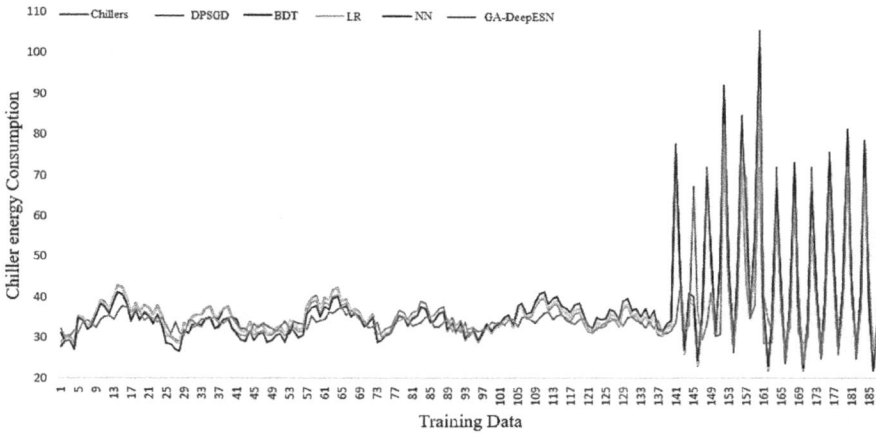

FIGURE 11.3 Experimental data vs predictions for training data.

TABLE 11.2
Comparison of different normalization functions for IHPC dataset

Normalizer	MAE	RMSE	RAE	RSE	COD
Binning	0.022080	0.038727	0.024056	0.001063	0.998937
MinMax	0.021893	0.037616	0.023852	0.001003	0.998997
Gaussian	0.030749	0.104393	0.033500	0.007723	0.992277
Without normalization	0.911138	1.188965	0.992659	1.001865	-0.001865

11.6 EXPERIMENTAL ANALYSIS OF INDIVIDUAL HOUSEHOLD POWER CONSUMPTION DATASET

Regression analysis is used to evaluate the network capability for active power prediction in a single household. The RMSE is used as a measure to evaluate how the trained network estimation is correlated to the experimental data. We conducted several experiments to choose the best normalization function for our data for a network containing 12 and 6 neurons in the first and second hidden layers respectively. The results with various normalizers are presented in Table 11.2. We observe that the MinMax normalization works better over other normalizers for our data.

Figure 11.4 and Figure 11.5 present the prediction results of the proposed model for test data and the training data respectively. These figures present the global active power prediction of typical household versus observed global active power with varying loads. In Figure 11.4 and Figure 11.5, it is shown that the predicted power has a similar pattern with the observed active power. Further, the proposed model has shown high accuracy in high and low power consumption conditions.

We compared the performance of our proposed model with linear regression, neural networks, boosted decision trees, and DPSGD. The proposed GA-DeepESN

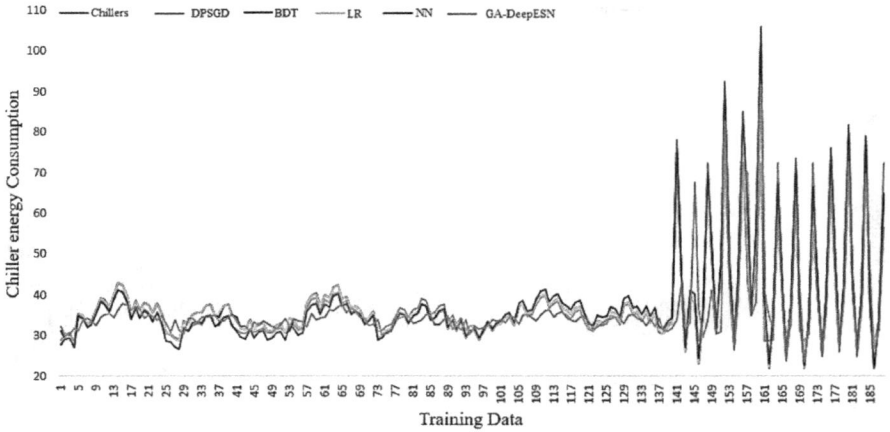

FIGURE 11.4   Experimental data vs predictions for test data.

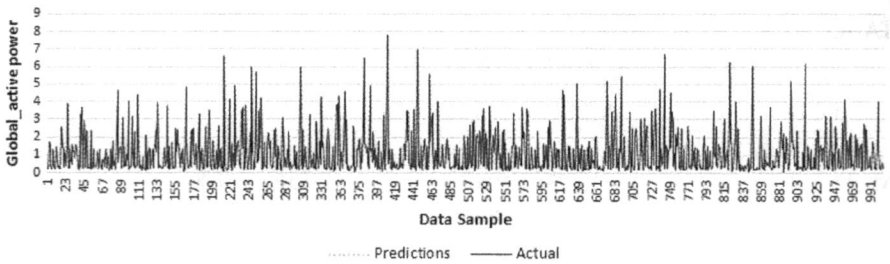

FIGURE 11.5   Experimental data vs predictions for training data.

TABLE 11.3
Performance comparison of various prediction models for household dataset

	Root mean squared error	Coefficient of determination
Linear regression	0.39298	0.91422
Neural network	0.03818	0.99896
Boosted decision trees	0.04748	0.99754
DPSGD	0.03583	0.99901
GA-DeepESN	0.02854	0.99926

method is a relatively advanced model and shows better performance over other methods. The performance results of the said models are given in Table 11.3. The GA-DeepESN model has low root mean square error (RMSE) and high COD. This prediction accuracy is promising and comparable with the reported results. We also present

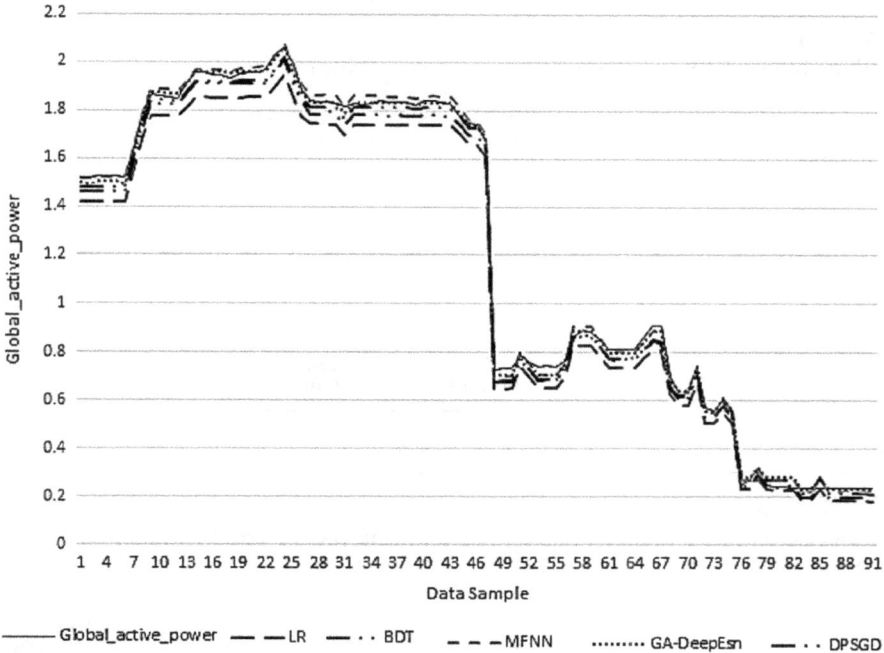

FIGURE 11.6 Predictions of the various models for the 90 minutes interval.

the predictions of the various models for the interval of 90 minutes for giving better visualization in Figure 11.6. Figure 11.6 presents the global active power prediction of typical household with different models versus observed global active power. It can be seen that the proposed model has the best performance to describe the realistic relationship between global active power and other variables in the dataset.

11.7 PERFORMANCE EVALUATION

Statistical significance testing is an essential procedure to evaluate which model is best supported by the sample data (Sirkin et al. 2005). Tests for statistical significance are used to check whether the prediction of the model is due only to random chance. These tests tell us the probability of making an error if we use the said model. As our data is interval based and we work with more than one sample with a repeated-measures (within subjects) design, we performed T-test (Sirkin 2005) and Wilcoxon signed rank test (Randles 1988) for our data. We compared our methods with each of other methods on both the datasets by performing these statistical tests. The results obtained by the T-test justify whether the predicted mean values of all approaches have a distinguished difference with 58 degrees of freedom at 0.05% level of significance. The statistical test values for all approaches are presented in Table 11.4 and Table 11.6. Each cell in the table represents the values of T/P. Based on Table 11.4 and 11.6, the prediction values are statistically significant (all P-values are less than

TABLE 11.4
Values for T and P for significance level of 0.05 for IHPC dataset

	DPSGD	GA-DeepESN	BDT	NN	LR
DPSGD	*	*	*	*	*
GA-DeepESN	1.9127	*	*	*	*
	0.030364				
BDT	2.124239	2.827	*	*	*
	0.0423	0.003219			
NN	5.453751	5.268	−0.866085	*	*
	<0.00001	<0.00001	0.393552		
LR	3.925519	4.587	−0.047383	1.470144	*
	0.00049	0.000012	0.962532	0.15229	

TABLE 11.5
Statistical analysis using Wilcoxon signed rank test (Z and P values) for IHPC dataset

	LR	NN	BDT	DPSGD	GA-DeepESN
LR	*	*	*	*	*
NN	Z = 4.7821	*	*	*	*
	P = 0				
BDT	Z = 2.3139	Z= 4.6587	*	*	*
	P = 0.02088	P = 0			
DPSGD	Z = 4.7821	Z = 4.7821	Z = 2.189	*	*
	P = 0	P = 0	P = 0.02859		
GA-DeepESN	Z = 5.2621	Z = 4.9621	Z = 4.2134	Z = 3.564	*
	P = 0	P = 0	P = 0	P = 0	

0.00001). Similarly, we performed Wilcoxon Signed-Rank test and the results are presented in Table 11.5 and Table 11.7. Based on Table 11.5 and Table 11.7, the obtained values are statistically significant (all Z-values are obtained by positive ranks and P values are mostly 0.000).

11.8 CONCLUSION

Minimizing the energy consumption is inevitable with the increasing of CO_2 emissions and rising energy prices in recent years. It is very important to predict the energy demands of the residential buildings and data centers. We developed an improved deep echo state network optimized with genetic algorithm (GA-DeepESN) for forecasting chiller energy consumption of a data center and a house hold active energy consumption. Based on our experiments and comparisons with other prediction models, the GA-DeepESN approach gives statistically significant results.

TABLE 11.6
Values for T and P for significance level of 0.05 for DCPC dataset

	GA-DeepESN	DPSGD	BDT	NN	LR
GA-DeepESN	*				
DPSGD	2.2153	*	*	*	*
	0.030680				
BDT	2.9532	2.7125	*	*	*
	0.002268	0.004387			
NN	3.1952	2.2153	1.9862	*	*
	0.001131	0.030680	0.025872		
LR	5.8536	4.7263	2.3139	1.9863	*
	<0.00001	<0.00001	0.02088	0.025866	

TABLE 11.7
Statistical analysis using Wilcoxon signed rank test (Z and P values) for DCPC dataset

	LR	NN	BDT	DPSGD	GA-Deep-ESN
LR	*	*	*	*	*
NN	Z = 1.9975	*	*	*	*
	P = 0.045771				
BDT	Z = 2.5816	Z = 4.6587	*	*	*
	P = 0.009834	P = 0			
DPSGD	Z = 3.9199	Z = 4.7821	Z = 2.189	*	*
	P = 0	P = 0	P = 0.02859		
GA-DeepESN	Z = 2.8524	Z = 4.7821	Z = 5.2484	Z = 3.564	*
	P = 0.004339	P = 0	P = 0	P = 0	

ACKNOWLEDGEMENTS

This work is partially supported by the Seed Grant, sponsored by the National Institute of Technology, Tiruchirappalli, India.

REFERENCES

Amber, K.P., Ahmad, R., Aslam, M.W., Kousar, A., Usman, M., Khan, M.S. 2018. Intelligent techniques for forecasting electricity consumption of buildings. *Energy*, 157:886–893.
Aydinalp-Koksal, M., Ugursal, V.I. 2008. Comparison of neural network, conditional demand analysis, and engineering approaches for modeling end-use energy consumption in the residential sector. *Applied Energy*, 85(4):271–296.
Bouktif, S., Fiaz, A., Ouni, A., Serhani, M.A. 2018. Optimal deep learning LSTM model for electric load forecasting using feature selection and genetic algorithm: Comparison with machine learning approaches. *Energies*, 11(7):1636.

Ekici, B.B., Aksoy, U.T. 2009. Prediction of building energy consumption by using artificial neural networks. *Advances in Engineering Software*, 40(5):356–362.

Gallicchio, C., Micheli, A., Pedrelli, L. 2017. Deep reservoir computing: A critical analysis. *Neurocomputing*. 268:87–99.

Gassar, A., Yun, G.Y., Kim, S. 2019. Data-driven approach to prediction of residential energy consumption at urban scales in London. *Energy*, 187:115973.

Georges, H., Alice, B. 2012. Individual household electric power consumption data set. UCI Machine Learning Repository. https://archive.ics.uci.edu/ml/datasets/Individual+household+electric+power+consumption

Guneralp, B., Zhou, Y., Urge-Vorsatz, D., Gupta, M., Yu, S., Patel, P.L., Fragkias, M., Li, X., Seto, K.C. 2017. Global scenarios of urban density and its impacts on building energy use through 2050. *Proceedings of the National Academy of Sciences*, 114(34):8945–8950.

Hazelwood, K., Bird, S., Brooks, D., Chintala, S., Diril, U., Dzhulgakov, D., Fawzy, M., Jia, B., Jia, Y., Kalro, A., Law, J., Lee, K., Lu, J., Noordhuis, P., Smelyanskiy, M., Xiong, L., Wang, X. 2018. Applied machine learning at Facebook: A datacenter infrastructure perspective. In *Proceedings of the IEEE International Symposium on High Performance Computer Architecture (HPCA)*, pp. 620–629. IEEE.

Holland, J.H. 1984. Genetic algorithms and adaptation. *Adaptive Control of Ill-Defined Systems*, 317–333.

Hu, H., Wang, L., Lv, S. 2020. Forecasting energy consumption and wind power generation using deep echo state network. *Renewable Energy*.

Jahani, E., Cetin, K., Cho, I.H. 2020. City-scale single family residential building energy consumption prediction using genetic algorithm-based numerical moment matching technique. *Building and Environment*, 172:106667.

Kialashaki, A., Reisel, J.R. 2013. Modeling of the energy demand of the residential sector in the United States using regression models and artificial neural networks. *Applied Energy*, 108:271–280.

LeCun, Y., Bengio, Y., Hinton, G. 2015. Deep learning. Nature, 521(7553):436–444.

Lee, E.W.M., Fung, I.W.H., Tam, V.W.Y., and Arashpour, M. 2014. A fully autonomous kernel-based online learning neural network model and its application to building cooling load prediction. *Soft Computing*, 18(10):1999–2014.

Liu, Z., Wu, D., Liu, Y., Han, Z., Lun, L., Gao, J., Jin, G., Cao, G. 2019. Accuracy analyses and model comparison of machine learning adopted in building energy consumption prediction. *Energy Exploration & Exploitation*, 37(4):1426–1451.

Malmodin, J., Lunden. D. 2018. *The electricity consumption and operational carbon emissions of ICT network operators 2010–2015*. Report from the KTH Centre for Sustainable Communications, Stockholm, Sweden.

Nishi, A., Chuan, S., Yanbing, S., Xiaogang, S., Abishai, D., Rahul, K., Tianyu, Z., Xiang, Z., Lifei, Z. 2017. Power variation trend prediction in modern datacenter. In *Proceedings of the 16th IEEE Intersociety Conference on Thermal and Thermomechanical Phenomena in Electronic Systems*, 977–980. IEEE.

Randles, R.H. 1988. Wilcoxon signed rank test. *Encyclopedia of statistical sciences*, Wiley.

Reddy, V.D., Gangadharan, G.R., Rao, G.S.V.R.K., Aiello, M. 2020. Energy-efficient resource allocation in data centers using a hybrid evolutionary algorithm. In *Machine Learning for Intelligent Decision Science*, 71–92. Springer.

Reddy, V.D., Setz, B., Rao, G.S.V.R.K., Gangadharan, G.R., and Aiello, M. 2018. Best practices for sustainable datacenters. *IT Professional*, 20(5):57–67.

Schmidhuber, J. 2015. Deep learning in neural networks: An overview. *Neural Networks*, 61:85–117, 2015.

Shalabi, L.A., Shaaban, Z., Kasasbeh, B. 2006. Data mining: A preprocessing engine. *Journal of Computer Science*, 2(9):735–739.

Sirkin, R.M. 2005. *Statistics for the social sciences*. Sage Publications.

U.S. Energy Information Administration. How much energy is consumed in U.S. buildings?. Online from www.eia.gov/totalenergy/data/monthly/. Retrieved on June 15, 2020.

Versick, D., Waßmann, I., Tavangarian, D. 2013. Power consumption estimation of cpu and peripheral components in virtual machines. *ACM SIGAPP Applied Computing Review*, 13(3):17–25.

Wong, S.L., Wan, K., Lam, T. 2010. Artificial neural networks for energy analysis of office buildings with daylighting. *Applied Energy*, 87(2):551–557, 2010.

Yu, Z., Haghighat, F., Fung, B., Yoshino, H. 2010. A decision tree method for building energy demand modeling. *Energy and Buildings*, 42(10):1637–1646.

Zhang, X, Cai, M, Wang, C., Gao, L., Fan, X. 2018. Research for SOC Prediction of Lithium Battery Based on GA-ESN. In *Proceedings of the 11th International Symposium on Computational Intelligence and Design*. 165–168.

Zhao, H.X., Magoules, F. 2010. Parallel support vector machines applied to the prediction of multiple buildings energy consumption. *Journal of Algorithms & Computational Technology*, 4(2):231–249, 2010.

12 State-of-the-Art Natural Language Processing Techniques

Tarun Jaiswal, Manju Pandey, and
Priyanka Tripathi

CONTENTS

12.1 INTRODUCTION

NLP is a branch of cognitive science (SC), machine/deep learning and linguistics concerned with the conversation among computers and human natural language for e.g. English language. Any language can include categories as the natural

DOI: 10.1201/9781003153405-12

language which accept and learn from the human system. Natural language is able to understand the emotions, express and deliver human feedback to another person in an environment. From birth, a child is able to understand and learn natural language. NLP accumulated the set of procedures to achieve that aim. As we know, the NLP area is very complex and varied, so NLP is constructed with some rules that bind in the system and extract meaningful information (i.e. linguistic construction and significance) from large datasets. There are numerous NLP applications, namely detection, classification, recognition, segmentation, speech recognition system, translation from language to language, text summarization, sentiment analysis, chatbots, and text classification, etc. Currently, in NLP, deep learning plays a very important role, the neural network grounded methods have achieved remarkable enhancement in numerous tasks, likewise syntactic parsing (Allahyari et al. 2017), language modelling (Arora et al. 2017) and machine translation (Bahdanau et al. 2015).

There are basically two categories of NLP – conventional NLP and DL based NLP. The statistical NLP method is currently in fashion (see Figure 12.1). However, current studies have shown that deep learning overcomes all former NLP techniques.

Conventional-NLP – There are three various prevalent NLP techniques in today's scenario: classical NLP, deep learning, and statics. The language detection system is the first stage of classical natural language processing. After this stage, the corresponding pipeline of preprocessing stage is performed, which consist of tokenization, named-entity recognition (NER) and part of speech-tagger. The consequence of these preprocessing stages is a human-designed feature, allowing a model to be developed and implications for the predicted work to be conducted.

Deep learning (Dl) – The technique of deep learning is based on an entirely altered style. After raw data preprocessing, the input is combined as an integral part into the dense vector, and this can be produced via various approaches such as doc2vec and glove,word2vec, etc. Thus it becomes the new input data of the NN, which feeds the hidden layer. By using this layer, the network acquires how to identify and touch the objective.

Deep learning uses representation learning where a machine automatically determines the representation of features from the raw data, without any human help, and uses them for accurate prediction. The architecture of the model is language independent.

Methods of NLP, NLP tasks are completed by using the two main techniques, i.e. semantic analysis and syntactic analysis.

The following represent their description:

12.1.1 Syntax

The syntax is the correct placing of words in the sentence in such a manner that they create sense grammatically. Syntactic analysis is used in NLP to evaluate the way a

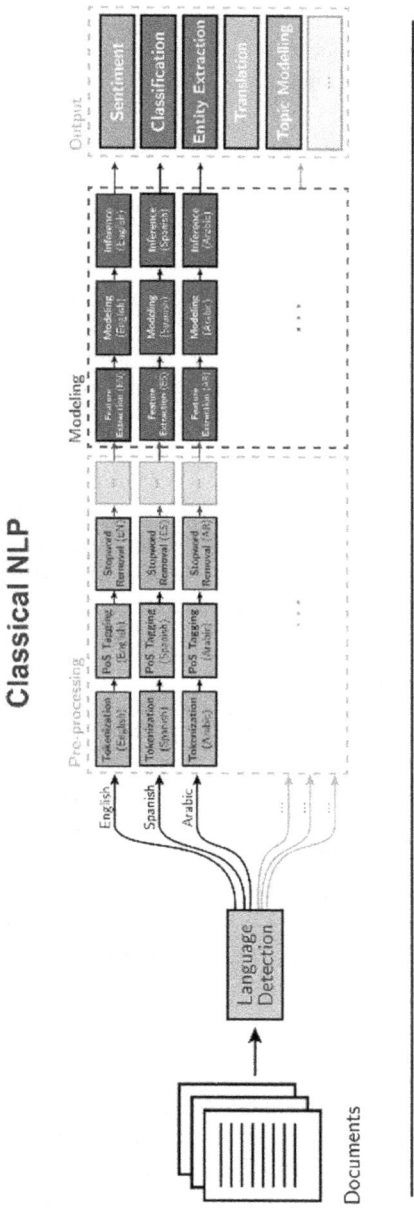

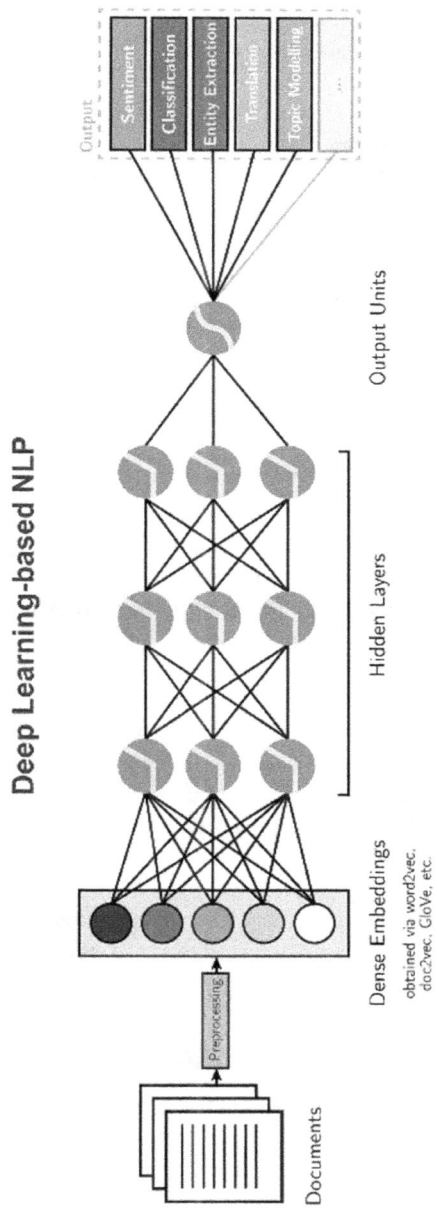

FIGURE 12.1 NLP schema.

natural language integrates with grammar rules. To apply specific grammar rules to words and extract their meaning, some essential algorithms are used.

Some unique techniques are further included in the syntax:

Lemmatization: Lemmatization is a process of minimizing the several modulated forms of a word to an atomic form for convenience.

Morphological-segmentation: Morphological segmentation splits words into atomic elements known as morphemes.

Word segmentation: Word segmentation performs the partition of a large portion of continuing text into various components.

Part-of-speech labeling: For every word, the part of speech is identified by this technique.

Parsing: Parsing perform grammatical analytics.

Sentence breaking: Sentence breaking sets boundaries positions on a large section of text.

Stemming: Stemming involves the cutting of changed words to their unique arrangement.

12.1.2 SEMANTICS

Semantics refers to the linguistics and logic that are conveyed through a text. The most important and intricate work of natural language processing is semantic analysis, and this part needs further research. Various computer algorithms are used in semantics to find the meaning and arrangement of the sentence. Specific methods that are used in semantic analysis are:

NER: It consists of determining the elements of the text that are recognized and categorized into predetermined groups. Some common examples include the names of places and people.

Word-sense-disambiguation: it defines a word's sense based on its perspective.

Natural-language generation: It uses the repository to identify the semantic meaning and convert it into a human-understandable form or language.

The structure of the chapter is as follows: Section 12.2 discusses recent literature of NLP, Section 12.3 describes the applications of NLP, and Section 12.4 describes the standard datasets.

12.2 RELATED WORK

This section designates the investigation of literature pertinent to the main aspects of this chapter. The essential aspects of this chapter are machine learning to natural language processing and classification methods. Over the last two decades, a lot of investigation has been done in the pitch of NLP. Some of the prevalent works have been discussed here (see Figure 12.2).

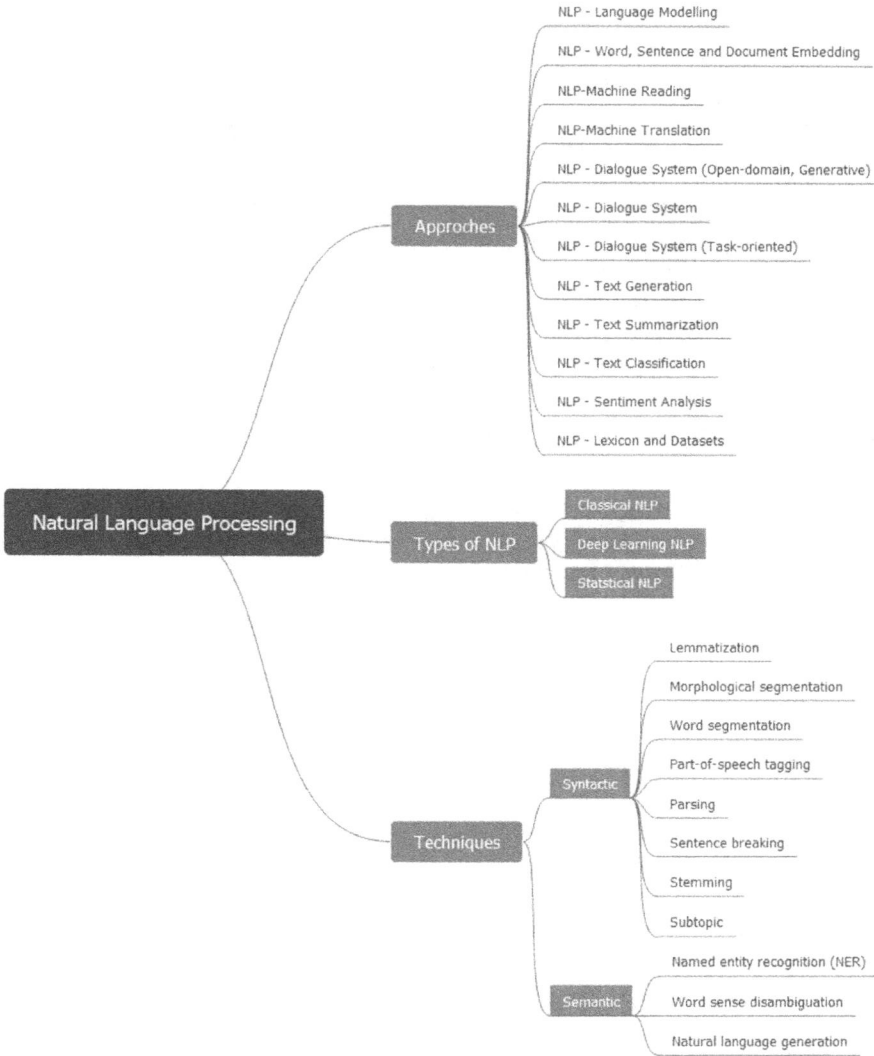

NLP - Language Modelling

NLP - Word, Sentence and Document Embedding

NLP-Machine Reading

NLP-Machine Translation

NLP - Dialogue System (Open-domain, Generative)

NLP - Dialogue System

NLP - Dialogue System (Task-oriented)

NLP - Text Generation

NLP - Text Summarization

NLP - Text Classification

NLP - Sentiment Analysis

NLP - Lexicon and Datasets

Approches

Natural Language Processing

Types of NLP

Classical NLP

Deep Learning NLP

Statstical NLP

Techniques

Syntactic

Lemmatization

Morphological segmentation

Word segmentation

Part-of-speech tagging

Parsing

Sentence breaking

Stemming

Subtopic

Semantic

Named entity recognition (NER)

Word sense disambiguation

Natural language generation

FIGURE 12.2 Natural language processing (NLP) and its techniques.

12.2.1 NLP – Language Modelling

The proposed novel model, Affect-LM, can rearrange the emotional context parameter with the framework (Lee et al. 2018). The Affect-LM advanced version of the LSTM (long-short-term memory) language model provides conversational content. The broad experiments validate the efficiency of the suggested model as used by Amazon Mechanical Turk. The proposed method is also able to learn discriminative feature representation.

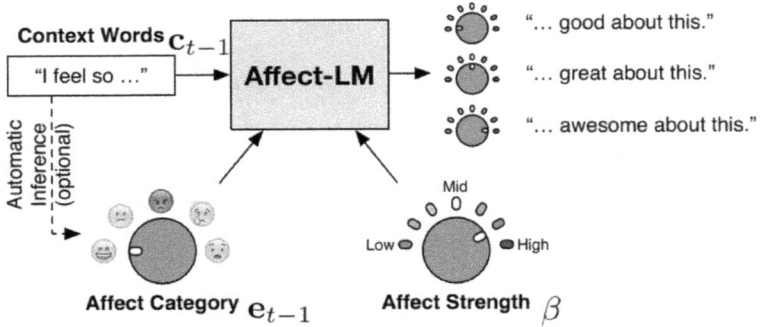

FIGURE 12.3   Affect-LM has five detailed groups (e_{t-1}) with fluctuating affect strengths (β) that show expressively shaded informal text. The three produced instance sentences for the happy affect category are shown in three distinct affect strengths (Lee et al. 2018).

Figure 12.3 shows Affect-LM and Affect-LM's capacity to create emotion colored conversational-text in various evaluation criteria with different strengths.

The suggested Affect-LM method has an extra energy term in the word estimation, and the following equation will explain it.

$$P\left(\omega_t = i \mid C_{t-1}, e_{t-1}\right) = \frac{exp\left(U_i^T f\left(C_{t-1}\right) + \beta V_i^T g\left(e_{t-1}\right) + b_i\right)}{\sum_{i=1}^{V} exp\left(U_i^T f\left(C_{t-1}\right) + \beta V_i^T g\left(e_{t-1}\right) + b_i\right)} \tag{12.1}$$

e_{t-1} is known as input vector, which describes the affect category information achieved from the words in the perspective through preparation, while $g(\cdot)$ shows the outcome of the operating network over e_{t-1}. V_i is known as embedding, which is learnt via the system for the $i-th$ word in the dictionary and the evaluative data acquired by each term is supposed to be discriminatory. The parameter β is given in Equation 12.3 that describes the impact of affect group information over the whole calculation of the target word ω_t.

The suggested approach is fast, effective and efficient to enhance the distributed word-embeddings (Bojanowski et al. 2017). The suggested method addresses this dispute and expands the advanced puzzlements on Penn-Treebank and Wiki-Text 02 to 47.69 and 40.68 correspondingly. The suggested method showed superiority on the large-scale 1B word dataset and outclassing the baseline via over 05.6 points.

The suggested technique makes a double contribution. First of all, it considers the Softmax reduction as a matrix factorization problem by framing the language model. Second, it offered a modest and impressive technique that significantly enhanced performance than the other prevalent methods.

The authors examined the functionality of the parametric model class $P_\theta\left(X \mid C\right)$ in this study. The Softmax function works on concealed state h_c & W_x (word embedding) to describe the restricted distribution $P_\theta\left(X \mid C\right)$. In addition, the model is basically described as distribution.

$$P_\theta(X \mid C) = \frac{exp\, h_c^\tau W_x}{\sum_{x'} exp\, h_c^\tau W_{x'}} \qquad (12.2)$$

Where W_x & h_c have a similar factor d. Logit is the dot product of $h_c^\tau W_x$. In a high-rank language model, in order to minimize the Softmax limiting factor problem, the authors suggested a high-rank language system called Mixture of Softmax (MoS). The mixture of Softmax expresses the conditional allocation as

$$P_\theta(X \mid C) = \sum_{k=1}^{K} \pi_{c,k} \frac{exp\, h_{c,k}^\tau W_x}{\sum_{x'} exp\, h_{c,k}^\tau W_{x'}}; s.t. \sum_{k=1}^{K} \pi_{c,k} = 1 \qquad (12.3)$$

Where $\pi_{c,k}$ describe the blend weight of kth element while the $h_{c,k}$ describes the context vector related to c. In other terms, MoS calculates variations of K Softmax and utilizes a weighted average of them as the distribution of next-token likelihood.

Mixture of Contexts: A Low-Rank Baseline, Another possible approach is to directly mix the context vectors (or logits) before taking the Softmax, rather than mixing the probabilities afterwards as in MoS. Specifically, the conditional distribution is parameterized as

$$P_\theta(X \mid C) = \frac{exp\left(\sum_{k=1}^{K} \pi_{c,k} h_{c,k}\right)^\tau W_x}{\sum_{x'} exp\left(\sum_{k=1}^{K} \pi_{c,k} h_{c,k}\right)^\tau W_{x'}} = \frac{exp\left(\sum_{k=1}^{K} \pi_{c,k} h_{c,k}^\tau w_x\right)}{\sum_{x'} exp\left(\sum_{k=1}^{K} \pi_{c,k} h_{c,k}^\tau w_{x'}\right)} \qquad (12.4)$$

Both $h_{c,k}$ & $\pi_{c,k}$ have a similar value, such as in MoS. This approach that we point to as the combination of perspectives (MoC), the same rank constraint issue as Softmax, considering its remarkable resemblance to MoS. Authors incorporate MoC as a reference point in an investigation and equate it scientifically with MoS.

The elaborated study on the prevalent method is based on advanced results of a natural language modelling benchmark standard; however, some challenges of the studied are based on different environment, different configuration and an unrestrained cradle of experimental disparity (Bordes et al. 2017). So authors used regularization approaches with large-scale automatic black-box hyperparameter tuning and restructured the several frameworks. The suggested approach outperformed the other advanced methods.

The weight-dropped LSTM models are able to tackle DropConnect on hidden to hidden weights for the form of recurrent regularization (Bowman et al. 2016). The main objectives of the authors were to optimize LSTM-based facsimiles. The authors also introduced NTAvSGD, a non-monotonically triggered (NT) modified by the averaged stochastic-gradient-technique (AvSGD) (see Algorithm 1). Comprehensive studies on 02 datasets, namely 52.8 on Penn-Treebank and 52.0 on Wiki-Text-02, demonstrated the validity and usefulness of the suggested approach.

Algorithm 1 Non-monotonically Triggered-AvSGD (NT-AvSGD)

Inputs: Initial point ω_0, learning rate γ logging interval L, non-monotone interval n.

  1. Initialize $k \leftarrow 0,\ t \leftarrow 0,\ T \leftarrow 0,\ logs \leftarrow [\ ]$
  2. *while stopping criterion not met do*
  3. *Calculate stochastic-gradient* $\widehat{\nabla}f\left(\omega_k\right)$ *and take SGD step (1).*
  4. If $mod\left(k,L\right) = 0\ and\ T = 0$ then
  5. *Calculate validation perplexity v.*
  6. If $t > n\ and\ v > \min\limits_{l\in\{0,...,t-n-1\}} logs\left[1\right]$ then
  7. Set $T \leftarrow k$
  8. *end if*
  9. Append $v\ to\ logs$
 10. $t \leftarrow t+1$
 11. *end if*
 12. $k \leftarrow k+1$
 13. *end while*

$$\text{return}\ \frac{\sum_{i=T}^{k}\omega_i}{\left(k-T+1\right)}$$

Novel neural architecture Transformer-XL was proposed by the authors (Chawla et al. 2018), which is able to work on learning enslavement of fixed scale. It contains the segment level recurrence + new position encode task. The proposed method solved the fragmentation problem that occurs in context. As extensive experiments showed the proposed method learns 80% more than the RNNs and 45% of vanilla transformers. The effectiveness showed on long/short sequences.

12.2.2 NLP – WORD, SENTENCE AND DOCUMENT EMBEDDING

Both supervised and unsupervised approaches are used and trained to learn word vector semantic data and sentiment data (Cheng et al. 2016). The suggested scheme influenced sequential and multidimensional sentiment data and non-sentiment data. The authors allowed the suggested system to exploit the document-level sentiment polarity annotations in the real scenario. The substantial studies showed the suggested system's usefulness, and the proposed system outperformed state-of-the-art techniques.

For checking the quality of the computational continuous vector representation that is achieved in a word similarity task the 02 new innovative designs are used. The suggested method outperforms based on numerous kinds of NN. For extensive experiments, the authors showed that the proposed method greatly increased accuracy at a considerable inferior computational cost. The proposed method showed that the high-quality word vectors that can learn within a short period of time provide advanced performance on a dataset designed for determining syntactic and semantic word likenesses (Dai et al. 2019).

The numerous accumulation can recover the quality of the vectors and the training time (Faruqui et al. 2014). Subsampling of the frequently used terms was employed for notable speed. The major problem challenge of word creations is to represent idiomatic sentences with order disparity. The authors showed a very easy method for discovering phrases in context, and vector representations used for lots of phrases were conceivable.

Several state-of-the-art techniques described only word representation rather than the sentiment of text, so authors focused on an approach that can learn word embedding designed for Twitter-sentiment classification (Ghosh et al. 2017). The suggested sentiment-specific-word-embedding (SSWE) task can encrypt sentiment data in the sequential representation of the word (see Figure 12.4). The authors designed the 03 NN to successfully integrate the sentiment polarity of text with loss functions. Extensive tests on the proposed task on a benchmark Twitter-sentiment classification dataset in SemEval in 2013 revealed that the proposed task feature performs comparably to manual features in the top-performing scheme, and that the performance was improved by concatenating SSWE by the prevailing feature set.

In the $SSWE_h$ model, the cross-entropy error of the Softmax layer is:

$$loss_h(t) = -\sum_{k=\{0,1\}} f_k^g(t) \cdot log\left(f_k^h(t)\right) \qquad (12.5)$$

Where $f^g(t)$ shows the distribution of sentiment while the $f^h(t)$ shows the predicted sentiment distribution.

The loss function $SSWE_u$ is the sequential blend of two hinge reductions in the $SSWE_u$ combined model.

$$loss_u(t,t^\tau) = \alpha \cdot loss_{c\omega}(t,t^\tau) + (1-\alpha) \cdot loss_{us}(t,t^\tau) \qquad (12.6)$$

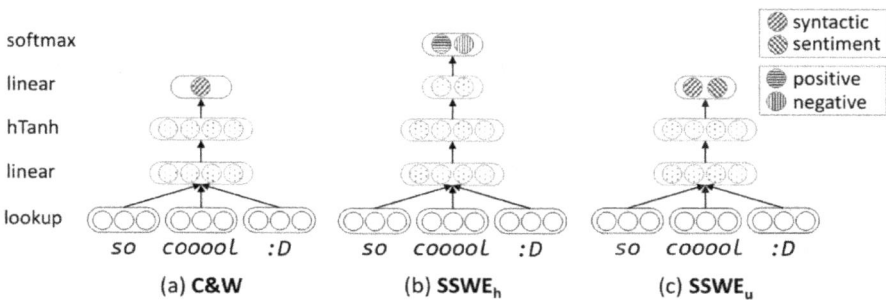

FIGURE 12.4 The traditional C & W model and proposed neural-networks ($SSWE_h$ and $SSWE_u$) for learning sentiment-specific word embedding (Ghosh et al. 2017).

where $loss_{c\omega}(t,t^\tau)$ is the syntactic loss, $loss_{us}(t,t^\tau)$ is the sentiment loss as described in Equation 12.7. The hyper-parameter α weighs the two parts.

$$loss_{us}(t,t^\tau) = \max\left(0, 1 - \delta_s(t)f_1^u(t) + \delta_s(t)f_1^u(t^\tau)\right) \tag{12.7}$$

With the identical NN and the similar parameter setting, the embedding of unigrams, bigrams and trigrams are performed distinctly.

The unigrams (bigram/trigram) contexts are the surrounding unigrams (bigrams/trigrams), respectively.

The introduced approach represents a similar vector demonstration through refining vector space representation as of semantic lexicons (Gonçalves et al. 2013). The extensive experiments validate the effectiveness tested on numerous languages and also attained enhancement of word vector models. The proposed method outperforms compared to the prevalent method.

The easy and efficient unsupervised method for embedding a sentence is grounded on the discourse vectors in the random walk prototypical designed for engendering text (Graves et al. 2014). The proposed method was superior to the baseline task arranged with the same tasks, and outperformed compared with the supervised approach, named as RNN, LSTM etc. sentence embedding, a feature achieved through downstream supervised scheme, and had superior performance to other advanced techniques.

Algorithm 2 Sentence Embedding

Input: Word embeddings $\{v_\omega : \omega \in \mathcal{V}\}$, a set of sentences S, parameter a and estimated probabilities $\{p(\omega) : \omega \in \mathcal{V}\}$ of the words.

Output: Sentence-embeddings $\{v_s : s \in S\}$

1. *for all sentence s in S do*
2. $v_s \leftarrow \dfrac{1}{|s|}\displaystyle\sum_{\omega \in s}\dfrac{a}{a + p(\omega)}v_\omega$
3. *end for*
4. *Form a matrix X whose columns are $\{v_s : s \in S\}$, and let u be its singular vector*
5. *for all sentence s in S do*
6. $v_s \leftarrow v_s - uu^T v_s$
7. *end for*

The learning of the word representation is enhanced through the affect lexical from an amount of data (Howard & Ruder 2018). Using a Word2Vec SkipGram, the fine-grained data were extracted, i.e. a word's psycholinguistic and expressive positioning during a training period. For accumulated learning both the Word2Vec CBOW and

GloVe approach were used. Unlabeled data control by Warriner's affect lexicon was used to regularize the vector representations learning. The extensive research showed the suggested methods' efficiency and the suggested method was superior on sentiment detection.

The authors have developed a Skipgram technique for each word denoted by a bag of character n-grams (Kim 2014). As generated in this design, each letter n-gram integrates with the vector and word connected with it. The proposed method is fast and delivers fast access to data during training, and permits word depictions. The authors tested the proposed approach on nine different languages for similarity and an analogous scheme. The proposed method outperformed, compared to recently proposed morphological word representation approaches.

The fast, easy and efficient method for learning the word embedding data on contextual text through the different platform has been given by the authors (Kiperwasser & Goldberg 2016). The extensive research showed the efficiency of the proposed approach on numerous down-stream NLP tasks.

The suggested scheme worked as follows:

First, let us describe the intent of the skipgram framework.

$$L_{\mathcal{D}} = \sum_{(\omega,c)\in\,\mathcal{D}} \#\left(\omega,c\right)\left(\left(log\,\sigma(\omega\cdot c)+\sum_{i=1}^{k}E_{c_i'\,\sim P(\omega)}\left[log\,\sigma\left(-\omega\cdot c_i'\right)\right]\right)\right) \quad (12.8)$$

Where $\mathcal{D}$ refers to the complete text corpus through which we learn the word embeddings. The word w is the recent word, c is the context word, and $\#\left(\omega,c\right)$ is the number of times they co-occur in $\mathcal{D}$. We use ω and c to designate the vector representations for ω and c, respectively. The symbol $\sigma(\cdot)$ denotes the sigmoid-function. The word c_i' is a "negative sample" sampled from the distribution $P(\omega)$ – typically selected as the unigram distribution $U(\omega)$ raised to the 3/4rd power.

The authors first learn for each word ω an embedding ws via the origin domain $\mathcal{D}_s$. Next, we learn the objective area embeddings as follows:

$$L'_{\mathcal{D}_t} = L_{\mathcal{D}_t} + \sum_{\omega\in\mathcal{D}_t\cap\mathcal{D}_s}\alpha_\omega\cdot\left\|\omega_t - \omega_s\right\|^2 \quad (12.9)$$

Where $\mathcal{D}_t$ refers to the target area, and ω_t is the target area representation for ω. Such a regularized objective can still be optimized using standard stochastic gradient descent.

$$\alpha_\omega = \sigma\left(\lambda\cdot\varnothing(\omega)\right) \quad (12.10)$$

Where λ is a hyper-parameter to decide the scaling factor of the significance function $\varnothing(\cdot)$. Define the normalized frequency for the word ω as follows-

$$\mathcal{F}_{\mathcal{D}}(\omega) = \frac{f_{\mathcal{D}}(\omega)}{\max_{\omega'\epsilon\mathcal{D}_k} f_{\mathcal{D}}(\omega')} \quad (12.11)$$

Function $\varnothing(\cdot)$ based on the following metric that is motivated by the well-known Sørensen-Dice coefficient calculated as:

$$\varnothing(\omega) = \frac{2 \cdot \mathcal{F}_{D_s}(\omega) \cdot \mathcal{F}_{D_t}(\omega)}{\mathcal{F}_{D_s}(\omega) + \mathcal{F}_{D_t}(\omega)} \tag{12.12}$$

note that the value of φ(w) would be high only if both $\mathcal{F}_{D_s}(\omega)$ and $\mathcal{F}_{D_t}(\omega)$ are high.

12.2.3 NLP-MACHINE READING

The long short-term memory construction is encompassed by a memory net instead of a solitary memory lockup. The presented m/c reading is based on a simulator that was used to test L-to-R and executed shallow-reasoning by memory and attention (Li et al. 2016a). The proposed method allowed adaptive cell usage throughout the recurrence via NN. They contributed a route to weekly induce relation midst tokens. It was used for a single sequence but also accumulated through the encode–decode framework (see Figure 12.5). The extensive studies showed the efficiency of the proposed approach.

Modeling Two Sequences with LSTMN, merge the LSTMN, which focuses on intra-relation reasoning, with the encoder–decoder network whose attention module learns the inter-alignment between two sequences.

The figure represents the two kinds of combinations.

Shallow Attention Fusion, Instead of a standard RNN or LSTM, shallow combination basically handles the LSTMN as a distinct part that can be simply used in an encoder and decoder design.

Deep Attention Fusion, Deep fusion, when performing state updates, merges inter- and intra-attention (initiated by the decoder) when calculating state updates.

Calculate inter attention among the input at time step t and tokens in the whole source sequence as follows:

$$b_j^t = u^t \ tanh\left(W_\gamma \gamma_j + W_x x_t + W_{\tilde{\gamma}} \tilde{\gamma}_{t-1}\right) \tag{12.13}$$

$$p_j^t = softmax\left(b_j^t\right) \tag{12.14}$$

We measure the adaptive representation of the source memory tape $\tilde{\alpha}_t$ and concealed tape $\tilde{\gamma}_t$ as:

$$\begin{bmatrix} \tilde{\gamma}_t \\ \tilde{\alpha}_t \end{bmatrix} = \sum_{j=1}^{m} p_j^t \cdot \begin{bmatrix} \gamma_j \\ \alpha_j \end{bmatrix} \tag{12.15}$$

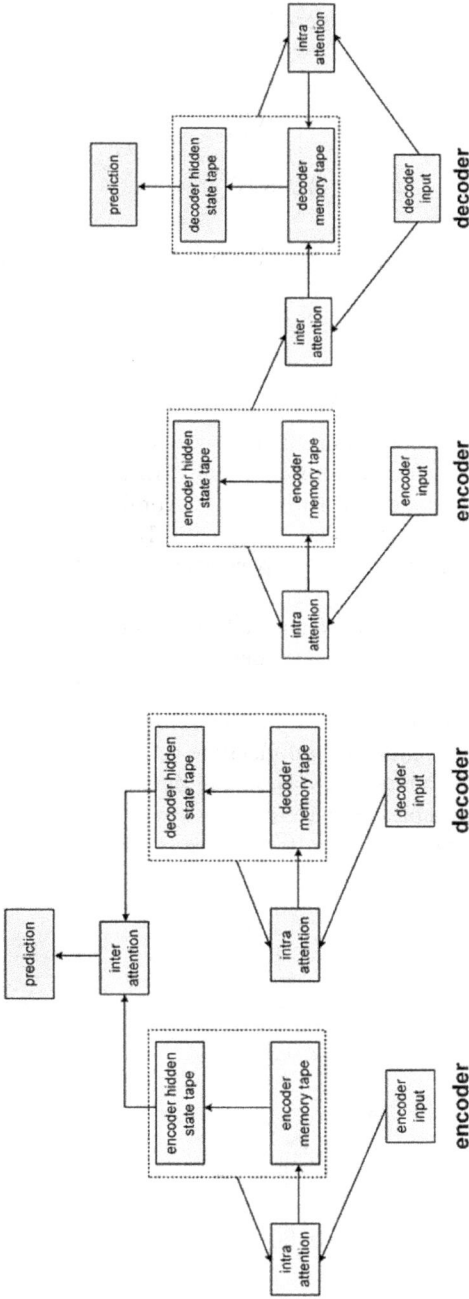

(a) Decoder with shallow attention fusion.

(b) Decoder with deep attention fusion.

FIGURE 12.5 The encoder utilizes the intra attention and the decoder utilizes intra as well as inter attention scheme. The above mentioned structure demonstrates the two approaches for merging the intra and inter attention in the decoder (Li et al. 2016a).

We can then shift the representation of the adaptive source $\tilde{\alpha}_t$ to the target memory with another gating operation r_t, similar to the gates in Equation.

The new target memory comprises inter-alignment $r_t \odot \tilde{\alpha}_t$, intra-relation $f_t \odot \tilde{C}_t$, and the novel input information $i_t \odot \hat{C}_t$:

$$C_t = r_t \odot \tilde{\alpha}_t + f_t \odot \tilde{C}_t + i_t \odot \hat{C}_t \tag{12.16}$$

$$h_t = o_t \odot \tanh\left(C_t\right) \tag{12.17}$$

The main modification of deep-fusion lies in the recurrent storage of the inter-alignment vector in the target memory net, as a way to help the target network analyze source information.

12.2.4 NLP-MACHINE TRANSLATION

The competencies of NN via connecting them to exterior remembrance resources are able to communicate via attentional progressions (Li et al. 2016b). The pooled arrangement is comparable towards the Turing m/c or Von-Neumann architecture but is differentiable endwise able to train the data effectively via gradient descent. The results show that Neural Turing Machines outperformed other algorithms by working on a guileless method.

The custom of a fixed length vector enhanced the encoder–decoder framework's effectiveness and suggested an automated model to explore the source sentence that is pertinent to predicting a target-word (Lowe et al. 2015). The authors tested the proposed method on an English-to-French translation and found it to be effective. The authors introduced an advanced neural machine interpretation architecture (see Figure 12.6). The advanced architecture contains bi-directional RNN as an encoder and a decoder that visualizes a basis phrase search when a translation is decoded.

Decoder: General Description – We describe each conditional probability in a new model framework in Equation 12.18 as:

$$p\left(y_i | y_1, \ldots, y_{i-1}, X\right) = g\left(y_{i-1}, s_i, c_i\right), \tag{12.18}$$

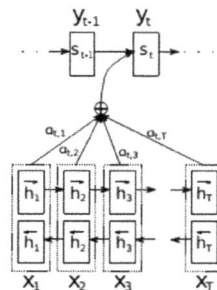

FIGURE 12.6 The graphical representation of the suggested framework trying to produce the t-th target word y_t given a source sentence $(x_1, x_2, \ldots, x_T)$.

Where s_i shows the RNN concealed state for time period i, calculated by

$$s_i = f\left(s_{i-1}, y_{i-1}, c_i\right) \tag{12.19}$$

The context vector c_i relies on a series of annotations $\left(h_1, \ldots, h_{T_x}\right)$ to which an encoder represents the input sentence. Each annotation hi comprises information about the entire input sequence with a robust concentration on the parts adjoining the i-th word of the input sequence.

The context vector c_i is, then, calculated as a total sum of weight of these annotations h_i:

$$c_i = \sum_{j=1}^{T_x} \alpha_{ij} h_j \tag{12.20}$$

The weight α_{ij} of each annotation h_j is calculated by

$$\alpha_{ij} = \frac{\exp\left(e_{ij}\right)}{\sum_{k=1}^{T_x} \exp\left(e_{ik}\right)} \tag{12.21}$$

Where

$$e_{ij} = a\left(s_{i-1}, h_j\right)$$

Represent the coordination scheme that evaluates how well the inputs around position j and the output at position match. The score is based on the RNN concealed state s_{i-1} (just before producing i, Equation 12.18) and the j-th annotation h_j of the input sentence.

Encoder: Bidirectional RNN for Annotating Sequences – A BiRNN comprises forward and backward RNNs. The forward RNN $\vec{f}$ reads the input structure as it is ordered (from x_1 to x_{T_x}) and estimates a sequence of forward concealed states $(\vec{h}_1, \ldots, \vec{h}_{T_x})$. The backward RNN f reads the sequence in the opposite order (from x_{T_x} to x_1), resulting in a sequence of backward hidden states ($\overleftarrow{h}_1, \ldots, \overleftarrow{h}_{T_x}$).

There are two key objectives, i.e. general objective and local objective, used for operative classes of attentional apparatus (Lowe et al. 2016). The general task constantly joins the entire source word, whereas local tasks merely take care of segments of words on a specific manner. The two tasks were tested on WMT translation for English and German. On the WMT 15 English-German translation task, the proposed ensembles technique received 25.9 BLEU points. The proposed method outperformed compared to the other advanced method.

The modeling coverage for neural machine translation (NMT) is integrated with the attention model, which is focused on untranslated source words. The broad research showed the suggested technique's efficiency compared to the translation and alignment quality (Luong et al. 2015).

12.2.5 NLP – Dialogue System (Open-Domain, Generative)

The unstructured Twitter conversations by the new comeback generation system are trained endwise (Maas et al. 2011). The authors used NN to handle sparsity issues that accumulate the text data and handle the dialogue system. The proposed method outperformed compared to other prevalent approaches and achieved consistent accuracy among context-sensitive and non-context-sensitive m/c paraphrase and data repossession standards.

The authors suggested the following context-sensitive-models:

Context-Sensitive Models, differentiate 03 linguistic entities in a conversa-tion among the 02 participant A and B: the context c, the Message m and Response r. The context c represents a series of earlier dialogue exchanges of any length; then B produces a message m to which A responds by formu-lating its reply r.

We use 03 context based generation models to evaluate a generation model of the response r, $r = r_1,....,r_T$, conditioned on past information c and m:

$$p(r|c,m) = \prod_{t=1}^{T} p(r_t|r_1,.....,r_{t-1},c,m) \tag{12.22}$$

Tripled Language Model – We concisely combine almost every expression m,c,r into a single sentence s in our first model, called RLMT, and develop and train the RLM to reduce $L(s)$. Provided c and m, measure the likelihood of the reply as follows: to acquire a concealed state encoding useful knowledge about prior utterances, the forward propagation was conducted over the recognized utterances c and m. Later, calculate the probability of the reaction from the secret condition.

Dynamic-Context Generative Model I – second model (DCGM-01), the meaning and the message are translated into a vector-representation of a fixed length. It is used by the RLM to decipher the answer (see Figure 12.7).

The model's parameters are $\Theta_{DCGM-I} = W_{in}, W_{hh}, W_{out}, \{W_f^l\}_{l=1}^{L}, where \{W_f^l\}_{l=1}^{L}$ represents weights for the L layers of the feedforward context networks. The fixed-length context vector K_L is achieved by forwarding propagation of the network:

DCGM-I   DCGM-II

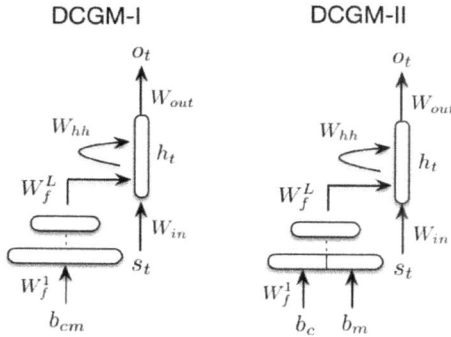

FIGURE 12.7   In the above scheme, the DCGM-01 is shown on the left, while DCGM-02 is shown on the right. The decoder RLM obtains a discrimination from the context encoder. In DCGM-01, the encoding of the bag of words describe c and m in a single vector b_{cm}. In DCGM-02, we combine the representations b_c and b_m on the first layer to preserve order information.

$$k_1 = b_{cm}^\mathsf{T} W_f^1$$
$$k_l = \sigma\left(k_{l-1}^\mathsf{T} W_f^l\right) for\, l = 2,\ldots,6 \qquad (12.23)$$

The rows of W_f^1 comprise the embeddings of the vocabulary.

Dynamic-Context Generative Model II- Since the DCGM-01 doesn't differentiate among $C\,\&\,M$, and that model has the tendency to misjudge the robust dependency among m and r. The DCGM-02 solves this problem via adding 02 direct mappings of bag of words illustrations b_c and b_m in the feed-forward network's input layer that demonstrates c and m.

The context–encoder forward equations are:

$$k_1 = \left[b_c^\mathsf{T} W, b_m^\mathsf{T} W_f^1\right]$$
$$k_l = \sigma\left(k_l^\mathsf{T} W_f^l\right) for\, l = 2,\ldots,L \qquad (12.24)$$

Where $[x, y]$ represents the concatenation of x and y vectors.

To tackle the problem of speaker consistency in neural response generation. The proposed Persona-Based Neural Conversation Model encrypts personas in distributed embeddings that extract innovative information (Melis et al. 2017). The extensive research showed the efficiency of the proposed model in both incomprehension and BLEU scores. The proposed method is superior in speaker steadiness as restrained thru human judiciaries.

The authors focused on the work, dialogue generation thru deep reinforcement learning (Merity et al. 2018). The proposed system pretends the dialogue through two effective representatives have three important communication schemes, i.e. informatively, coherence, and ease of answering in forward-looking function. The proposed method superior to long-term dialogues. The authors evaluated the three factors, i.e. diversity, length, and human judges, and showed that the proposed method gained innovative comeback.

12.2.6 NLP – Dialogue System

The innovative datasets named Ubuntu Dialogue Corpus contain one million multi-turn dialogue as of entire over 7 million utterances and 100 million words (Mikolov et al. 2013a). They provided the novel direction for investigators, a model built with NN and stored huge unlabeled information. The proposed datasets in two variations multi-turn property and unstructured nature. The authors presented the benchmark performance on the datasets and showed the proposed work superior.

The inspection of the Next Utterance Classification (NUC) is done through the evaluation of dialogue systems (Mikolov et al. 2010). Through the extensive experiments, the outcome generated on the foremost three discoveries – first, achieved superior classification responses with high-rate; second, the performance phases vary on the task and expertise manner; third, through automation by used prevalent ML method, achieved good accuracy.

12.2.7 NLP – Dialogue System (Task-Oriented)

This chapter represented the testbed towards break down the métiers and inadequacies of endways dialogue systems in goal-oriented solicitations (Mikolov et al. 2013b). In the scenario of restaurant reservation text the API system generated the resultant calls and need operative sentence and symbols throughout the communications. The authors also represented the endwise dialogue system constructed with a memory net, not given the good response whenever the non-trivial operations conditions. The authors showed the effectiveness of the proposed system.

The endwise, goal-oriented and fast novel pipe-lined Wizard-of-Oz framework made by NN in text in and out manner through this system dialogue is generated in a simple manner without the extra conditions. The extensive experiments showed the proposed system's effectiveness, which worked naturally in a restaurant search domain with rapidity (Musto et al. 2014).

12.2.8 NLP – Text Generation

To resolve the text generation problem, the sequence generation framework is used, known as SeqGAN, which worked on a stochastic strategy in reinforcement learning and solved the generator differentiation situation by applying the updated gradient strategy (Nallapati et al. 2017). The GAN discriminator generates the signal and passes it to RL, and forward it to the intermediate state-action phase by using the

Monte Carlo search. The experiments showed the efficiency of the suggested method on synthetic data and real world tasks.

For getting each and every sentence with the policy of RNN-based variational autoencoder generative prototypical worked on distributed latent representations (Preoţiuc-Pietro et al. 2016). By factorization the optimistic sentences were achieved, i.e. style, topic and syntactic features. By latent space the coherent novel sentences were generated, and hence the proposed framework gained the learning solution. The experiments showed the effectiveness of the proposed system.

12.2.9   NLP – TEXT SUMMARIZATION

The new framework supplements the customary sequence-to-sequence attentional prototypical. There are two important parts given in these arrangements, i.e., fusion-based pointer generator net, that regenerate the data and deliver new words by generator, extract the words from its main sources, treatment of the produced words. The proposed method worked with CNN and the Daily Mail summarization scheme and was superior to the other most advanced prevalent methods (Rojas-Barahona et al. 2017).

The following algorithm depicts an advanced definition and back-propagation algorithm flow in a CNN as it passes through numerous epochs before reaching either the maximum iterations or the cost function's objective.

Algorithm 3 Pseudo Code of the CNN Backpropagation Algorithm

1: Weight initialization for randomly produced value(small)
2: Set learning level to a small value (+ve)
3: Iteration n = 1;Begin
4: **for** n < max *iteration OR Cost function criteria met*, **do**
5: for image x_i *to* x_i do
6: a. Forward propagation via convolution, pooling, and then entirely opposing layers
7: b. Obtain cost function value for the image
8: c. Compute error term $\delta^{(l)}$ with respect to weights for each type of layers
9: Note that the error gets propagated from layer to layer in the following sequence
10. i. fully connected layer
11: ii. Pooling layer
13: iii. Convolution layer
14: Gradient computed as follows
15: i. convolution-layer
16: ii. Pooling-layer
17: iii. Convolution-layer
18: d. Update-weights

19: $w_{ji}^{(l)} \leftarrow w_{ji}^{(l)} + \Delta w_{ji}^{(l)}$

20: e. Update bias

21: $b_j^{(l)} \leftarrow b_j^{(l)} + \Delta b_j^{(l)}$

The proposed Recurrent Neural Network (RNN) built a prototypical sequence to summarize documents named as SummaRuNNer (See et al. 2017). The proposed system is effective because it is easy to understand and visualize the results of its predictions. The proposed method gave a superior performance compared to its prevalent method. The proposed method eliminated the complexity and generated the text smoothly without struggle.

The brief review of the text summarization method shows very limited data and information (Sordoni et al. 2015). This chapter also defined the advantages and disadvantages of the numerous methods based on text summarization.

12.2.10 NLP – Text Classification

The sentence-level classification strategy is designed using the sequence of operations via CNN by which the model is trained (Tang et al. 2014). By improving the CNN, i.e. slight hyperparameter regulation, it achieved effectiveness. The proposed model is superior to the prevalent method on four out of seven tasks based on question classification and sentiment analysis. The novel universal language framework fine-tuning (ULMFiT) for efficient transfer-learning approach is used in NLP and it is used for fine-tuning (Tu et al. 2016). Extensive experiments showed the suggested system's efficiency on the six text classification schemes, and the proposed system decreases the rate by 18 to 24% on the mainstream of datasets.

12.2.11 NLP – Sentiment Analysis

The extensive work on Twitter posts by a lexicon-based method is used for sentiment classification (Warriner et al. 2013). The proposed method is grounded on the enchanting improvement of extensive lexical possessions, likewise SentiWordNet, WordNet-Affect, MPQA, and SenticNet. The extensive experiments is validated on two datasets. The integration of the novel approach with the prevalent method and delivered superior outcome and competitive analysis. The extensive research showed the suggested method's efficiency compared to the eight widespread sentiment analysis methods in relation to coverage and agreement (Yang et al. 2017). It also introduced an unrestricted Web service called iFeel that was able to compare the API for given contexts.

12.2.12 NLP – Lexicon and Datasets

In this section, the enhanced datasets over 14,000 English lemma are able to deliver smart information, i.e. age, sex and education alteration in emotion value (Yang et al.

2018). The enhanced dataset provides a greater chance to represent different norms, such as different disease types, occupations, and taboo words. The proposed dataset contains 2895 social media posts and is preferred on two parameters, i.e., psychologically trained annotators on two different nine-point scales, and describes the sentiment and intensity of post-position (Yu et al. 2017). The two measurements were used to train the forecast model and achieved superior performances correlated at r =:65 with valence and r =:85 with arousal annotations. The practical evaluation authenticates the effectiveness of the suggested approach.

12.3 APPLICATIONS OF NLP

It provides the ability to system or computer to read, recognize, and derive the meaning of the human language. Several corporate companies have identified the advantage of this innovative technology and have also tested and deployed several natural language processing application for their business growth and intelligence. In today's scenario, using the natural language processing knowledge, industries obtain the data and information, and then use them to study the market trend, understand the client and economic benefit.

NLP has made a huge leap in the past couple of years, both in terms of theory and functional adoption into different business-based solutions.

Some commonly used natural language processing industries applications are described below (see Figure 12.8).

Sentiment Analysis – Identifying natural language is especially difficult for computers when it comes to decision-making, considering that people mostly use feelings. However, the study of sentiment may understand subtle thoughts and beliefs and assess how true or false they are.

Chatbots and Virtual Assistants – The NLP approach is used for virtual assistance and chatbots developments this NLP based application automatically provides the answer of questions by understanding their context. Typical response systems follow the predetermined procedures whereas artificial intelligence-based applications can acquire the knowledge from each and every communication. The best part is that they discover from interaction and developed over time. These smart computers are constantly at the forefront of customer service.

Text Classification – Text categorization, a task of text processing often requires emotion analysis, includes understanding, converting, and classification of unordered text automatically.

Text Extraction – Extracting text or extracting information automatically identifies unique information in the text, such as person's names, corporations, locations and more. This process is also known as the identification of the entity. In a text, one can also extract phrases and certain predefined characteristics like item details and style. Text extraction applications involve sifting through incoming service tickets and defining relevant details without opening and reading the document, such as business names, order numbers, and mailing addresses.

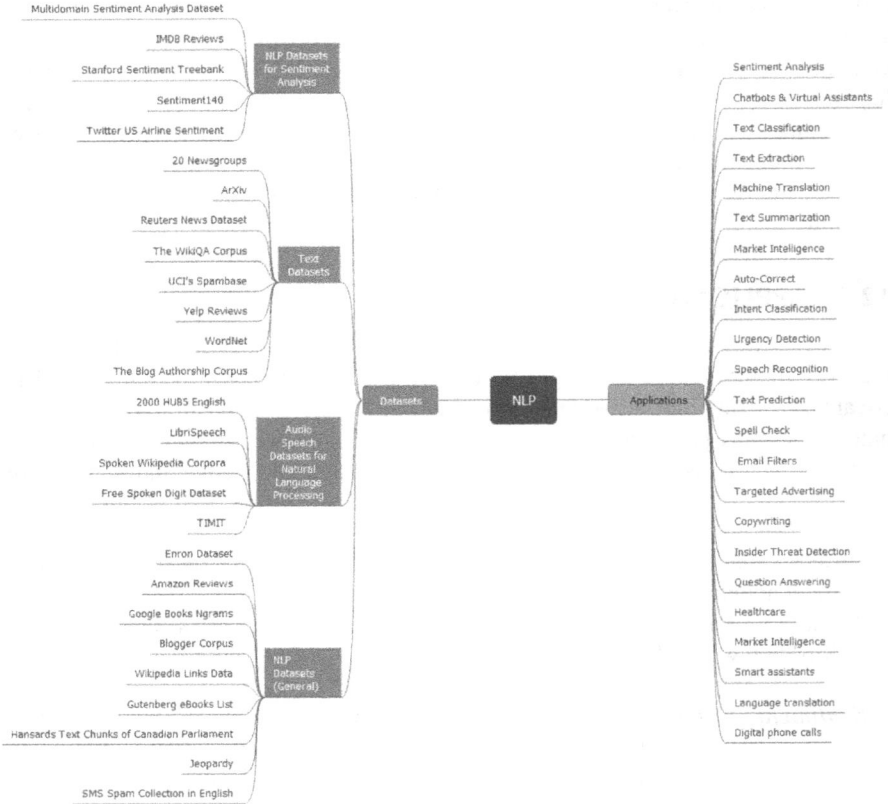

FIGURE 12.8 Different applications and datasets of NLP.

Machine Translation – NLP's first application, is a machine translation. Although the translation from face-books has been designated as the perfect one, this translation still has some context understanding issues. This interpretation is especially important for the business since it promotes contact, helps businesses reach wider markets, and interprets international documents easily.

Text Summarization – Automatic description is pretty understandable. By gathering the most relevant info, the text is summarized. This technique's key purpose is to simplify the process by which one pass through various content such as research article, media content, or legal papers. There are two approaches of data summarization using NLP. The first one is extraction-based-summarization, and the second is abstraction-based-summarization. The former approach extracts key-phrases and produce the summary while the latter approach generates fresh phrases, and gives improved performance.

Market Intelligence – To understand their clients and use those experiences to develop more productive campaigns, sellers will benefit from natural language analysis. The study of subjects, emotions, keywords and purpose in unstructured data will improve one's consumer research. To find consumer-specific problems and keep an eye on rivals, we can even evaluate data (seeing what things go better for them and which are not good for them).

Auto-Correct – NLP is very important for grammar correcting application such as Ginger and Grammarly. These applications use natural language processing and correct the writer's spelling mistakes, grammar, and sentence structure.

Intent Classification – This approach uses natural language processing and recognizes the intention of the given underlined manuscript. Intent recognition is very beneficial in the case of trades and consumer support.

Urgency Detection – NLP strategies could also help to spot urgency in a manuscript. Using our own parameters, we may develop an urgency recognition system to identify specific terms and phrases that represent importance or dissatisfaction. It can also help to select the more significant demands to make sure that under a mountain of unanswered questions they are not lost. Urgency detection lets you boost reaction times and productivity, which has a positive impact on consumer loyalty.

Speech Recognition – Natural language plays a significant part in the speech recognition approach. In speech recognition, the human spoken sentence is transformed into computer understandable form. Apple's Siri and Amazon's Alexa, are two of the best examples of speech recognition.

Text Prediction – Uses the natural language processing text prediction technique to predict the subsequent word in the sentence. Google search is one of the most prominent and general examples of text prediction. Google uses the BERT and NLP combination that uses NN to develop the pertained model. A large volume of unannotated text existing on the internet is used to train this system. The search engine uses the BERT algorithm to recognize the questions as human recognize them. Several other applications like Google docs and Gmail compose and utilize natural language processing techniques that predict text.

Spell Check – A computer program that detects grammar mistakes or spelling errors in a document and rectifies them. Grammarly is one of the common examples of spell checker programs. It is an application that interfaces with a number of text documents while the user starts to type, immediately scans for spelling mistakes, and recommends modifications.

Email Filters – This is another most often used NLP program. In this approach, the email text is analyzed and email provider after analyzing the text prevents the unwanted message from entering into the mailbox. In addition, it also adds an extra security layer and saves the user time.

Targeted Advertising – Several corporate companies always try to reach the maximum number of customers to increase their business. In that case natural language processing helps corporate companies to place their ad at the right place at the right time to promote their product. This is accomplished by observing and analyzing the search keyword, social media platform and browsing behavior for finding the potential consumer on-line. Targeted ads operate primarily on identical keywords. Data analytics and text mining software are mainly utilized for this, e.g., Apache OpenNLP.

Copywriting – NLP is able to increase companies' growth by enhancing their approach for content marketing. This can create marketing material that suits the voice of your product effectively and can offer insights into which messages are more attractive to your intended audience.

Insider Threat Detection – Illegitimate and untrustworthy intentions inside the communication can be detected by using the NLP based model. This model can also identify the hazardous pattern so that quick solution of that situation can be achieved. This is crucial because security breaches could result in big damages for both firms and consumers.

Question-answering – Questions by human beings using natural language can have answers provided by question answering wherein this NLP application recognizes the speech delivered and establishes a replies in return. Excellent examples of question answering applications are OK Google, Apple's Siri, and virtual assistants.

Healthcare – Natural language processing is very helpful in the medical and healthcare sectors since it simplifies the various complex processes such as patient history or record retrieval and provides insight information and suggestion to the doctor and caretaker so that prompt decisions can be taken.

Market Intelligence – It can be difficult to stay competitive in the industry, but industries can filter via various blogs, websites, and social media posts using NLP-based, specialized tools to update the latest trend and what is happening in the different sectors. Interesting and informative statistics may also be collected to reinforce the company's business and minimize future consumer frustration.

Smart assistants – Several smart assistants such as Amazon's Alexa and Apple's Siri provide a useful response by knowing the pattern of speech. With these smart devices, we can ask any question, and these smart devices understand the context of the questions asked and answer it. Throughout our house and everyday life, we are getting used to having Siri or Alexa show up as we speak to them via things such as the thermostat, light switches, vehicle, and much more.

Language translation – Several languages do not permit direct conversion but have diverse sentence orders. NLP helps online interpreters to more reliably interpret languages and show grammatically truthful outcomes. This is particularly useful when attempting to communicate in a different language with the others. Not just that, but software now understand the language based on given input text and convert it while translating from another language to your own.

Digital phone calls – Whenever we call customer care, we hear a well-known voice that your call can be recorded for quality and training purposes, then the technology that works behind that voice is NLP. Most of these recordings are used for training and monitoring, and all these recordings are stored in the database where they are used to train NLP based models. The automated system uses these recordings for the training of chat boat, or the system understands these calls and directs those calls to the customer representative.

12.4   NLP DATASETS

NLP is a vast area of research. It can often be hard to know where to start, let alone start looking for NLP datasets with too many places to explore. We have done our best to collect datasets for a wide variety of NLP research fields, from sentiment analysis to audio and voice recognition projects, while it is difficult to cover any field of interest.

12.4.1 NLP Datasets for Sentiment Analysis

Machine learning (ML) based models need to be prepared with massive, sophisticated datasets for sentiment analysis. The following list should hint at some of the ways that you can enhance your sentiment analysis algorithm.

- **Multidomain-Sentiment-Analysis Dataset**. This dataset describes the diverse product review obtained from Amazon, and this dataset is quite old.
- **Reviews.** This dataset is comparatively small compared to the other dataset. This dataset contains 25,000 reviews of movies that are basically used for binary sentiment categorization.
- **Stanford Sentiment Treebank.** Stanford's dataset was also developed from evaluation of movie reviews, configured to build the classifier to recognize emotion in extensive sentences. About 10,000 snippets obtained from Rotten Tomatoes are included.
- **Sentiment140**: There are 160,000 tweets that are arranged in six fields in this famous dataset, namely: polarity, ID, tweet period, query, reader, and messages.
- **Twitter US Airline Sentiment**: This dataset contains tweets about United States airlines, and these tweets are divided into three categories: negative, positive, and neutral. In this dataset, negative tweets are included for complaint identification purpose.

12.4.2 Text Datasets

A huge area of study is natural language processing, but the following list contains a wide variety of databases for various activities of natural-language processing, such as speech recognition and chatbots.

- **20-Newsgroups**: This dataset contains about 20,000 text files that encompass 20 newsgroups from games to spirituality.
- **ArXiv**: ArXiv dataset contains the whole arXiv research paper repository as full text, and its size is around 270 gigabytes.
- **Reuters-News-Dataset**: In 1987, the records in this dataset emerged on Reuters. Since then, they have been accumulated and indexed for machine learning use.
- **The WikiQA Corpus**: This dataset contains various queries and their response. Basically, this dataset was generated for open field research.
- **UCI's Spambase**: This dataset is very beneficial for the development of the spam filter for personal use and it was developed by HP Corporation.
- **Yelp Reviews**: This dataset contains five million reviews and was developed by Yelp.
- **WordNet**: Implemented by researchers at the University of Princeton, this dataset is basically a huge lexical repository of English 'synsets' or collections of alternative expression.

- **The Blog Authorship Corpus**: More than 681,000 posts by 19,320 separate writers are included in this dataset, and around 140 million words are included in this.

12.4.3 AUDIO SPEECH DATASETS FOR NLP

For the development of NLP systems, audio speech datasets are most important, and their application includes virtual assistance, vehicle navigation, voice-activated system etc.

- **2000 HUB5 English**: This dataset includes recordings that are compiled from 40 English telephone conversations. The related speech files can also be accessed via this tab.
- **LibriSpeech**: This repository includes nearly 1000 hours of conversation in English, consisting of audio files read by different speakers. The information or data is grouped according to the chapter of the book.
- **Spoken Wikipedia Corpora**: This dataset comprises spoken articles from Wikipedia in English, Dutch, and German, and their duration is of several hours. According to its nature, it also involves a large number of reader and related subject.
- **Free Spoken Digit Dataset**: This dataset contains 1500 soundtracks in English.
- **TIMIT**: This information is intended for research in acoustic-phonetic analyses and the construction of automated tools for voice recognition. It comprises videos of 630 American English speakers reading ten 'phonetically intensive' sentences.
- **Enron Dataset:** This dataset contains 500,000 messages of Enron management, and this dataset was created as a source for those individuals who want to enhance and study the prevalent email tools.
- **Amazon Reviews:** This dataset comprises about 35 million Amazon reviews over a period of 18 years. This dataset also covers items and customer details, comments, and the analysis of plaintext.
- **Google Books Ngrams**: Collected from the blogger.com is the aggregation of 681,288 post that comprises 140 million words and every single blog incorporated here consist of at least 200 occurrences of common daily English words.
- **Wikipedia Links Data:** This dataset is given by the Google and it contains near about 13 million text document and it also contain the web pages that has hyperlink that point to the English encyclopedia/Wikipedia.
- **Gutenberg eBooks List:** The annotated list of e-books from program Gutenberg comprises elementary information about each eBook, structured according to the year.
- **Hansards Text Chunks of Canadian Parliament:** This dataset contains around 1.3 million text documents from the Canadian parliament record.
- **Jeopardy:** This dataset comprises 200,000 queries and their response from the famous game show jeopardy.it also has other relevant information such as number of show, date of show and class of queries.

- **SMS Spam Collection in English**: This dataset contains 5574 messages in English language which is marked either as appropriate/valid or junk. In this dataset near about 425 texts are spam which were separated from the grumble text website.

12.5 CONCLUSION

NLP is a prominent pitch of cognitive science that covenants with smearing linguistic and geometric procedures to context/text in edict to extract meaning in a way that is comparable to how the human brain understands language. Consideration of the NN in the field of NLP is very attractive. This chapter has been written keeping in mind the current study of NLP, and it covers all the content related to it.

NLP provides several sets of techniques and tools that can be used in several aspects of life. By learning and utilizing in our everyday interaction, the quality of life is highly enhanced. This chapter presents the various NLP approaches and related datasets. The main aim of this chapter is that all the related research that has been done so far in the NLP field is made readily available to researchers.

REFERENCES

Allahyari, M., Pouriyeh, S.A., Assefi, M., Safaei, S., Trippe, E.D., Gutierrez, J.B., & Kochut, K. (2017). Text Summarization Techniques: A Brief Survey. *ArXiv, abs/1707.02268.*

Arora, S., Liang, Y., & Ma, T. (2017). A Simple but Tough-to-Beat Baseline for Sentence Embeddings. *ICLR.*

Bahdanau, D., Cho, K., & Bengio, Y. (2015). Neural Machine Translation by Jointly Learning to Align and Translate. *CoRR, abs/1409.0473.*

Bojanowski, P., Grave, E., Joulin, A., & Mikolov, T. (2017). Enriching Word Vectors with Subword Information. *Transactions of the Association for Computational Linguistics, 5*, 135–146.

Bordes, A., & Weston, J. (2017). Learning End-to-End Goal-Oriented Dialog. *ArXiv, abs/1605.07683.*

Bowman, S.R., Vilnis, L., Vinyals, O., Dai, A.M., Józefowicz, R., & Bengio, S. (2016). Generating Sentences from a Continuous Space. *CoNLL.*

Chawla, K., Khosla, S., Chhaya, N., & Jaidka, K. (2018). Towards Building Affect sensitive Word Distributions.

Cheng, J., Dong, L., & Lapata, M. (2016). Long Short-Term Memory-Networks for Machine Reading. *ArXiv, abs/1601.06733.*

Dai, Z., Yang, Z., Yang, Y., Carbonell, J., Le, Q.V., & Salakhutdinov, R. (2019). Transformer-XL: Attentive Language Models Beyond a Fixed-Length Context. *ACL.*

Faruqui, M., Dodge, J., Jauhar, S. K., Dyer, C., Hovy, E., & Smith, N. A. (2014). Retrofitting word vectors to semantic lexicons. arXiv preprint arXiv:1411.4166.

Ghosh, S., Chollet, M., Laksana, E., Morency, L., & Scherer, S. (2017). Affect-LM: A Neural Language Model for Customizable Affective Text Generation. *ACL.*

Gonçalves, P., Araújo, M., Benevenuto, F., & Cha, M. (2013). Comparing and combining sentiment analysis methods. *ArXiv, abs/1406.0032.*

Graves, A., Wayne, G., & Danihelka, I. (2014). Neural Turing Machines. *ArXiv, abs/1410.5401.*

Howard, J., & Ruder, S. (2018). Universal Language Model Fine-tuning for Text Classification. *ACL.*

Kim, Y. (2014). Convolutional Neural Networks for Sentence Classification. *EMNLP*.

Kiperwasser, E., & Goldberg, Y. (2016). Simple and Accurate Dependency Parsing Using Bidirectional LSTM Feature Representations. *Transactions of the Association for Computational Linguistics, 4*, 313–327.

Lee, Y., & Kim, Y. (2018). Coverage Modeling in Neural Machine Translation Using Orthogonal Regularization, 561–566.

Li, J., Galley, M., Brockett, C., Spithourakis, G.P., Gao, J., & Dolan, W. (2016a). A Persona-Based Neural Conversation Model. *ArXiv, abs/1603.06155*.

Li, J., Monroe, W., Ritter, A., Jurafsky, D., Galley, M., & Gao, J. (2016b). Deep Reinforcement Learning for Dialogue Generation. *ArXiv, abs/1606.01541*.

Lowe, R., Pow, N., Serban, I., & Pineau, J. (2015). The Ubuntu Dialogue Corpus: A Large Dataset for Research in Unstructured Multi-Turn Dialogue Systems. *ArXiv, abs/1506.08909*.

Lowe, R., Serban, I., Noseworthy, M., Charlin, L., & Pineau, J. (2016). On the Evaluation of Dialogue Systems with Next Utterance Classification. *ArXiv, abs/1605.05414*.

Luong, T., Pham, H., & Manning, C.D. (2015). Effective Approaches to Attention-based Neural Machine Translation. *ArXiv, abs/1508.04025*.

Maas, A.L., Daly, R.E., Pham, P.T., Huang, D., Ng, A., & Potts, C. (2011). Learning Word Vectors for Sentiment Analysis. *ACL*.

Melis, G., Dyer, C., & Blunsom, P. (2018). On the State of the Art of Evaluation in Neural Language Models. *ArXiv, abs/1707.05589*.

Merity, S., Keskar, N., & Socher, R. (2018). Regularizing and Optimizing LSTM Language Models. *ArXiv, abs/1708.02182*.

Mikolov, T., Chen, K., Corrado, G.S., & Dean, J. (2013a). Efficient Estimation of Word Representations in Vector Space. *ICLR*.

Mikolov, T., Karafiát, M., Burget, L., Černocký, J., & Khudanpur, S. (2010). Recurrent neural network based language model. *INTERSPEECH*.

Mikolov, T., Sutskever, I., Chen, K., Corrado, G.S., & Dean, J. (2013b). Distributed Representations of Words and Phrases and their Compositionality. *NIPS*.

Musto, C., Semeraro, G., & Polignano, M. (2014). A Comparison of Lexicon-based Approaches for Sentiment Analysis of Microblog Posts. *DART@AI*IA*.

Nallapati, R., Zhai, F., & Zhou, B. (2017). SummaRuNNer: A Recurrent Neural Network Based Sequence Model for Extractive Summarization of Documents. *AAAI*.

Preoţiuc-Pietro, D., Schwartz, H. A., Park, G., Eichstaedt, J., Kern, M., Ungar, L., & Shulman, E. (2016, June). Modelling valence and arousal in facebook posts. In Proceedings of the 7th workshop on computational approaches to subjectivity, sentiment and social media analysis (pp. 9–15).

Rojas-Barahona, L., Gašić, M., Mrksic, N., Su, P., Ultes, S., Wen, T., Young, S., & Vandyke, D. (2017). A Network-based End-to-End Trainable Task-oriented Dialogue System. *EACL*.

See, A., Liu, P.J., & Manning, C.D. (2017). Get To The Point: Summarization with Pointer-Generator Networks. *ACL*.

Sordoni, A., Galley, M., Auli, M., Brockett, C., Ji, Y., Mitchell, M., Nie, J., Gao, J., & Dolan, W. (2015). A Neural Network Approach to Context-Sensitive Generation of Conversational Responses. *HLT-NAACL*.

Tang, D., Wei, F., Yang, N., Zhou, M., Liu, T., & Qin, B. (2014). Learning Sentiment-Specific Word Embedding for Twitter Sentiment Classification. *ACL*.

Tu, Z., Lu, Z., Liu, Y., Liu, X., & Li, H. (2016). Modeling Coverage for Neural Machine Translation. *arXiv: Computation and Language*.

Warriner, A.B., Kuperman, V., & Brysbaert, M. (2013). Norms of valence, arousal, and dominance for 13,915 English lemmas. *Behavior Research Methods, 45*, 1191–1207.

Yang, W., Lu, W., & Zheng, V. (2017). A Simple Regularization-based Algorithm for Learning Cross-Domain Word Embeddings. *EMNLP*.

Yang, Z., Dai, Z., Salakhutdinov, R., & Cohen, W.W. (2018). Breaking the Softmax Bottleneck: A High-Rank RNN Language Model. *ArXiv, abs/1711.03953*.

Yu, L., Zhang, W., Wang, J., & Yu, Y. (2017). SeqGAN: Sequence Generative Adversarial Nets with Policy Gradient. *AAAI*.

13 Modern Approaches in HR Analytics towards Predictive Decision-Making for Competitive Advantage

K. Guru, S. Raja, A. Umadevi, M. Ashok, and Kumar Ramasamy

CONTENTS

13.1 INTRODUCTION

The technology in human resources management (HRM) is gradually coupled with the related changes in data and information processing that restructure the environment. The area of human resource analytics is increasingly an integral part of the organizational system, which can be understood as a data-centered and analytical

DOI: 10.1201/9781003153405-13

thinking approach. This thesis explores new literatures and their consequences for predictive decision-making in the area of HR analytics. It also includes a comprehensive analysis of the literatures on incorporating HR analytics into organizational set-ups by implementing sufficient IT infrastructure and provisions. This qualitative, descriptive case study approach has the aim of exploring the strategies in HRA to enhance the performance of their organization. The methods used for data collection include semi-structured interviews, analysis of company document, and company website materials. It is essential, in a competitive business environment, to leverage the potentials of an employee to the fulfil the needs for organizational performance. Hence human resources always remain as one of the key distinguishing factors for a company that can generate the requisite organizational value for competitive growth.

The effective use of an organization's human resources capital is an ongoing process; continuous steps in the direction will ensure that an organization's human resources remain an asset rather than a liability. Management of human resources has to be carried out taking into account the interests of the company on the whole. It is understood that this area of study is aimed at discovering certain behaviors and methods that can be introduced with workers in order to achieve organizational objectives. It should nevertheless seek to obtain greater insight into the interpersonal traits of employees in order to ensure sufficiently efficient control of human capital and to support changes and introductions that yield good outcomes or have beneficial effects. HRM aims to provide tools and measures in personnel management and focuses on the basic premise that managers and employees can work together to accomplish shared goals within the organizational space of the hierarchies. In order to accomplish these objectives, HRM needs a range of well-established techniques and practices as well as innovative ones which are specific to the organizational context.

Significant issues are dealt with and accepted as part of the HRM work of a company. Decisions were defined as one of the most critical organizational processes, such as employee behavior, efficiency of jobs, productivity standards and employee stress levels (Griffin and Moorland, 2010). It is essential to harmonize the nature of HRM practices with broader criteria and recommendations for employee performance and strategic objectives. In order to strengthen the role of the employee in critical areas such as decision-making, which is the requisite capacity, knowledge and expertise, different strategic business strategies should be closely aligned with the organizational conditions (Pereira, 2013). Human resources analysis is a comparatively recent approach in the broader sense of the HRM which concerns the use of scientific tools, strategies and procedures to implement and disguise the most successful actions, such as policies and practices of the HRM. It is also called individual intelligence, ability intelligence or research. HR analytics can be viewed by providing scientifically related facts and statistics that can be used to develop new policies and enforce existing HR techniques and other behavior. The HRM resources provided in analytics have been used by employers and organizations, but there is still a large room for sector development and for the analysis of the various categories covered by HRM.

The aim of this thesis is to examine the recent literature on the relationship between research into human resources and its approach to improving the current set of management and human resources activities. The field of human resource analytics is becoming an important part of the organizational system, and can be interpreted as a data-centered and critical thought approach. This chapter explores new literatures and their consequences for predictive decision-making in the area of HR analysis. This overview will provide an understanding of how critical human analysis is for decision-making and how businesses can expect strong returns on investments. This chapter aims to explore, if integrated into the enterprise, the ability of HR analytics to support managers in making strategic decision-making using statistical and relevant HR analytical data and literature.

The emphasis also included reviewing the IT infrastructure and technical changes, including those influencing the manner in which data is processed and collected, in order to efficiently enforce the HR analysis and to ensure efficient data storage in order to become important for HR analysis. This will provide critical review of the comprehensive steps taken to incorporate HR analytics in an organizational structure, the methods and techniques used for data management, and the approach taken as analytics are being used in industrial decision-making. HR analytics thus provide tremendous opportunities and has an immense potential to enhance the processes of HR and management decision-making to be analyzed during this report.

13.2 REVIEW OF LITERATURE

13.2.1 HR ANALYTICS: A MODERN HR DECISION-MAKING TOOL

13.2.1.1 Concept of HR Analytics

For an organization, employees are incomparable assets. Effective means of gaining a strategic edge in a diverse corporate world and a major obstacle for firms to manage and map employees with varying expertise according to their approach can be focused through HR analytics. This demands large amounts of information for decision-making to be produced, evaluated and processed. Management of human resources comprises instruments for managers to learn about patterns from various HR positions that enable companies to eliminate stars from the database of big employees. Research on the analytical and rational control of employee data and organizational success offers the response of data which have to be analyzed. HR analytics involves the use of mathematical methods, analysis and algorithms to analyze employee data and to explain conclusions in evocative reports (Levenson 2005). HR analytics uses predictive algorithms to inspect employee records, and data show patterns that enable employers to forecast trends in behavior including turnover, training expenses, and employee engagement. This is also called mathematical analysis.

A traditional HR processing framework collects HRIS, archives of outcomes, mobile equipment and data storage for convergence of the social network. It uses big data, statistical analysis and data mining strategies to understand the data's hidden trends, associations, probability and forecasting. Data storage networks are used for capturing, processing, transforming and storing data in various databases. HR analysis in the larger human resources consulting world is a comparatively new activity. It is sometimes called study of individuals or skills or analysis of the workforce. HR analytics are more accurate and they provide objectively relevant data and information that can be used in developing and implementing future techniques and existing HR strategies and controls. Employers and companies have recognized HRM's promise in analytics. However, there is still tremendous space for growth in the field and for the study of the importance of analytics among the numerous groups that are subject to human resources management.

Efficient HR analytics allows HR administrators to execute HR activities such as estimation of individual demand and supply, appropriate working tests, the measurement of requirements of preparation for workers, pay implementation and retention of good employee knowledge in order to measure benefits and monitor the discipline of their employees. In general, it helps HR managers to choose from details on recruitment, hiring, training, promotions, job planning and corporate success. Work can be categorized as descriptive, predictive and optimization analytics (Watson 2010; Narula 2015). Descriptive analysis is a first stage of study involving the understanding of historical experience, behavior and outcome and only explains the relationship (Fitz-enz 2009). This includes data panels, ad hoc reports, boxes, dashboards / ranking sheets and SQL queries. Revenue ratios, recruit costs and absentee rates can be determined by descriptive analysis. The second stage of study involves forecasting potential behavior and outcomes on the basis of previous data. Requires statistical research related to: The use of data mining (data correlation), decision-making, pattern recognition, forecasting, study of root causes and statistical modeling (what is next). Predictive modeling can assist HR managers in predicting turnover rates and job results through recruiting approaches. The third level of study is optimization analysis, which deals not just with minimal capital. In order to build the right alternative education approach to achieve organizational efficiency, it includes the use of linear programming, simulation, application and deployment of mathematical modeling (Narula, 2015).

13.2.1.2   Analytical Methods for HR

Various business intelligence vendors are Microsoft, SAP, Oracle and IBM. BI (Business Intelligence) HR modules and data analysis tools apply for BI applications (Kapoor & Sherif 2012). R-Studio is a data collection and visualization tool that can be used for large datasets. Python is a favorite programming language for all data scientists for computational data and visualization. Microsoft Excel is traditionally an important data-processing method for the compilation, analysis and transformation of data by means of formulas, pivot tables, scenario setup and graphical methods. The software will gather data from various sources and make analyzing, compiling and displaying data simple for us; the most important aspect of the program is that successful implementation (Erik Van Vulpen, n.d).

Mondore, Douthitt and Carson (2011), in their HR research paper, have prepared a semester summary on how HR analytics should be used to ensure it aligns with HR targets and desired organizational results. The study phases involved and identified include identifying the most significant findings; establishing an inter-functional data strategy; taking necessary steps in order to produce key findings; preparing and implementing the plan; evaluating, revising and adapting the plan after execution, and making further changes (pages 23–24). The research also looked at how HR analysis can be used to identify and protect a company's talent and thus develop an organization's performance and talent strategy. The suggested HR research strategy referred to in the study is designed to incorporate such organizations and thus, with minimal guidelines such as the provision of HR analyzes for representatives of an organization, may be viewed as a holistic viewpoint.

The core of the point, that HR analytics should add value to decision-making by providing backed and validated statistical evidence, is discussed in Rasmussen and Ulrich's (2015) study. To avoid HR analytics becoming yet another "management fad," the study suggests that progress is being made in adjusting the traditional methodology adopted by HR and associated efforts to ensure tangible and material results are reached successfully. In place of the traditional inside-out approach based on HR, the study proposes to incorporate a shift to the "outside-in" approach with an emphasis on specific measures. This suggestion for a changed solution can be integrated in the application of relevant technical interventions. The Ulrich and Rasmussen report explores two case studies on the application of HR analysis and studies its positive effect on productivity and organizational talent development. The research offers HR analytics as an insights into how HR's actual effect on the accomplishment of company goals can be enhanced.

Levenson stresses the importance of HR analysis in its 2011 study by offering a comparative cost-benefit/impact analysis and a ROI analysis. Based on his 2005 report, Levenson (2011) estimates the high quality capacity of decision making on human capital and human resources challenges in companies through the use of analytics and metrics. The analysis reveals that the HR function examination was described as a skill held by the analytical specialist. The Levenson study indicates that the measurements are required to raise the standard of analytical decisions in HR in detail. Furthermore, the report postulates that measures and research endorse a much needed in-depth study to discuss the modeling of labor markets and business architecture that have both a diagnostic component and a reasoning exercise. Thus, the study shows that the time, attention, and money of HR practitioners must typically be based on good analysis that improves the assessment of HR problems and offers better advice to establish business models and behavior.

The research also involved the development of an HR-analytics COE and a basis for a number of operational activities based on HR analysis. The study by Rasmussen and Ulrich (2015) examines the substance of the case that HR analytics aim to add value in the management and office decision-making, including substantive and validated statistical data. The study proposes modifying the conventional methodology and associated policies adopted by HR to ensure that tangible and material outcomes are influenced to prevent the development of HR analytics into yet another "technology fad." The research proposed a shift in the "outside-in" approach to concrete behavior, instead of utilizing the traditional HR-oriented "inside-out" approach. Relevant technical interventions may incorporate this proposal for a changed methodology. In two cases analyses of the implementation of HR analytics, Ulrich and Rasmussen's study examines the beneficial effect this has on the optimization of success and organizational talent. The research identifies HR analytics as an informative way to maximize the actual impact of HR on company goals.

Mondore, Douthitt and Carson (2011) have demonstrated in their research into human resources measurement that they must also consider how HR analytics can be used in combination with HR strategies and desired organizational outcomes by including a six-phase guide on how HR analytics can be used in order to achieve HR strategies and desired organizational results. In the course of the study, the key findings are identified, a cross-functional data strategy developed, appropriate initiatives to assess crucial results introduced, the plan drafted and implemented, the post-enforcing plan evaluated, updated and further modified. The research also examined how HR analytics can be used to recognize and protect talent for the company, and thus to develop succession and talent for the company. The suggested HR analysis approach presented in the report focuses on the execution of such organizations. Thus, minor advice such as including HR analytics for individuals in a company's management roles can be seen as giving detailed analysis.

13.2.1.3   IT Technology and Central Storage of Data

Technology has guided the development of HR analytical instruments and has increasingly improved the research landscape. This segment examines the technology use literature for the progress of HR research in decision-making. The Fairhurst (2014) research explains the nature of measures to incorporate data-driven architecture in HR analytics and assimilates HRM analytics. The study constricted five steps to turn market problems into issues of data understanding, i.e. questions that can be addressed with data. The other steps included taking the requisite measures to systematizing and archive the data collected; interpreting the data using the corresponding statistical instruments and measures; further analyzing the data using analytics, computer training and neural networking; and reporting or finding the findings to the organization or entity concerned in a consistent manner. A study by Angrave, Charlwood, Kirkpatrick, Lawrence and Stuart (2016) explored the truthfulness of the disruptive claims of HR analytics, which are rapidly becoming a necessity to review organizational outcomes. The findings of the report are cautionary and seek to analyze the implementation of rational thinking in research into human capital. Angrave et al. explored the need for metric research focused on sophisticated mathematical analyzes and parallel tests. The proposal to develop the current paradigm focuses on the use of HR-related variables in constructing analysis models, but also advises against the uncontrolled use of analytical technologies and instruments. In a case study and critical assessment, King's research (2016) discussed the role of HRM data analysis. The study also indicates that changes in the region have contributed to a degree of confusion as to the willingness of HR employees to use their data or knowledge to generate positive results. The thesis reviewed literature both to analyze the inconveniences of human resources research, and to contend that the implementation of analytics and similar practices is included in the academic debate. The usage of case studies in research gives constructive recommendations when analyzing with an emphasis on a more effective and action-oriented approach. The analysis thus defined core analytical goals and proposed practices, such as outsourcing, maintaining security and accessibility of the available data, and the creation of models for using available data.

Jasmit Kaur and Alexis A. Fink (2017) based on 22 interviews with 16 firms, concluded that most R Studio businesses use a standard visualization and dashboard creation platform for analysis and presentation of data accompanied by a tableau (83%). Some of the latest technical capabilities are IBM's Watson Talent, HireVue's video processing tool and Intel's Saffron. For quality polls, surveyGizmo, sirota, and market statistics, and analyzes, Cognos Visier is used. This list of new technology opportunities reveals that HR analysis is still developing and there is still much to be done in the future.

13.2.1.4   Make Insightful Judgment

Predictive analytics are used in the prediction of unpredictable events using predictive information, modeling, computer training, data mining and AI (artificial intelligence) to analyze past and existing data to forecast future information. HR predictive analysis is rising increasingly and is important. HR predictive analytics was used by HR administrators in organizations to anticipate and optimize human conduct and to maximize investment gains for enterprises using predictive computational approaches through decision-making. The related literature is discussed in this section to explain the method of HR prediction research. The study of Rich, Lepine and Crawford (2010) suggests that employee engagement is more important than other factors that reflect comparatively limited facets of the personality of an employee. Their study was based on an evaluation by 245 burners and their superiors to obtain the optimal findings. The study states that engagement mediates relationships through basic dimensions such as mission success and the organization of workers.

Oogle argues in his book *Career Rules* that Google uses statistics for applicants, that interview matters are thoroughly programmed and specifically suited to the profiles of applicants in order to choose the right candidates. Google also notes that employees who retire from the HR predictive research association are expected to leave the business because they receive raises within the first four years. Kluemper, Rosen and Mossholder (2012) considered the potential to forecast job efficiency, IQ-led intervention, personality testing, and organized interviews with the applicant's Facebook profile to serve as an additional method to pick their workers following legal and ethical considerations. Best Buy used HR analytics to estimate the company's results on a quarterly basis, estimating that a 0.1% rise in workers' engagement contributed to a profit increase of $100,000 (Erik Van Vulpen n.d). In the 2011 HR data review Hewlett-Packard (HP) measured the amount of employee revenues known as "Flight Risk Rating." They found that higher pay and bonuses and positive results are adversely associated and someone who is promoted but does not earn a salary raise will definitely leave the job. HP has developed a dashboard with critical matrices for all its HR managers. By predicting the employee's behavior and their commitment to organizational performance, companies will save millions of money.

Ballinger, Cross and Holtom analyzed the impact that the network system has on voluntary purchases in their 2016 paper. The study focuses on employee social relations as a resource that impacts the organization's labor capital costs. The study indicates that the reputation of networks as connectivity for well-connected persons is a major factor in a company's turnover evaluation. Ballinger, Cross and Holtom

(2016) noted that the brokerage represents a moderating force for productivity, creation of ideas and performance. The major impact of workforce turnover is to improve brokerage and qualifications for employees with a more established social network and to exit businesses more likely. According to the HR prediction review, the concept and suggested measures that could be used for prediction modeling had a significant impact on all fields concerned with human resources management. The role of the field of prediction analytics in HR research has been examined by Mishra, Lama and Pal (2016). It helps businesses, through HR-related interventions, to reduce implementation costs and enhance company performance and increase employee participation. This research seeks to explain how issues such as low retention and high turnover rates correlate. The HR predictive review reveals, "The quickly emerging technology that has 100 percent consistency in HR decision-making."

Reddy and Lakshmikeerthi (2017), in their study, found that HR analytics are an efficient method for assimilating data gathered, so that workers within an enterprise can recognize, view, predict and optimize capacity. The study also reduced variables such as organizational processes, strategic processes, organization structure and architecture, which are significant in HR analysis. In total, 22 interviews were carried out between Jasmit Kaur and Alexis A. Fink (2017) and 16 firms. They find that HR analytics was used by organizations to build predictable models of recruitment, acquisition, sales and profit design on the basis of employee profiles and sales patterns. Companies use HR analytics to involve employees, through satisfaction polls, inputs from employees about best or worst management practices based on data from HR's vacancies, talent migration (transfer, promotional) and labor force data, to help managers identify their style and growth needs. Measuring training programs success is also a key application of HR analytics.

13.2.1.5 Problems of Data Governance in HR Analytics

Data management uses systematic processes to ensure the compilation and the ethical and legal use of the quality data. The field of HR analytics has the potential for ethical and legal concerns as it uses new technology to produce data and analyze vast quantities to promote decision-making by management. Toyama (2015) proposed that the first sin of AI is to make ethically based decisions by computer systems (machines), and suggested that human decision makers should gain feedback from HR analytics. Predictive models can be helpful for alerting decision makers, but not for making or reaching decisions. Decisions do not infringe essential labor laws such as discrimination and disability. Owing to HR data collection and the use of clean data for decision making, employee confidentiality, trust and privacy are at risk.

13.2.1.6 Strategic Study

There are not sufficiently analyses of the strategic approach mentioned in the analysis by Mondore, Douthitt and Carson, which can provide sufficient evidence of its applicability in the world, in the reviewed literature on the implementation of HR analytics. The model of the analysis cannot however be assumed to be consistently applicable and contextualized in line with the criteria of a specific organization. Rasmussen and Ulrich (2015) conducted the study based on the assumption that the strategic response of the HR companies was to revolutionize analysis in relation to the company's corporate priorities, although they saw inadequate execution to demonstrate the extent to which HR analytics is a return in order to help achieve organizational goals. In order to illustrate their fundamental value for HR analysis, Levenson's study explored various approaches and mathematical techniques in the area of HR analysis. The thesis also focused on developing business models and practical plans rather than standard execution and deployment of data mining software. However, research does not define accurate time frames and resources for production of the proposed HR analytical COE and the development of models required for HR analysis.

Angrave et al. (2016) provides statistically based evidence, figures from quantitative statistics and the integration into the literature described in this section of the HR-related variables. The methodology suggested in the thesis, while supporting the idea of a research methodological model, is beyond the framework of a quantitative analysis and can only be carried out within an organization. Results and recommendations cannot be errorless and thus cannot be relied on blindly. A week's analysis reveals that the online and online inclusion of HRIS makes it easier to collect Hour-based variable results depends on machinery, and decreases human resources needs; at this stage, however, costs and other HRIS systems inclusion variables can only be approximated to be feasible. There will be other obstacles to implementing HRIS, such as expensive electronic components and the need for personnel to operate them. In the optimistic comments contained in the article this is not taken into account. Fairhurst's (2014) study gives an example of how data obtained in HR analytics can be integrated into an organizational structure of the data-driven architecture. The paradigm introduced by the analysis remains a cursory formulation and does not explain each of these actions in depth. As a result, the functional feasibility of the concept is minimal and considerable attempts will be made to implement it into an enterprise.

In order to legitimize those postulations relating to HR analytics' costs-sustainability prospects the arguments of Mishra, Lama and Pal's research on predictive decision-making must be supported by appropriate case studies in organizations. Moreover, the forecasts of future growth in the sector posed by the study remain factory importance and the sustainability of these estimates must be seen in real-world companies. The Fitzenz and Mattox reports, on the other hand, provide a broader roadmap to provide HRM analysis that is thoroughly familiar with the context of the shift in approach required to adopt a quantitative paradigm that favors policy-making to help leverage human resources. The study of Ballinger, Cross and Holtom and Rich was helpful in presenting criteria based on predictive policy criteria, such as productivity and personnel turnover. The recommendations of the studies were, however, isolated; there was no action plan or blueprint for adoption. Similarly, research into the implementation of a training curriculum in Puhakainen and Siponen is suffering from an absence of an adequate role model for assessing practical feasibility in organizations.

13.3 MEASURING THE BUSINESS ADVANTAGE THROUGH HR PREDICTIVE ANALYTICS

The analysis of Fitzenz and Mattox (2014) showed that predictive analytical companies in the HR sector have at least 4% higher efficiency. More generally, mathematical modeling will highlight the economic value of HR research investments. Levenson (2005) says it is especially helpful to compute causal relationships rather than just a ROI or cost-effectiveness analysis. The process is so condensed for ROI that it combines all costs and benefits into a single number. The root factors or conditions that could lead to poor decisions are not therefore taken into account. Alternatively, if all the parameters, complicated to explain the stock market performance in HR, should be combined it took too long to slow down the decision-making process. Cost-benefit analysis adds the ability to change the Levenson equation (2005). He says that the cost-benefit study is not a conclusive statistic but an opportunity to take a good look at the risks and benefits. However, numerical weight cannot be assigned to all rewards such as better innovation or consumer satisfaction. Unquantified profits often are left out, but the cost-benefit analysis is insufficient because, after all, they are crucial for the business' sustainability.

Levenson (2005) adds that an impact analysis should first be undertaken rather than closely summarizing monetarized benefits. Impact analysis needs to detect causalities between the input and output variables. The HR leaders will list these causes very easily, but it is difficult to grasp the relationships and causes. The behavioral simulation of personal impacts, group dynamics, incentives and behaviors must be paired with the same model and then lead to success on the market. The first half of the problem is behavioral modeling or the analysis of causal relations, such that the predictive outcome models are developed. The analysis of triggers can be the biggest challenge for people who begin predictive human resources analytics. Ashton et al. (2004) continue that leaders today need to improve the analytical skills of HR teams to make the best of HR people understand what's important to HR and how processes are working. However, they add that the best choice, depending on the situation, may be to learn those expertise from outside sources. HR is traditionally evaluated as its own capsular area, how HR behaves itself. This measurement method is also important to optimize internal processes.

Chester's research (2012) shows that several variables are used in HR and can be used as indicators like achievement ratings, employee rotation or headcount. It is therefore necessary to decide precisely which data are available and what can be measured. In fact, the true importance of measuring something in HR over some other element of HR should be taken into consideration. Then again, you have to go past HR features when measuring the HR results. The HR advanced field experiments are doubled by Fitz-enz and Mattox (2014) and will realize financial and economic benefits. Financial gains are preferred adjustments that can be seen in the balance sheet, the declaration of revenue or, for example, the asset prices. Good non-monetary changes, including the credibility of customers or employees, client satisfaction and public relations, are often known as non-balance sheet reservations. In the ideal case, these economic gains will all transform into positive cash flows. Since research is mainly focused on identifying possible consumer advantages, it is beneficial to quantify the business benefits in future research with financial and economic meaning under the above terms. Presenting quantified reward figures provides greater corporate sponsorship and encourages change management.

Predictive modeling is based on algorithm testing in order to show the potential of an event. Predictive modeling may be known as cognitive computing, which refers to a system that can be educated and developed after learning new information. The aim of cognitive systems is to automatically interact with people. This will make people feel insecure or afraid of their jobs; in the future the robot can overtake people's work. Nevertheless, neither cognitive nor predictive analysis are designed to replace HR specialists, but are designed to help them by increasing their human ability to understand and interpret the data. Cognitive computing will assist you in speeding up everyday operations. Each organization should have a comparative advantage over those who do not have the capacity to master this business.

IBM Corporation indicates that if human characteristics of common sense, critical reasoning and independent thought are paired with computer-driven mathematical analysis and statistical estimates, the highest economic benefit would be accomplished. They also tried this in the chess game, where people can use machines as advisors, and enemies can be human or computers. Computers were definitely the most effective players when paired with their instincts. In more scientific terminology, and as described by Naasz and Nadel (2015), the following advantages are obtained through predictive modelling:

- Capacity to identify the best predictors that affect the given outputs most
- Observable and quantifiable details on the effect of multiple predictors
- A measurable statistical model of the present situation and future expectations

All the above statements are broad enough to be applied in every field of industry and every business operation. Although the possibilities are vast, it makes sense to identify which uses are more possibility where predictive modelling is to be sought. The following chapters discuss potentially useful areas for implementation. The options for using predictive analytics in the HR are again explored in the following categories: acquisition of personnel, management and retention of employees.

13.3.1 EMPLOYEE ACQUISITION OPPORTUNITIES OF PREDICTIVE ANALYTICS

As mentioned in previous pages, the process consists in the assessment, recruitment and incorporation of employees. Traditional metrics tend to depend on summarizing employee numbers and, for example, how quickly employees should be onboard. These reviews are valuable for creating a perspective on the current situation, but they lack insight into how systems can be enhanced. The subsequent sections discuss what more can be done beyond the overview papers. Oehler and Falletta (2015) have shown an indication of the financial gains achieved by predictive analysis in terms of identifying the best candidates for the company. In a logistics firm, workers had a 74% voluntary attrition rate, resulting in costs of $68 million. Then the company discovered that by applying predictive analytical techniques, it could minimize annual churn by 28% by increasing the value of entrepreneurship, flexibility, adaptability and ability to overcome uncertainty in the candidate evaluation level. This has helped them to lower sales costs by $27 million, raise employee satisfaction by 41% and customer satisfaction by 52%.

In the employee acquisition sector (Fitz-enz and Mattox 2014), the need for hiring efficiencies to be increased, calculated by the pace and the efficiency of hires. As the result of good jobs, one achieves improved participation of the employee, efficiency, profitability and managed retention. In other words, recruit efficiency must be increased and the process must be quicker.

Naasz and Nadel (2015) have identified the key possibility for the use of computerized research in the employee acquisition period; to make smarter recruiting decisions. This involves the hunt for work applicants across new networks of masses of details, which are impossible to do through human interpretation alone. In particular, they state that social networks should be searched, databases resumed or applications obtained for matching the appropriate candidates to roles by projecting the potential expertise and results of candidates in the target role. One example, as Farley et al. (1989) pointed out in their recruitment process, is the hiring channel: informal recruitment networks appear to deliver better candidates. In other words, predictive analytics can be used to filter candidates from the volumes of data on the basis of an open place description. It does not fully substitute hiring professionals, but can be used as a method to assist with this process.

The big benefit of using machine power in the recruiting process is that if historically HR experts have made a subjective estimate of their abilities, a robot will objectively evaluate the potential match. This removes the choice of recruiters to exclude applicants depending on their review of appropriate data. Personal preference can also be reduced when using computers in interpreting programs, rather than human beings, which appear to favor identical personalities (Naasz and Nadel 2015). The forecasting of demand by workers and skills is another field worth exploring in predictive research. Cappelli (2009) further illustrates the volatility of workforce movement in the modern market world, which includes the opportunity to respond to transition. Where one wants to be working next is hard to predict; in which divisions or fields of talent. He suggests that conventional methods of employee prediction have failed, but competitiveness and externalization of workers have improved. This poses the need for improved estimation tools and ways to forecast the movement of workers. In other words, the current employee condition may be regulated more accurately by risk management and more precise forecasts would be helpful for this purpose.

The identification and successful management of the skills available in the organization is necessarily where predictive capability should be used. This is addressed in part with the planned workforce proposed by Cappelli (2009), which suggests that one has to realize what is expected in the future of the industry and prepare for it. Bhattacharya et al. (2005) provide a further insight. They note that the HR versatility of the organization, generated by individual expertise of its employees, has a huge positive impact on its success. In this sense, HR versatility relates to the capacity of HR to respond to evolving external demands, by adjusting employee capacities and managing the expertise of the entire workforce. This is a recent field of study since there is historically very little substantial scientific data on the effects on the business results. Bhattacharya et al. (2005) recommend a different strategy if you cannot foresee what talents will quickly go or which will have greater demand; you may also seek to recruit individuals with high interpersonal resilience, desire and enthusiasm to learn new skills. This question is therefore multifaceted, but all abilities are carefully handled to have a positive impact on the success of the organization.

However, Guinn (2012) gives further insight into the dilemma by specifying which knowledge is important to the desired result when recruiting an employee. Unlike traditional views, he claims that career experience, job knowledge, knowledge of the right people, qualifications or doing the right things do not predict progress at work. It notes instead that spending time with the applicant, long observation and psychometric profiling shows the characteristics of employees which often contribute to a poor performance in the job: poor hearing abilities, lack of organizational skills, lack of leadership ability, incapacity to respond to change, a tunnel vision linked to work or a lack of follow-up in work. Therefore, predictive regression in the recruiting process cannot fix anything. On behalf of HR specialists, it cannot determine the conditions for achievement or replace the value of meeting the applicant directly. Specialists still have to determine what the job is specifically desired, how it is calculated and how the individual interacts with the staff and community of the organization.

13.3.2 Effects of Workforce Performance Predictive Analytics

In the field of workforce growth, analytics can be used to organize unstructured data for performance assessments and satisfaction surveys to allow more data-based decision-making and action planning (Naasz and Nadel, 2015). This section addresses both HR aspects for growth and governance to be addressed after the employee enters the company; such as expertise, talent, apprenticeship, health, pay, profit and success management. Predicting the success of workers is one aspect businesses want to monitor and forecast. Anschober et al. (2010) analyze this field and state that successful competence-based management is a characteristic differentiating high-performing organizations from the others. In today's dynamic economies, the freedom to develop, preserve and use employee capacity is a tremendous asset. In this resource-based viewpoint the business can be distinguished from others by its willingness to produce better than its rivals what the consumer needs. In order to achieve this, the organization must first define its core competences. The organization must then ensure that it has staff and a capability base that will ensure the continuum of its work. The main competencies contribute to the company's inherent resources and talents within its staff, which can lead either to differentiated, inimitable goods or to perceptible cost advantages. Whatever the core competencies of the businesses, companies also depend heavily on individuals who generate value according to the priorities of the business. Thus, it helps to forecast the best performers who help the mission of the business, whatever the objectives are.

Brown (2014) has fascinating concepts in the sense of incentives for predictive analytics. He says the overall compensation scheme that provides fair rewards for jobs in the company is obsolete. Instead of a set and inflexible scheme, it proposes an "intelligent rewards" system which has been developed worldwide. The intelligent incentives are based on a simpler emphasis on the key values, a clear reliance on facts and measures, greater interaction of workers by rewards and more transparent dialogue on rewards. Simply put, workers should know from what incentives they receive, which should be adequately rewarding for any person to participate for multiple purposes in order to achieve superior results. So it can be inferred that projections can help to find the right recompense scheme for any employee. According to Brown's (2014) reports, relatively few businesses have assessed the viability but have merely benchmarked the other companies throughout the business in market. The biggest problem was the lack of data and nowadays data dispersal on various locations; retaliation data in an external payroll supplier, financial system expenses and HR department engagement data and so on. The need to combine these data has been recognized to handle the incentives more efficiently. In a more operational aspect of HR, talent acquisition and are also advantages education planning analytics. The industrial age, Fitz-enz and Mattox (2014) noted that a career planning and competencies management perspective is obsolete as a detailed list of expertise in the rapidly evolving knowledge landscape will be virtually impossible to create. Predictive analytics can be used by forecasting which training courses produce meaningful returns on investment. According to Fitz-enz and Mattox (2014) the greater gain has been gained by delivering instruction that has the greatest applicability to employee perceptions. Predictions could be used at the employee level, or in general for the long-term advantage of preparation. Many businesses do not have unrestricted budgets to offer training, in which the amount of training per employee has to be determined.

13.3.3 EFFECTS OF EMPLOYEE RETENTION PREDICTIVE ANALYTICS

The company benefits from its management, listens to employee requests and establishes a high degree of commitment to employees in many ways. Committed people work harder, deliver more efficiently, provide customers with superior offerings and influence the company's success. In other words, employers would actually increase business productivity by increasing employees' commitment. Although there are more and more studies of employee happiness in companies, the full view of the employee's preferences is not provided, and it must be further expanded. If involuntary resignations are to be stopped, jobs must stay, that is, statistical modeling should be used in the area of staffing. Baron (2012) insists that any organization must view the commitment in accordance with the organizational culture, so that the study content will vary accordingly. There are common nominators, however, and disengagement begins with one absent. Satisfaction, motivation and efficiency must be considered in the role of the workforce. Furthermore, the commitment varies with time and the goal of engagement differs between employees; others are more involved in employees or administrators, while some are "tasks" that make the missions more driven. There are two types of commitment, related and transactional. In the short term, transactional players do well because the organization needs them to be rewarded – fame or money for example. This type of dedication, however, still contributes to burnout and does not really promote the employee's common sense of purpose. Emotionally committed employees have faith in the goal of the business and become more committed and well-being. Because increasing incentive for workers has also been shown to increase productivity, it is valuable to find out how to improve it and predict the effect of each employee category. Oehler and Falletta (2015) offer an example of how statistical analysis of employees' retention has achieved real-life benefits. A transport company has started assessing productivity and predicting success. They were urged to both develop a fairer reward model but also to acknowledge future leading players who should directly focus on job planning and motivating management, and potentially contribute to lower employee turnover. The average productivity increase was 1.7 million pounds. Fitzenz and Mattox (2014) plan to investigate the influx and displacement of jobs at a more general basis. They show the complexity of the retention study by means of a comparison to the principle of chaos. Employees quitting and entering the company at entirely irregular intervals can first look mysterious and unpredictable. Theoreticians and mathematicians of the disorderly process prove otherwise. An unexpected situation should be described in full; research could only require more complex calculations as measures can be added to the equation. Tymon et al. (2011) suggest the following methods that management can affect employee turnover: experience with inherent rewards; good feeling of the worker is associated positively with retention; meaning, growth, preference and expertise; the personal contribution of the organization, the more dedicated the company is to be. The more ambitious the employee's perspective is for the good future that is formed by the boss' inspiration and underlying awards, the less likely the employee is retained. This is a collection of assumptions that you might use at the outset of your predictive research, collect data about certain areas and examine correlations in your company background. But there is no ready-made blueprint or pattern that may be said to be right for company employee data, but data must be collected and investigated. If a pattern can then be found in the seemingly random material, it can in the best case be explained and controlled. Finally, it increases efficiency, competitiveness, profitability and cost-control capability by analyzing patterns in outcomes. The person has not only left, but his knowledge and experience has simultaneously gone. Shaw (2011) argues that high turnover rates affect organizational performance directly. The losses include losses in place of the outcome and losses of social resources, including the loss of the skills and knowledge of the person in his/her field of work. When the attrition rate is high, total competence is lost instantly. Shaw (2011) continues that if the turnover rate were constantly high, it would not be as detrimental as in the case of limited turnover and employees would like to exit. The employee has gained a more qualitative job expertise, which cannot be replaced. He also says that if the company leaves the employee, it has a greater effect on business productivity. It is up to management to decide the necessary type of turnover, but predictive research would also be needed in the field of skill management to effectively address this capacity deficit. The following characteristics are also used in the classification of data within the working groups, according to Fitzenz and Mattox (2014). The four most common reasons for employee propensity to quit are also referred to as four nominators. The disengagement of staff starts as all these simple human needs disappear; the need for faith, optimism, a sense of worth and a sense of competence. There are aspects where the manager can influence, of course, other places, for example moving overseas, are not controllable. A list of the seven simple reasons why employees are leaving is given below (Fitzenz and Mattox, 2014).

- Disengagement
- Employee confidence in job/workplace was not planned
- Mismatch of tenure hope between work and individual

- Very little mentoring or training
- Don't feel respected or remembered
- Stress from overwork and harmony between work and life
- Loss of trust in senior management

This points of departure can be used in "what happens" analysis or informative analysis. The study of the relationships between the individual variables and inclusion of data such as outcomes of employee retention surveys showing involvement helps the development of statistical models that can be used in pharmaceutical action planning.

13.4 HR ANALYTICS AND PREDICTIVE DECISION-MAKING MODEL

After studying the value of HR for the enterprise, the metrics needed in HR and the desirability of predictive analytics to achieve those goals, one might ask what keeps organizations from doing predictive analytics. Each organization needs all important elements on the way to the necessary elements. These elements are described in many models of analysis maturity, which concentrate the research on its overall yet holistic nature. The model outlines the methodological elements required to use the data for business interviews. In this analysis, the analytical measures used in the analytical maturity model are used. The analysis in the models below uses the measurements as the basis for the analysis complexity assessment tool, which allows organizations to define where they are currently and what further steps are required to accomplish advanced analytics, for example predictive analysis. After reviewing an associated current HR research and statistical decision-making literature, the following model has been developed to explain the value of effective decision-making. The proposed model is based on the literature previously studied.

The illustrated model, Figure 13.1, is based on studies conducted by Ballinger, Cross and Holtom (2016), which display the indicator for the turnover of employees

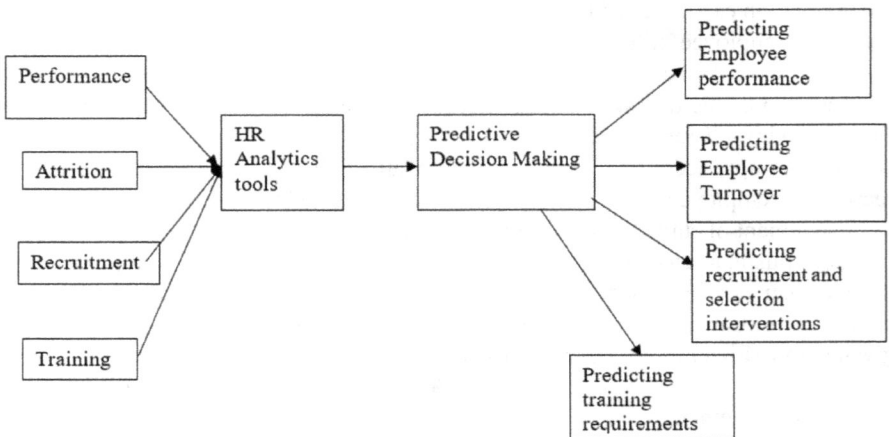

FIGURE 13.1 HR analytics and predictive decision-making model.

of data collected from the employee network structure. The model proposed also discusses, as explained in the study by Rich et al. (2010), how data is collected from different criteria will affect employee efficiency and be measured by analytical methods. The Puhakainen and Siponen research explored the need for training employees to ensure conformity with the stated policies in the organization's ISS policies, as previously discussed in the literature. The necessary training-related interventions in this regard can be understood through data collect in connection with the present study on existing levels of training and awareness among employees. After collecting this material, it can be analyzed using computational methods in order to draw conclusions on the predictive decisions to be made about training standards and employee policies. The model shows how computer methods can be used to decide quantitatively on various parameters necessary for organizational activities.

13.5 FINDINGS AND CONCLUSION

The objective of this chapter was to review existing literature to explain the relationship between human resources and analysis and to understand its importance in the improvement of the current area of management and human resources studies. This chapter aimed to explore, if integrated into the enterprise, the ability of HR analytics to support managers in making strategic decision-making using statistical and relevant HR analytical data and literature. The focus was also on reviewing of IT technology and technological improvements, especially those that affect the way data are stored and retrieved, to successfully incorporate HR analyzes and ensure reliable data storage to become relevant for HR analyses. A major conclusion from a literature review of the incorporation of research in the resource management area indicates that, while there are proposals for theoretical models to adopt, there is a lack of literature which examines their applicability in practice and examines the efficacy (or failure) of these models.

Predictive behavior literature exploration indicates that field data analytics can be used to identify certain parameters which can support HRM and HRD in an organization. Predictive modeling can be utilized to provide the raw data helps draw critical lessons and real observations that promote organizational growth, similar to the model discussed in this article. However, in case of predictive models studies the usable literary instruments should be more enhanced in order to accurately assess the validity and viability of the model designed for such industries. This is critical because quantitative decision-making and modeling cannot be used homogenously around the industry and organizations, since they are subject to numerous parameters, such as their corporate form, the field in which they operate, the number of staff and the amount of money that they will spend in human resources analytics. The findings show possible results of scientific intervention and how statistical decision-making impacts main project parameters. Inside the company. The organization. The organization. The literature review will draw more insights and related findings on the existence of large-scale deployment and potential implications of HR analysis. The results reached may also be helpful to consider the variety of opportunities for activities such as predictive decision-making and restrictions opened up by HR analysis.

13.6   LIMITATION AND SCOPE

The key focus of the chapter is on the availability of literature in this field. Predictive modeling and organizational research are still void in the literature. Therefore, existing literature should be better elaborated and concentrated on case studies in mathematical modeling and operational analysis to guarantee that the model and roadmaps are validated, tracked and dangerous. The selection of literature eligible for review has been limited as a result of this study. In the future, studies on the case studies of statistical modeling and organizational studies should be focused on modeling and roadmaps from different validated, well-established and successful organizations. The research may also focus on a theoretical approach to the study to analyze prediction modeling as a field within an organization's human resources studies.

REFERENCES

Angrave, D., Charlwood, A., Kirkpatrick, I., Lawrence, M., & Stuart, M. 2016. HR and analytics: why HR is set to fail the big data challenge. *Human Resource Management Journal*, 26, 1–11.

Anschober, M., & Richardson, S. 2010. Sustaining corporate success: what drives the top performers? *Journal of Business Strategy*, 31(5), 4–13.

Ashton, M. C., Lee, K., et al. 2004. A six-factor structure of personality-descriptive adjectives: solutions from psychological studies in seven languages. *Soc. Psychol.*, 86, 356–366.

Ballinger, G. A., Cross, R., & Holtom, B. C. 2016. The right friends in the right places: understanding network structure as a predictor of voluntary turnover, *Journal of Applied Psychology*, 101(4), 535.

Baron 2012. Organisational culture and dynamics. *International Journal of Scientific Research and Management*, 6(1), 2321–3418.

Bhattacharya, M., Gibson, D. E., & Doty, D. H. 2005. The effects of flexibility in employee skills, employee behaviors, and HR practices on firm performance. *Journal of Management*, 31, 622–640.

Brown, U. J. 2014. The influence of internet usage on academic performance and face-to-face communication. *Journal of Psychology and Behavioral Science*, 2, 163–186.

Cappelli, P. 2009. Talent on demand – managing talent in an age of uncertainty. *Strategic Direction*, 25(3).

Chesters, G. S. 2012. Social movements and the ethics of knowledge production. *Social Movement Studies*, 11(2), 1–16.

Fairhurst. 2014. *Big data and analytics of human capital*. Jobs Research Institute.

Farley, J. et al. 1989. The relationship between recruiting source, applicant quality, and hire performance: An analysis by sex, ethnicity, and age. *Personnel Psychology*, 42(2), 293–308.

Fitz-Enz, J. 2009. *The ROI of human capital: measuring the economic value of employee performance*. Amacom.

Griffin, R. W., & Moorhead, G. 2010. Organizational conduct. Commitment Learning.

Guinn, D. G. 2012. Recruitment is the psychological contract: An employer perspective. *Human Resource Management Journal*, 10(2), 32–41.

Kapoor, B. & Sherif, J. 2012. Human resources in an enriched environment of business intelligence, *Kybernetes*, 41(10), 212.

Kaur, J., & Fink, A. A. 2017. Trends and practices in talent analytics. Society for Human Resource Management (SHRM)–Society for Industrial Organizational Psychology (SIOP) Science of HR White Paper Series.

Kirkpatrick, I., Angrave, D., Charlwood, A. & Stuart, M. 2016. HR and analytics: why the big data problem is for HR to fail. *Journal of Human Resource Management*, 26(1), 1–11.

King, K. G. & King, K. 2016. Human resource data analysis: case study and critical evaluation. *Examination of Human Resource Development*, 15(4), 32–43.

Kluemper, D. H., Rosen, P. A., & Mossholder, K. W. 2012. Social networking websites, personality ratings, and the organizational context. *Journal of Applied Social Psychology*, 42 (5), 1143–1172.

Levenson, A. 2005. *Survey on HR analytics and HR transformation: feedback report.* Center for Effective Organizations, University of Southern California. 131–142.

Levenson, A. 2011. Targeted analytics was used to enhance talent decision making. *Human Rights and Policy*, 34(2), 34.

Mishra, S. N., Lama, R., & Pal, Y. Mishra, S. N. 2016. HRPA (Human Resource Predictive Analytics) in companies for HR management. *Science Foreign Days & Technology Study*, 5(5), 56–60.

Mondore, S. & Carson, M. 2011. Maximize the effect and productivity of HR analytics to achieve business performance. *Human Resources and Policy*, 34(2), 20.

Mattox, J. & Fitz-enz, J. 2014. *Predictive human resources analytics.* Hoboken, NJ: Wiley.

Naasz, K. & Nadel, S. 2015. Advances in big data and analytics can unlock insights and drive HR actions. *HR Focus*, 92 (5), 1–4.

Narula, S. 2015. HR analytics: its use, techniques and impact. *Business International Journal of Research in Commerce and Management.*

Pereira, V. E. 2013. A longitudinal case-study analysis of HRM high performance company activities (unpublished master's thesis) in Indian HRO / BPO industries. Portsmouth University. Recall from the https:/resourcesportal.port.ac.uk / portal / files/6033330 / Vijay Phd Oct 2013 Revised.pdf February 22, 2018

Oehler, K. H. & Falletta, S. V. 2015. Point/counterpoint: should companies have free rein to use predictive analytics. *HR Magazine*, 60(5), 26–27.

Radtke, P. H. 2003. Online teams: results of team mediation. *Group Dynamics: Theory, Study & Practice*, 7, 297–323.

Rasmussen, T., & Ulrich, D. 2015. Practice learning: how HR analytics stops management fading. *Dynamics of Organization*, 44(3), 236–242.

Reddy, P. R. & Lakshmikeerthi, P. 2017. HR analytics – an effective evidence based HRM tool, *International Journal of Business and Management Invention*, 6(7), 2017, 23–34.

Rich, B. L., Lepine, J. A., & Crawford, E. R. 2010. Job engagement: antecedents and effects on job performance. *Academy of Management Journal*, 53, 617–635.

Shaw 2011. *Evaluating in practice.* Cambridge UK: Cambridge University Press.

Siponen, M., 2010. Improving enforcement by employees through security training in information systems: a review of action analysis. *Quarterly, MIS*, 34(4), 757–778.

Toyama, K. 2015. *Geek heresy: rescuing social change from the cult of technology.* New York: PublicAffairs.

Tymon Jr, W. G., Stumpf, S. A., & Smith, R. R. 2011. Manager support predicts turnover of professionals in India. *Career Development International*, 16, 293–312.

Watson, H. J. 2010. Business analytics insight: hype or here to stay? *Business Intelligence Journal*, 16(1), 4–8.

14 Digital Transformation

Utilization of Analytics and Machine Learning in Smart Manufacturing

Iman Raeesi Vanani and Akram Shaabani

CONTENTS

14.1 INTRODUCTION

Digital transformation is recognized as a change in businesses, manufacturing processes, culture and organizational aspects to meet the needs of the market (Nasiri et al., 2020). Rapid changes and the introduction of emerging technologies into industries have led companies to use new technologies in their manufacturing processes to be present in the market and maintain a competitive advantage, and to achieve that it is necessary to have the appropriate knowledge and infrastructure. The application of new technologies and their optimal use in industries can create value, because the information and data that industries and managers obtain from these technologies can be considered as a valuable resource (Barbosa et al., 2018).

In addition, because of the rapid development of the Internet, data is growing and transmitting at an uncontrollable rate in organizations and industries. The important point is the optimal use of manufacturing data obtained through these technologies that managers can use them in decision-making, so analytics and machine learning are more important.

The remainder of this chapter is organized as follows. Section 14.2 discusses digital transformation, machine learning and analytics. Section 14.3 analyzes the importance of digital transformation and smart manufacturing, Section 14.4 covers analytics and smart manufacturing, Section 14.5 discusses machine learning and smart manufacturing, and the chapter concludes with Section 14.6.

DOI: 10.1201/9781003153405-14

14.2 DIGITAL TRANSFORMATION, ANALYTICS, AND MACHINE LEARNING

Digitalization and digital transformation started in organizations when computers were used for accounting, decision support, and transaction processing in the 1940–1950s. Gradually, the developments increased. In the 1960s, production robots and online transaction processing were used. By the mid-1970s, the personal computer revolution was beginning. Throughout the 1980s adoption of computing technology accelerated. In the 1990s and 2000s saw the expansion of local area networks, the global Internet, digital telephony, data warehouses, and digital data storage that brought more possibilities to the organizations. Then, the ability to store and compute digital data was significantly increased in early 2010 (Heavin & Power, 2018). After that, there was the emergence of artificial intelligence (AI), data science, machine learning, analytics (Miklosik & Evans, 2020) 3D printing, and the blockchain technology (Paschek et al., 2017), which have created many effects and transformed organizations and industries.

AI is "Science and engineering of building smart machines, especially smart computer applications" and its application is when a machine simulates functions related to the human mind, such as learning and problem solving (Cioffi et al., 2020). As a result of digital transformation and technologies, every day a lot of raw data are generated. This availability of data in large sizes has drawn attention to machine learning (Dogan & Birant, 2021). Machine learning, a subset of AI focused on knowledge acquisition, is a practical computational method with the ability to process big, intricate, and multidimensional data. In addition, machine learning is able to learn from patterns in data and create models to predict a specific process that will ultimately be effective in making advanced and informed decisions (Suvarna et al., 2020). Machine learning is known as a way to access the concepts behind the data (Lenz et al., 2020; Raeesi Vanani & Kheiri, 2018).

Machine learning use varies algorithm for learning patterns using data (Blier-Wong et al., 2020). Machine learning algorithms are classified into four types of learning: supervised, unsupervised, semi-supervised and reinforcement (Kang et al., 2020), which are shown in Figure 14.1. Supervised learning uses more classification, while unsupervised learning uses more clustering. Classification include techniques

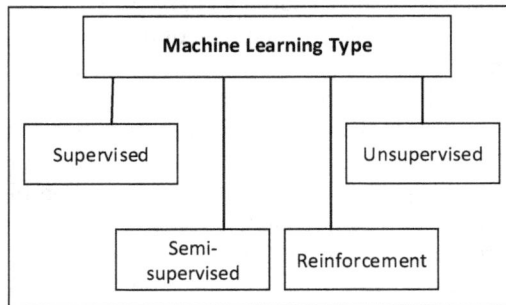

FIGURE 14.1 Machine learning type.

such as neural networks, support vector machines, decision trees, and the most common clustering technique used is k-means (Dogan & Birant, 2021).

Another point in digital transformation is analytics, which is very important to improve business performance. Analytics is becoming the foundation for businesses. Due to the large amount of data in business, data can be used for strategic decisions using analytics (Bordeleau et al., 2020). Analytics involves collecting, processing, analyzing, and interpreting data to gain insight (Tavana et al., 2020). In general, analytics is the science of converting data into applicable measures and is defined as "the solution to identify meaningful patterns and relationships between data and support decision making (Jun et al., 2020).

Data analytics refers to investigating data to obtain important information using analytics methods that are used to identify trends and support decision-making (Raut et al., 2021). It is also defined as advanced analytic methods that work on a big data-base (Ungermann et al., 2019) and is very useful to identify the features of a product (Filz et al., 2020). It increases the agility of organizations for rapid change in the face of fluctuations (Jain et al., 2017), and also shows the trend of future changes (Maruthamuthu et al., 2020). As shown in Figure 14.2, analytics involves AI, machine learning, deep learning, data science, statistics science and also is very useful and practical for business.

Over time, developments and progress have been made in analytics, leading to increased complexity and new methods being introduced. Analytics methods are shown in Figure 14.3 and are described below (Bag et al., 2020; Meister et al., 2019; Menezes et al., 2019):

- **Descriptive analytics**: This method is a series of historical data analysis, which is used in savings and seasonal demand forecasting methods.
- **Diagnostic analytics:** Investigating the main and root reason.
- **Predictive analytics:** Past data is considered to forecast future events, biases, trends, and behaviors through causation and correlation.
- **Prescriptive analytics**: Based on the data and models given (inputs), finds or suggests the best mode, path, method, or movement for performance (outputs).

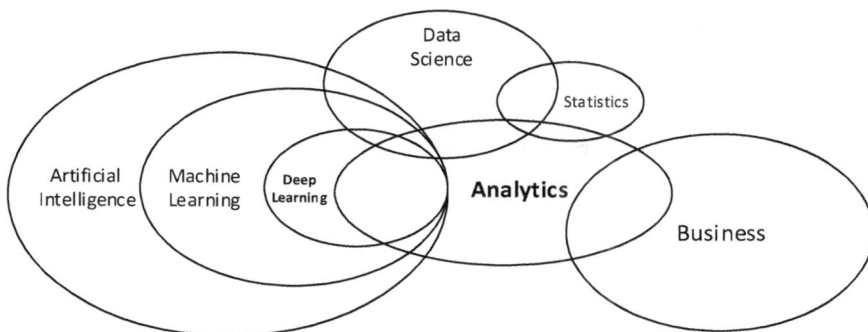

FIGURE 14.2 Analytics.

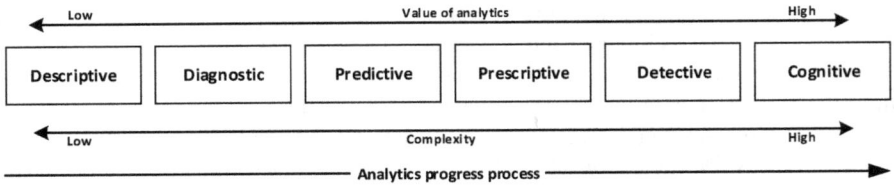

FIGURE 14.3   Progress process of the Analytics methods.

- **Detective analytics**: This method corrects and detects the collected data that has inappropriate values, which are mostly used for predictive and prescriptive analytics, and allows to achieve the desired results.
- **Cognitive analytics**: Used in predicting, prescribing and automating diagnostics for smarter decisions by considering adaptation and learning processes.

As shown is Figure 14.3, cognitive analytics has high complexity and is one of the advanced analytics methods.

The important features and structures are determined by data analytics, and effective models are constructed by machine learning to train the behaviour of a manufacturing system (Dogan & Birant, 2021). Effective data analytics with the help of machine learning to predict sales, disease diagnosis, process optimization, economic analytics, performance improvement, business management, social network analytics, etc. is very effective and ultimately leads to a competitive advantage (Kaur et al., 2018).

14.3   DIGITAL TRANSFORMATION IN SMART MANUFACTURING

One of the main factors in the prosperity of countries' economies is industries, whose high performance leads to greater economic efficiency. Digital transformation and smart technologies are known as the main pillars for industries that aim to strengthen manufacturing systems. To maintain a competitive position in the market, companies must seek to use different technologies and techniques to optimize their products and processes. Smart technologies can increase production by up to 20%. So, manufacturers have turned to smart manufacturing and digitalization of manufacturing processes to respond quickly to the market, produce low-cost products, respond quickly to final consumer demand, increase flexibility, and stay competitive (Phuyal et al., 2020).

Smart manufacturing has several definitions. A few of the definition are provided as follows: It means to use the interconnected machines and tools to improve production performance and optimize the energy and workforce required through the implementation of big data processing, AI, robotics, and mutual connection among such technologies (Phuyal et al., 2020). In another definition, smart manufacturing refers to the application of AI methods in advanced big data analytics, data-driven automation, machine learning, and decision support systems in the realm of manufacturing (Majeed et al., 2021). In a third definition, smart manufacturing is considered an IT-based empowerment to revolutionize traditional manufacturing systems through the integration of the systems with digital technologies, such as Internet of Things, big data, cloud computing, and CPS (Lin et al., 2020), which aims at producing higher quality products and reducing the production and service costs (Wu et al., 2017).

Thus, smart manufacturing is a new form of smart, autonomous, customizable, and flexible manufacturing system (Ghobakhloo, 2020) in which ICT plays an important role (Wang et al., 2020). Digitization operations in the field of smart manufacturing refer to the application of information and digital technologies in the manufacturing process. These technologies include IoT, augmented reality, blockchain, big data analytics, additive manufacturing, industrial simulation technologies (Ghobakhloo, 2020), cyber physical systems, cloud computing, digital twins, and next-generation AI (Bueno et al., 2020; Wang et al., 2020).

The application of these technologies in smart manufacturing has significant effects on increasing the sustainability, productivity and profitability of industries (Majeed et al., 2021). The important point is data analytics and the tools used to analyze the data.

14.4 ANALYTICS AND SMART MANUFACTURING

Smart manufacturing emphasizes the importance of human ingenuity and the manufacturing knowledge creation from data (Lenz et al., 2020). Equipping manufacturing systems with technologies is the best opportunity for significant industries advancement, as this data can be used to make smart decisions (Shah et al., 2020; Zheng et al., 2018). The most important challenge for the industry is to understand information and extract insights from data, which is why advanced data analytics is an important factor in smart manufacturing (Lepenioti, Pertselakis, et al., 2020).

As shown Figure 14.4, smart industry uses different methods to data analytics (Tang & Meng, 2020). Analytics benefits from various data and techniques such as machine learning, mathematical statistical algorithms to support timely decision making (Sadat Mosavi & Filipe Santos, 2020). Data mining is also widely used in the manufacturing industry to make decisions about existing data (Majeed et al., 2021). In the application fields of data analytics, issues such as the ability to predict,

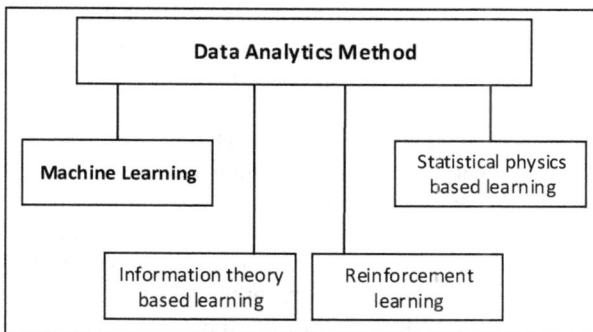

FIGURE 14.4 Data analytics method to smart industry (Tang & Meng, 2020).

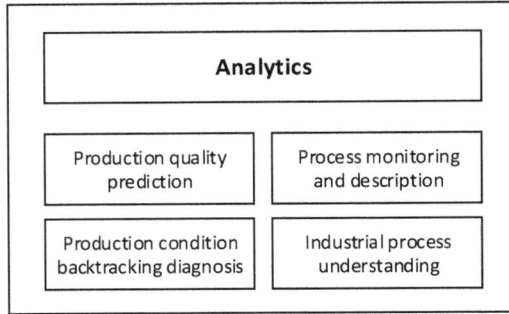

FIGURE 14.5   Analytics in smart industry.

monitor and control the product quality and process in real time are considered (Lenz et al., 2020).

Analytics is widely used in the smart industry, for example, for production quality prediction, industrial process understanding and some other cases that are shown in Figure 14.5 and are very important in smart manufacturing (Tang & Meng, 2020).

Industrial process understanding generally includes understanding the industrial sound, image and video. Image and video recognition plays an important role in detecting and monitoring production processes. Image data analytics on a large scale is very complex for industries. For image processing, monitoring, and fault detection in industrial production processes, analytics has a very important role because it leads to the identification of defects in equipment and production process. **Process monitoring and description** is used to save energy and ensure production in complex industrial production activities also to measure energy and resources. This can be done by analytics and achieve good results. Due to the complexity of **production condition**, recognizing product quality and productivity are difficult. Production condition can be analyzed by machine learning and statistical learning. Analytics leads to improved safety and increases the reliability of the production process. Analytics of quality data in the production process involves measuring the **production quality**. By measuring and identifying problems in product quality, changes can be made in the production process and change the quality parameters according to product performance indicators. This optimizes the production process, reduces production costs, and increases product quality (Tang & Meng, 2020).

Industries usually use some analytics methods to data analysis. For example, predictive and prescriptive analytics are very common and new in smart manufacturing, which often uses machine learning approaches and data mining tools.

Predictive analytics is used in smart manufacturing to monitor the health of machines and to predict failure, and as part of a predictive maintenance strategy (Lepenioti, Pertselakis, et al., 2020), and records and evaluates data on machinery and equipment to estimate the life cycle of equipment. One of the applications of the machine learning algorithm is error detection and the main cause of error, and this is an advanced analytics to optimize operation planning, real-time data monitoring and model pattern flow and data identification (E et al., 2019; Sohrabi et al., 2017). Also, the application of predictive analytics using advanced machine learning is invaluable in production, because predictive analytics leads to production improvement with respect to the cost, quantity, quality and sustainability of manufactured products (Lechevalier et al., 2014).

Prescriptive analytics uses mathematical programs and operations research techniques to make automated decisions based on the data generated and the results of predictive analytics (Menezes et al., 2019). Prescriptive analytics is one of the most sophisticated types of analytics in the business and is highly valued by businesses. The aim of predictive analytics is to select the best decision options using the large amounts of data and analytics, which uses AI, optimization algorithms, and expert systems (Lepenioti, Bousdekis, et al., 2020).

Predictive and prescriptive analytics in smart manufacturing have challenges such as:

- Lack of determination and uncertainty in prediction, particularly when the available data are not enough in terms of quantity or quality.
- Existence of contradictory, imperfect, or lost data with small dimensions, which can lead to misleading results.
- Lack of a common "language" between data scientists and experts prevents the extraction of appropriate hypotheses and the correct interpretation and explanation of results.
- Challenges of prescriptive analytics include combining machine learning and data mining methods with the "engineering technology" of specialists and the development of prescriptive analytics methods and the use of AI and machine learning algorithms (Lepenioti, Pertselakis, et al., 2020).

14.5 MACHINE LEARNING AND SMART MANUFACTURING

The machine learning function in manufacturing is such that the machine detects production processes and intelligently optimizes product quality (E et al., 2019) and is used in the design and new products development (de Carvalho et al., 2020) process monitoring, control, and supply chain planning (Suvarna et al., 2020). But in general, machine learning is used to support effective decision-making for manufacturing. In other words, one of the main applications of machine learning in the production process is "decision support" (Sharp et al., 2018).

In this way, smart factories use powerful systems to collect data create by technologies in all organizational processes. Many production operations are measured continuously and their data store in databases. The data is related to product specifications, the line of production, human resources, raw materials, failure of the machine, maintenance of the machine, product quality, and more (Dogan & Birant, 2021).

Machine learning algorithms detect errors by analyzing data in a timely manner and by making strategic decisions, prevent failure, increase maintenance, and durability of machines (Suvarna et al., 2020). Some machine learning methods that are used in manufacturing are classification and clustering, which are shown in Figure 14.6 (Dogan & Birant, 2021). For example, classification is used for fault detection, defect

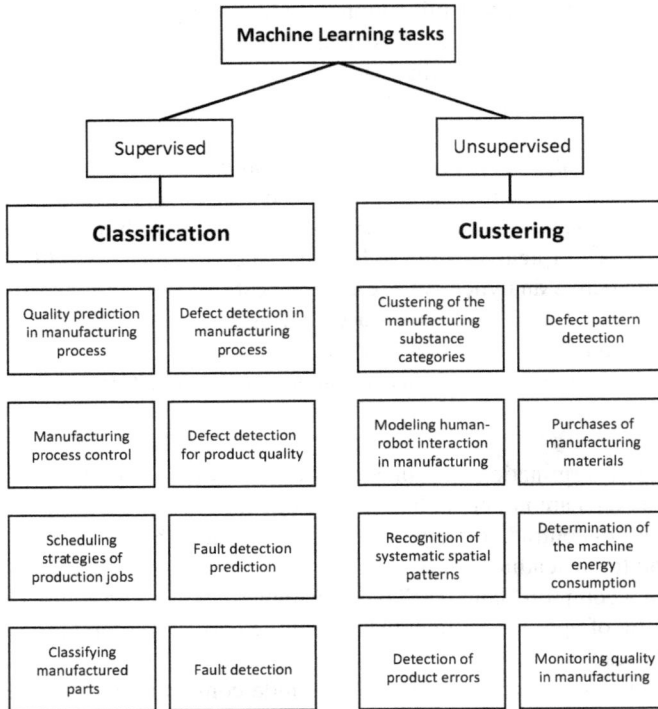

FIGURE 14.6 Machine learning tasks.

detection and prediction; also, clustering is used for error detection and monitoring quality.

Generally, we have shown the impact of machine learning and analytics in smart manufacturing with regard to digital transformation (Figure 14.7), and smart manufacturing needs to integrate all concepts.

14.6 CONCLUSION

New digital technologies are evolving and the process of digitization has led to a broader phenomenon of "digital transformation" in most industries (Kretschmer & Khashabi, 2020; Steiber et al., 2020). Digital transformation is one of the main priorities of many organizations active in various business fields. To compete and survive, large economic organizations and companies must consider appropriate digital transformation as a key strategy, such as using technology to fundamentally improve performance (Heavin & Power, 2018).

Moreover, the manufacturing industry is evolving digitally, commonly referred to as Industry 4.0, due to the advent of advanced information and digital technologies. Industry 4.0 is known as the digital transformation of industrial markets in which both smart manufacturing and using information and digital technologies in every aspect

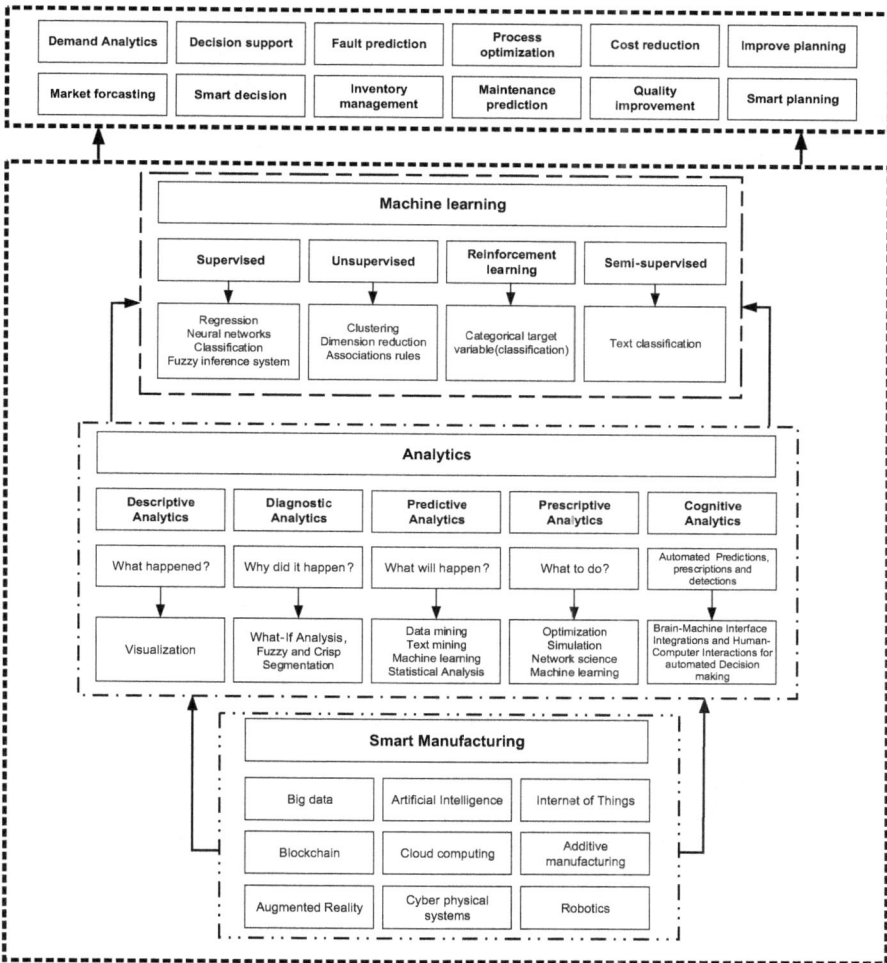

FIGURE 14.7 Smart manufacturing and digital transformation.

of production is a strategic priority for manufacturing companies (Ghobakhloo, 2020). Equipping manufacturing systems with technologies and producing massive data by them and ultimately analyzing data to support timely decision-making can lead to competitiveness, productivity growth, and industry innovation (Shah et al., 2020; Zheng et al., 2018).

Data analytics is one of the main components for using large volumes of data. By systematically analyzing the data generated in industries, more informed decisions can be made that increase the effectiveness of smart manufacturing. Data-based manufacturing is a prerequisite for smart manufacturing (Tao et al., 2018).

Thus, data is becoming a major factor in increasing competition in manufacturing, followed by using technologies such as IoT and sensors in the manufacturing process to collect real data is helpful and effective. A network of devices, machines, computers, and smart objects by collecting and sharing large volumes of data, increase and improve the state of automation in many industries. Analytics data collected with the help of machine learning can lead to better awareness of the possible malfunction of machinery and equipment, or predict the probable time for failure of a machine, and this will increase operational efficiency and economy. It saves time and money, which ultimately leads to smarter and faster decisions and better maintenance planning, and minimizes downtime in the production process. Another important application of data analytics in industries is waste reduction and product quality improvement, and with data analytics, deviations in product quality can be identified and eliminated as soon as possible.

REFERENCES

Bag, S., Pretorius, J. H. C., Gupta, S., & Dwivedi, Y. K. (2020). Role of institutional pressures and resources in the adoption of big data analytics powered artificial intelligence, sustainable manufacturing practices and circular economy capabilities. *Technological Forecasting and Social Change*, 120420. https://doi.org/10.1016/j.techfore.2020.120420

Barbosa, M. W., Vicente, A. de la C., Ladeira, M. B., & Oliveira, M. P. V. de. (2018). Managing supply chain resources with big data analytics: a systematic review. *International Journal of Logistics Research and Applications*, *21*(3), 177–200. https://doi.org/10.1080/13675567.2017.1369501

Blier-Wong, C., Cossette, H., Lamontagne, L., & Marceau, E. (2020). Machine learning in P&C insurance: a review for pricing and reserving. *Risks*, *9*(1), 4. https://doi.org/10.3390/risks9010004

Bordeleau, F.-E., Mosconi, E., & de Santa-Eulalia, L. A. (2020). Business intelligence and analytics value creation in Industry 4.0: a multiple case study in manufacturing medium enterprises. *Production Planning & Control*, *31*(2–3), 173–185. https://doi.org/10.1080/09537287.2019.1631458

Bueno, A., Godinho Filho, M., & Frank, A. G. (2020). Smart production planning and control in the Industry 4.0 context: A systematic literature review. *Computers & Industrial Engineering*, *149*, 106774. https://doi.org/10.1016/j.cie.2020.106774

Cioffi, R., Travaglioni, M., Piscitelli, G., Petrillo, A., & De Felice, F. (2020). Artificial intelligence and machine learning applications in smart production: progress, trends, and directions. *Sustainability*, *12*(2), 492. https://doi.org/10.3390/su12020492

de Carvalho, A. M., Sampaio, P., Rebentisch, E., & Oehmen, J. (2020). Technology and quality management: A review of concepts and opportunities in the digital transformation. *International Conference on Quality Engineering and Management*, 698–714.

Dogan, A., & Birant, D. (2021). Machine learning and data mining in manufacturing. *Expert Systems with Applications*, *166*, 114060. https://doi.org/10.1016/j.eswa.2020.114060

E, B., Flaih, L. R., Yuvaraj, D., K, S., Jayanthiladevi, A., & Kumar, T. S. (2019). Use case of artificial intelligence in machine learning manufacturing 4.0. *2019 International Conference on Computational Intelligence and Knowledge Economy (ICCIKE)*, 656–659. https://doi.org/10.1109/ICCIKE47802.2019.9004327

Filz, M.-A., Gellrich, S., Herrmann, C., & Thiede, S. (2020). Data-driven analysis of product state propagation in manufacturing systems using visual analytics and machine learning. *Procedia CIRP*, *93*, 449–454. https://doi.org/10.1016/j.procir.2020.03.065

Ghobakhloo, M. (2020). Determinants of information and digital technology implementation for smart manufacturing. *International Journal of Production Research*, *58*(8), 2384–2405. https://doi.org/10.1080/00207543.2019.1630775

Heavin, C., & Power, D. J. (2018). Challenges for digital transformation – towards a conceptual decision support guide for managers. *Journal of Decision Systems*, *27*(sup1), 38–45. https://doi.org/10.1080/12460125.2018.1468697

Jain, S., Shao, G., & Shin, S.-J. (2017). Manufacturing data analytics using a virtual factory representation. *International Journal of Production Research*, *55*(18), 5450–5464. https://doi.org/10.1080/00207543.2017.1321799

Jun, J., Chang, T.-W., & Jun, S. (2020). Quality prediction and yield improvement in process manufacturing based on data analytics. *Processes*, *8*(9), 1068. https://doi.org/10.3390/pr8091068

Kang, Z., Catal, C., & Tekinerdogan, B. (2020). Machine learning applications in production lines: a systematic literature review. *Computers & Industrial Engineering*, *149*, 106773. https://doi.org/10.1016/j.cie.2020.106773

Kaur, P., Sharma, M., & Mittal, M. (2018). Big data and machine learning based secure healthcare framework. *Procedia Computer Science*, *132*, 1049–1059. https://doi.org/10.1016/j.procs.2018.05.020

Kretschmer, T., & Khashabi, P. (2020). Digital transformation and organization design: an integrated approach. *California Management Review*, *62*(4), 86–104. https://doi.org/10.1177/0008125620940296

Lechevalier, D., Narayanan, A., & Rachuri, S. (2014). Towards a domain-specific framework for predictive analytics in manufacturing. *2014 IEEE International Conference on Big Data (Big Data)*, 987–995. https://doi.org/10.1109/BigData.2014.7004332

Lenz, J., MacDonald, E., Harik, R., & Wuest, T. (2020). Optimizing smart manufacturing systems by extending the smart products paradigm to the beginning of life. *Journal of Manufacturing Systems*, *57*, 274–286. https://doi.org/10.1016/j.jmsy.2020.10.001

Lepenioti, K., Bousdekis, A., Apostolou, D., & Mentzas, G. (2020). Prescriptive analytics: literature review and research challenges. *International Journal of Information Management*, *50*, 57–70. https://doi.org/10.1016/j.ijinfomgt.2019.04.003

Lepenioti, K., Pertselakis, M., Bousdekis, A., Louca, A., Lampathaki, F., Apostolou, D., Mentzas, G., & Anastasiou, S. (2020). *Machine Learning for Predictive and Prescriptive Analytics of Operational Data in Smart Manufacturing* (pp. 5–16). https://doi.org/10.1007/978-3-030-49165-9_1

Lin, T.-C., Sheng, M. L., & Jeng Wang, K. (2020). Dynamic capabilities for smart manufacturing transformation by manufacturing enterprises. *Asian Journal of Technology Innovation*, *28*(3), 403–426. https://doi.org/10.1080/19761597.2020.1769486

Majeed, A., Zhang, Y., Ren, S., Lv, J., Peng, T., Waqar, S., & Yin, E. (2021). A big data-driven framework for sustainable and smart additive manufacturing. *Robotics and Computer-Integrated Manufacturing*, *67*, 102026. https://doi.org/10.1016/j.rcim.2020.102026

Maruthamuthu, M. K., Rudge, S. R., Ardekani, A. M., Ladisch, M. R., & Verma, M. S. (2020). Process analytical technologies and data analytics for the manufacture of monoclonal antibodies. *Trends in Biotechnology*, *38*(10), 1169–1186. https://doi.org/10.1016/j.tibtech.2020.07.004

Meister, M., Beßle, J., Cviko, A., Böing, T., & Mett, J. (2019). Manufacturing analytics for problem-solving processes in production. *28th CIRP Design Conference,. Procedia CIRP*.

Menezes, B. C., Kelly, J. D., Leal, A. G., & Le Roux, G. C. (2019). Predictive, prescriptive and detective analytics for smart manufacturing in the information age. *IFAC-PapersOnLine*, *52*(1), 568–573.

Miklosik, A., & Evans, N. (2020). Impact of big data and machine learning on digital transformation in marketing: a literature review. *IEEE Access*, *8*, 101284–101292. https://doi.org/10.1109/ACCESS.2020.2998754

Nasiri, M., Ukko, J., Saunila, M., & Rantala, T. (2020). Managing the digital supply chain: The role of smart technologies. *Technovation*, 102121. https://doi.org/https://doi.org/10.1016/j.technovation.2020.102121

Paschek, D., Luminosu, C. T., & Draghici, A. (2017). Automated business process management–in times of digital transformation using machine learning or artificial intelligence. *MATEC Web of Conferences*, *121*, 4007.

Phuyal, S., Bista, D., & Bista, R. (2020). Challenges, opportunities and future directions of smart manufacturing: a state of art review. *Sustainable Futures*, *2*, 100023. https://doi.org/10.1016/j.sftr.2020.100023

Raeesi Vanani, I., & Kheiri, M. S. (2018). Big data analytics and visualization of performance of stock exchange companies based on balanced scorecard indicators. *Handbook of Research on Big Data Storage and Visualization Techniques*, 853–872. https://doi.org/10.4018/978-1-5225-3142-5.ch029

Raut, R. D., Yadav, V. S., Cheikhrouhou, N., Narwane, V. S., & Narkhede, B. E. (2021). Big data analytics: implementation challenges in Indian manufacturing supply chains. *Computers in Industry*, *125*, 103368. https://doi.org/10.1016/j.compind.2020.103368

Sadat Mosavi, N., & Filipe Santos, M. (2020). How prescriptive analytics influences decision making in precision medicine. *Procedia Computer Science*, *177*, 528–533. https://doi.org/10.1016/j.procs.2020.10.073

Shah, D., Wang, J., & He, Q. P. (2020). Feature engineering in big data analytics for IoT-enabled smart manufacturing – comparison between deep learning and statistical learning. *Computers & Chemical Engineering*, *141*, 106970. https://doi.org/10.1016/j.compchemeng.2020.106970

Sharp, M., Ak, R., & Hedberg, T. (2018). A survey of the advancing use and development of machine learning in smart manufacturing. *Journal of Manufacturing Systems*, *48*, 170–179. https://doi.org/10.1016/j.jmsy.2018.02.004

Sohrabi, B., Raeesi Vanani, I., & Shineh, M. B. (2017). Designing a predictive analytics solution for evaluating the scientific trends in information systems domain. *Webology*, *14*(1).

Steiber, A., Alänge, S., Ghosh, S., & Goncalves, D. (2020). Digital transformation of industrial firms: an innovation diffusion perspective. *European Journal of Innovation Management*, *ahead-of-p*(ahead-of-print). https://doi.org/10.1108/EJIM-01-2020-0018

Suvarna, M., Büth, L., Hejny, J., Mennenga, M., Li, J., Ng, Y. T., Herrmann, C., & Wang, X. (2020). Smart manufacturing for smart cities—overview, insights, and future directions. *Advanced Intelligent Systems*, *2*(10), 2000043. https://doi.org/10.1002/aisy.202000043

Tang, L., & Meng, Y. (2020). Data analytics and optimization for smart industry. *Frontiers of Engineering Management*. https://doi.org/10.1007/s42524-020-0126-0

Tao, F., Qi, Q., Liu, A., & Kusiak, A. (2018). Data-driven smart manufacturing. *Journal of Manufacturing Systems*, *48*, 157–169. https://doi.org/10.1016/j.jmsy.2018.01.006

Tavana, M., Shaabani, A., Javier Santos-Arteaga, F., & Raeesi Vanani, I. (2020). A review of uncertain decision-making methods in energy management using text mining and data analytics. *Energies*, *13*(15), 3947. https://doi.org/10.3390/en13153947

Ungermann, F., Kuhnle, A., Stricker, N., & Lanza, G. (2019). Data analytics for manufacturing systems – a data-driven approach for process optimization. *Procedia CIRP*, *81*, 369–374. https://doi.org/10.1016/j.procir.2019.03.064

Wang, B., Tao, F., Fang, X., Liu, C., Liu, Y., & Freiheit, T. (2020). Smart manufacturing and intelligent manufacturing: a comparative review. *Engineering*. https://doi.org/10.1016/j.eng.2020.07.017

Wu, D., Jennings, C., Terpenny, J., Gao, R. X., & Kumara, S. (2017). A comparative study on machine learning algorithms for smart manufacturing: tool wear prediction using random forests. *Journal of Manufacturing Science and Engineering, 139*(7). https://doi.org/10.1115/1.4036350

Zheng, P., Wang, H., Sang, Z., Zhong, R. Y., Liu, Y., Liu, C., Mubarok, K., Yu, S., & Xu, X. (2018). Smart manufacturing systems for Industry 4.0: Conceptual framework, scenarios, and future perspectives. *Frontiers of Mechanical Engineering, 13*(2), 137–150. https://doi.org/10.1007/s11465-018-0499-5

15 A Robust Cyber Security
Challenges and Opportunities

Glimpse Salwan, Priyanka Kaushal,
Surbhi Gupta, and Padmalaya Nayak

CONTENTS

15.1 INTRODUCTION

Cyber security today has become one of the major pillars on which the most intelligent species of all times, i.e. humans, have become dependent on. Cyber security is a term that in layperson's language defines the protection of the technological resources from the term cyber crime.

India's cyber law defines cyber crime as any kind of crime in which an electronic magnetic, optical, or other high-speed data processing device or system is used which performs logical, arithmetic, and memory functions by manipulations of electronic,

DOI: 10.1201/9781003153405-15

magnetic or optical impulses, and includes all input, output, processing, storage, computer software, or communication facilities which are connected or related to the computer in a computer system or computer network.

The concept of cyber security further stands on the three pillars, i.e. the CIA triad. This stands for Confidentiality, Integrity, and Availability of the data that is being transmitted through the various mediums.

- **Confidentiality** is the term used to prevent the disclosure of information to unauthorized individuals or systems (Julian Jang-Jaccard, 2014).
- **Integrity** is the term used to prevent any modification/deletion in an unauthorized manner (Julian Jang-Jaccard, 2014).
- **Availability** is the term used to assure that the systems responsible for delivering, storing, and processing information are accessible when needed and by those who need them (Julian Jang-Jaccard, 2014).

The one thing that every country agrees is that cyber security is rapidly needed for the protection of the modern-day infrastructure such as banking systems, Multi-National Companies' (MNC's) systems, clouds systems where most of the data is stored today.

Therefore, there is an urgent need for the improvement of these security infrastructure for every country in this planet.

15.2 LITERATURE SURVEY

Our literature survey has spanned a variety of sources including a broad range of all the cyber security related articles from various policies, tools currently used in the field, methodology of attacks used by the attackers/hackers, and also thinking about why a cyber security oriented infrastructure is needed for our upcoming generations and technologies.

Craigen (2014) states that the world of cyber security has many interlocking discourses. Deconstructing the word cyber security tends to situate the debate across both "cyber" and "security" realms and exposes some of the issues of legacy. "Cyber" is a cyberspace prefix that refers to computer networks of networking and augmented reality (Oxford, 2014). Then came the term "cybernetics" which clarifies the meaning as "field of control and communication theory, whether in machine or in the animal" (Wiener, 1948; Craigen, 2014).

William Gibson's 1984 novel, *Neuromancer*, popularized the word "cyberspace" in which he explains his view of a three-dimensional space of pure knowledge, traveling between machine and computer clusters in which individuals are knowledge generators and consumers (Kizza, 2011). We just know that the cyberspace was invented as an informational environment (Singer, 2014).

Now, forwarding to the term "security," in context to the literature on reviewing, Friedman and West in 2010 stated that this term is notoriously hard to define in a general sense. Security discourses inevitably involve and aim to explain who securitizes, on what issues (threats), with whom (the reference object), when, with what consequences, and under what circumstances, according to Buzan, Wæver, and Wilde (1998) (the structure).

It continues to be a disputed word, but a fundamental tenet of protection is free from risks or challenges (Oxford University, 2014). Furthermore, although we have indicated that security is a disputed subject, Baldwin (1997) states that this designation cannot be used as "a justification for not formulating as simply and specifically as conceivable one's own conception of security."

So, for finalizing the definition of cyber security for our literature survey, we have referred to the following definitions which we think provided the material perspective of cyber security:

1. "Cybersecurity consists largely of defensive methods used to detect and thwart would-be intruders" (Kemmerer, 2003).
2. "Cybersecurity entails the safeguarding of computer networks and the information they contain from penetration and from malicious damage or disruption" (Lewis, 2006).
3. "Cyber Security involves reducing the risk of malicious attack to software, computers and networks. This includes tools used to detect break-ins, stop viruses, block malicious access, enforce authentication, enable encrypted communications, and on and on" (Amoroso, 2006).
4. "Cybersecurity is the collection of tools, policies, security concepts, security safeguards, guidelines, risk management approaches, actions, training, best practices, assurance and technologies that can be used to protect the cyber environment and organization and user's assets" (ITU, 2009).
5. "The art of ensuring the existence and continuity of the information society of a nation, guaranteeing and protecting, in Cyberspace, its information, assets and critical infrastructure" (Canongia & Mandarino, 2014).
6. "The state of being protected against the criminal or unauthorized use of electronic data, or the measures taken to achieve this" (Oxford University, 2014).
7. "The activity or process, ability or capability, or state whereby information and communications systems and the information contained therein are protected from and/or defended against damage, unauthorized use or modification, or exploitation" (DHS, 2014).

A dynamic problem involving interdisciplinary reasoning is cyber security; thus, the subsequent concept must draw currently disparate definitions. However, being transparent may attract cyber security stakeholders, and being relevant should be profoundly beneficial.

However, the definition by Dan Craigen, Nadia Diakun-Thibault, Randy Purse was 4 "Cybersecurity is the organization and collection of resources, processes, and structures used to protect cyberspace and cyberspace-enabled systems from occurrences that misalign de jure from de facto property rights." Holds all the aspect nicely (Craigen et al., 2014).

Some of the important information regarding cyber security and the emerging technologies have been mentioned here, which we found throughout absorbing for the field of cyber security are mentioned below.

The United States even found out in 2000 that, rather than in independent state communication systems, data defense must be handled analytically (GAO-05-434, 2005; Council, 2008; NATO, 2010; Lithuania National Security Strategy, 2012, Johnson, 2015.

Investigators from around the globe are constantly looking for an effective approach of cyber defense. As the essential infrastructure components are so strongly related, it is recognized that a cyber-attack can extend very broadly and affect other networks as a disruption in one region can rapidly spread towards the other (Bulakh, 2016).

Cyber intruder groups are classified by the US Government Accountability Office by the goals that the attacker is seeking to accomplish by his activity:

Bot network operators, criminal groups, foreign intelligence agencies, hackers, insiders, phishers, spammers, writers of spyware/malware, terrorists.

For different reasons, all these groups have threatened critical infrastructure or other information technology properties, but all of them are very dangerous and any group can be recognized as innocuous (GAO-05-434, 2005).

All the above-mentioned attacks clearly states that attackers/hackers if they have enough opportunities could possibly take on the complete internet via an advanced exploitation method such as an automated vulnerability scanner.

The exploitation probability has been increased with the coming of the IoT field, which has direct access to our homes, personal data, and is today not yet developed up to the mark that is capable of securing virtual assets.

15.3   GETTING FAMILIAR WITH HACKING

According to *The Economic Times*, "**Hacking** is an attempt to exploit a computer system or a private network inside a computer. Simply put, it is the unauthorized access to or control over computer network security systems for some illicit purpose."

In layperson's term, the process of finding loopholes in the system of the networks or even in the person's mind that leads to unauthorized access to anything that the attacker shouldn't have access to can be termed as hacking.

In the police forces worldwide, it is always said if you want to catch a thief you must think like a thief; similarly, to protect yourself or catch a hacker you must think like one. After this statement one question in our minds consciously or subconsciously rises, which is "How do hackers think?" Let's get to that now.

15.3.1   How Do Hackers Think?

Eric Steven Raymond, who is a software developer and open-source contributor, as well as an author in his online journal titled *How To Be a Hacker* states some qualities that he over the years has experienced or seen in hackers. These qualities include:

- Problem-solving perspective.
- Not solving the previously done problems.
- Boredom is evil.

- Freedom is valuable.
- Attitude isn't a substitute for competence.

As the name suggests, those with a problem-solving perspective clearly define their work as a constant challenge when they start by breaking the layers of security, using problem solving to clear each and every path from finding open ports to privilege escalation to the root/administrative directory of the victim.

What Eric Steven Raymond meant by "Not solving the previously done problems" is that since the creative power of the brains is a limited resource for a period of time, it shouldn't be used to reinvent the wheel in today's world when it has already been done it many years ago. It should rather be applied to unsolved problems of modern-day technology.

Boredom and freedom these words here are self-explanatory because boredom in a hacker's work means lack of affection to solve problems, contradicting the above statement of problem-solving perspective. Attitude here refers to the respect of competence in the field; a learning attitude is necessary in order to keep improving ourselves in this mysterious field.

So, to sum up, we can say that hackers have a problem-solving mind, constantly seeing through the backend of application; that is, finding the loopholes and exploiting them or patching them.

15.3.2 Types of Hackers

Though hackers can be classified into many types, here we'll only talk about the main types of hackers, which include white hat hacker, gray hat hacker, and black hat hackers. Let's gain some brief information about them.

White hat: These are also known as "ethical hackers." They have the same skills of any professional hacker and instead of using these skills in a wrong manner, they use them to protect the organization they are working for. That's right, they have permission to break into the company and are provided with a handsome amount for protecting the company.

Black hat: These are the typical hackers we think of when we hear the term "hacker." They have an evil intent and have the aim of exploiting the target. They are the guys who have the skills and use them for their own means for stealing, harming an organizations and much more.

Gray hat: These are the ambiguous kind of hackers who actually look for their own mischievous intent as well report a company's vulnerabilities to them getting twice the benefit. But these guys also get into trouble due to breaking a company's bug bounty policies.

So, this was a brief introduction to the major types of hackers.

15.3.3 What Path Does a Hacker/Attacker Follow?

An attacker usually starts by studying the mechanism of the APIs, web architecture, devices, that is how does these things work, e.g. if we want to exploit a website then

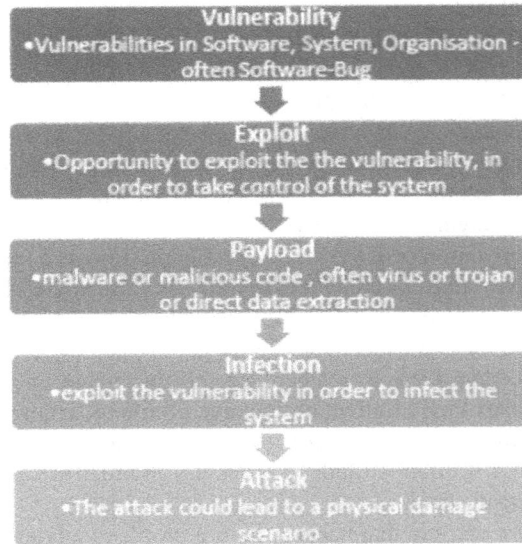

FIGURE 15.1 A typical cyber-attack scenario.

Source: (Limba, Plėta, Agafonov, & Damkus, 2017)

we start the procedure by studying the structure of the website. In most cases today, the attacker starts with studying some of the previous and some newly starred vulnerabilities which the company might be fixing right now (Keith Stouffer, 2013).

Usually, a cyber-attack that is completed successfully goes through a specific five-stage scenario that certainly includes finding a vulnerability, exploiting the vulnerability, using some payloads such as malicious software, infecting the system, and attacking either the complete system or a part of it (Limba, Plėta, Agafonov, & Damkus, 2017).

All these stages are shown in Figure 15.1

15.4 DEVELOPMENT OF CYBER SECURITY MANAGEMENT MODEL

Today, the important thing is that the cyber-attacks are becoming incredibly difficult to predict and anticipate. Taking preventive measures in time is the need of the hour (Limba, Plėta, Agafonov, & Damkus, 2017). The success rate of every cyber-attack has been constantly increasing over the time without noticing a downfall in the growth graph. Presently, there is no cyber security management model that has the proper procedure in respect to giving the response to the variety of the cyber-attacks grown within a short span of time. With a little bit of distortion in an unexpected scenario the cyber security model would be unable to work properly. It should be observed that a standalone model such as a cyber security management model is a technical feature that will not be valid throughout the upcoming technologies (WaterISAC, 2016; Limba, Plėta, Agafonov, & Damkus, 2017).

15.4.1 Mistakes in Cyber Security

Some of the common mistakes that organizations make knowing or unknowingly, which are a serious threat to their assets, are:

- Falsely assuming that it is possible to keep any infrastructure secure from every vulnerability. Any agency that has a notable amount of infrastructure in particular should recognize that maximum security is only a dream. The most critical part of protection is to consider what the most sensitive areas are and what should one do to prevent risks, and in which systems you need to identify irregular facilities. Following a proper procedure and a concrete roadmap that explains how to minimize losses and recover your daily activity is the need of the hour (Limba, Plèta, Agafonov, & Damkus, 2017). In addition to the context detection and response should be considered urgently implacable since these things are capable of reducing the losses (Limba, Plèta, Agafonov, & Damkus, 2017).
- Imaginable opinions such as hiring the best cyber security professional will save or protect you from all kinds of threats have proven to be just mere myths of the field. Organizations need to make one thing clear in each and every mind of the people working in the institute that cyber security is not just a mere department but the organization's approach (Singer, 2014). Thinking of cyber security as a department and professional association forms a strong and non-repellent sense of vulnerabilities or loopholes. Cyber security among the organizations and the institutions must be considered as a fundamental objective, not just an objective but it should become an urgent priority in the aims of the company leaders as well as the employees. The above statement holds up to date as we all know the human factor in an organization is the most vulnerable part of the security field (Dong Wei et al., 2010). Cyber security must be treated as an organization's policy that is influencing earnings.
- Blindly trusting the security technologies and tools that are used to secure the organization has always been proven a myth by itself.

  Companies that create security softwares never guarantee 100 percent protection or safety from any kind of attacks. Technology used in the modern world helps in prevention as well as detection of things like the unwanted attacks where it came from, what was the kind of attack etc. But we all cannot rely on just the technology because it can't give us the total security (Dong Wei, 2010; Limba, Plèta, Agafonov, & Damkus, 2017). Although as we know tools and various software play an important role in securing the data and the privacy of the organization, they alone aren't enough to provide us with the best security as most vulnerable link in the defense is the human factor itself (WaterISAC, 2016). We should only invest in tools if we are sure that employees can't be manipulated. A simple but solid example of the whole scenario can be social engineering attacks. In order to be an organization with enhanced security, the people of the organization should be trained well for any kind of attacks using social protocols.
- Considering monitoring as a limited field related to resources, networks, etc. is a common mistake among companies. Monitoring has a wider sense

in this context than just monitoring of the properties of equipment and resources. Monitoring is not a limited technological viewpoint that is related to particular digital information, information structures and monitoring of databases, but a broad perspective that puts together the whole enterprise of the surrounding world and the tracking of emerging cybercrime patterns. Monitoring proves to be worthless to those who are not able to learn from it (Limba, Plėta, Agafonov, & Damkus, 2017). In the long-term monitoring has proven to be beneficial to those who have constantly observed the difference in the emerging patterns of attacks are less vulnerable as compared to those who just ignore the monitoring. A company must ensure that security vulnerability information is exchanged with others, since only effective communication can provide an overall impression of the actual security situation in the region, state or worldwide (Govindarasu, 2017; Limba, Plėta, Agafonov, & Damkus, 2017).

- Lastly, most organizations think the security they are using is superior or not inferior to any other technology on the planet. It's like competing in the Olympic marathon to build productive data protection and attempt to stop cyber threats, but security is not victory vs. cybercriminals. New approaches and tactics are developed by the attackers, and defenders are still one stage behind. Investing in more advanced security systems to eliminate attacks seems to be useful, but the truth is very different.

Cyber security policies must give priority to investments in national infrastructure, instead of the emerging innovations or systems that can detect any hazard and assets (WaterISAC, 2016).

One of the main aspects of cyber security is to provide a strong foundation for building the new technological development solutions and systems, it has only been observed that systems and projects are relatively used for the connectivity across the globe, and shall further be used to promote security instead (Moslehpour, 2016; Limba, Plėta, Agafonov, & Damkus, 2017).

15.5 CHALLENGES IN CYBER SECURITY

With the advancement of technologies such as IoT, edge computing etc. it is necessary to protect these technology as the new technology comes with new vulnerabilities that have always been challenging the cyber security technology (Zhicheng, 2018). In this section we shall explore the present and the upcoming challenges.

15.5.1 Large Number of IoT Devices

Due to the lack of computational, memory, and battery resources it has become increasingly difficult to provide better protection. A large number of unused IoT devices could be at high risk for all future malicious attacks of IoT. Such exposure can lead to serious security breaches and can lead to economic losses (CID, 2016).

IoT systems tend to have limited capabilities to use standard encryption algorithms, authentication, and access control. When faced with the onslaught of a violent Denial of Service (DoS) acquisition, they are at high risk (Zhicheng, 2018). In general, IoT devices have limited power for standard encryption, authentication, and access control algorithms. There are many types of attacks that can be used to exploit IoT devices, some of which are:

- It is always possible to physically interrupt these devices and steal data and all sorts of codes and keys (Zhicheng, 2018).
- The type of attack carried out, i.e. identity building, can be used against these resources, which could lead to the disclosure of sensitive data.
- Microphones or speakers can interfere with devices also known as attentive listening via shared wireless channels.
- One can attack and use these devices using non-malicious nodes to set up links in the setup network between IoT devices.

The edge computing platform can ultimately play an important role in helping IoT systems, coordinating security operations, and tracking and protecting against cyber-external attacks. However, unlike standard mid-level cloud computing, the edge computing paradigm is relatively modern compared to its various security measures (Zhicheng, 2018).

15.5.2 NFV-SDN Integrated Edge Cloud Platform

Network Function Virtualization (NFV) and Software Defined Networking (SDN) are two new cloud-based and integrated cloud platform solutions that can match each other.

Using NFV, Virtual Machines (VMs) can be operated seamlessly by a horizontal cloud to perform device-specific processing or to provide security-related protection such as firewall VMs or application access. SDN can be configured to work with NFV on the network, install, monitor, and host VMs with embedded separation for data management and transmission.

NFV integration and SDN integration can allow for high-quality cloud infrastructure for future IoT applications to be optimized and managed (Pan et al., 2016). What is worse, however, is that both NFV and SDN are emerging and changing respectively, and many cyber security issues for both of them are yet to be resolved, not to mention that because of their merger, other security threats and problems may arise.

For example, prevention of DoS attacks, spoofing attacks, and dangerous injections in the visual environment remain an unresolved issue of SDN (Akhunzada, 2015; Zhicheng, 2018).

For NFV, hypervisor protection threats often occur by dissecting Virtualized Network Functions (VNF) and managing imaginary network topology, crossing the phase limit to VNFs, and avoiding DoS attacks and malicious internal attacks (Lal et al., 2017; Zhicheng, 2018).

15.5.3 Data Privacy and Security

Huge volumes of data at a very high speed can also be generated by a much greater number of IoT devices accessing the internet. As it is difficult to store all the information in consolidated locations for

Processing in the present paradigm of cloud computing as what happened. In some of the clustered edge storage clusters or edge clouds, similar data would have to be stored and analyzed. The treatment of data protection and confidentiality would also be distinct. In the one hand, it could be beneficial for application consumers (such as users with medical or health IoT applications) who choose to own, store and manage their own data entirely instead of having it in the hands of cloud providers where they are unable to provide absolute control and are unable to promise that cloud providers may not use their data or use the data in all accepted ways (Zhicheng, 2018).

It is still an open and difficult challenge to make data privacy and protection properly protected in a far more decentralized edge computing environment (Zhicheng, 2018).

15.5.4 Offloading and Interaction between Edge and IoT Devices

One of the characteristics of the open edge cloud era is that it would be very widespread for resource-poor IoT computers to offload tasks for fast processing to the resource-rich edge computing infrastructure (Pan, 2016; McElhannon, 2018; Zhicheng, 2018).

The IoT modules will also assist one another with multiple activities based on their accessibility to services and the policy for benefits. Offloading and collaboration with such duties may also raise new security issues such as:

- The first one is about the safety of code. It is important to write and create the task codes in such a manner that they can be seamlessly scheduled to run on various platforms, such as edge computing and IoT computers. Migrating cross-platform code and complex scheduling may also be a daunting job that involves APIs or interfaces to be secured (Zhicheng, 2018).
- Second, to manage the interface between mobile/wireless IoT devices and edge cloud institutions such as VMs, the facilitator of the edge cloud is also expected to provide appropriate support from the edge cloud side. To protect the traveling codes between both the edge cloud and IoT devices from targeted hackers, adequate access control mechanisms are needed. In addition, primarily wireless and telephone connections are the contact links between the IoT devices and the edge cloud (Zhicheng, 2018).
- All the security issues relevant to the wireless/mobile network still occur here. Suitable encryption, for example, is necessary for all interactions over these wireless connections. However, it is important to find some successful but lightweight methods for authentication, encryption, and access control since IoT devices are largely resource-poor. The resource-rich edge cloud would help make these IoTs simpler for the functions of defense (Zhicheng, 2018).

15.6   OPPORTUNITIES IN CYBER SECURITY

Since the cyber security field has always been developing hence the opportunities in this field have always been growing in a constant pace. In respect to the development of new technologies such neuro networks, blockchain, machine learning, deep learning, and, of course, artificial intelligence the opportunities have so far been explored and in this section we shall talk about our vision in the following emergence in following fields:

15.6.1   BLOCKCHAIN AND ZERO-TRUST SECURITY

Recently, blockchain technology has drawn broad interest from business as well as academia. It is considered a tool that will eventually transform not just how people use currencies (Bitcoin or other cryptocurrencies) but also how people and organizations they don't trust would interact with all sorts of transactions.

The transactions are highly open and visible to all interested parties.

The blockchain structure makes it nearly impossible to intentionally mess with the transactions reported (Zhicheng, 2018).

It has been commonly accepted that blockchain could have tremendous potential for enterprises to develop their own cyber security networks for logging, communications, user verification, identification and user access management activities.

Private blockchains should be developed in the sense of "edge computing + IoT" to allow intelligent contracts to redefine communications between individuals without a pre-defined relationship of trust (Zhicheng, 2018).

A zero-confidence framework generally means that the members never trust each other by association and instead validate each other's identities (Zhicheng, 2018). To create a more stable "edge computing + IoT" environment, a modern zero-trust architecture using blockchain may technically be very useful.

15.6.2   ARTIFICIAL INTELLIGENCE (AI) AND MACHINE LEARNING (ML)

Over the last few years, the resurgence of AI and machine learning (deep learning in particular) technology has seen considerable advances, notably with the success of AlphaGo and autonomous driving applications.

When there is a broad data scale available for model creation and parameter tuning, these technologies perform well and provide the best predictive performance (Zhicheng, 2018).

It is also a huge opportunity from the cyber security point of view to apply AI and deep learning to edge computing + IoT to help evaluate various cyber activities, recognize possible risks and weaknesses for repair or patching, and detect malicious attacks.

Hence, they opportunity for the cyber security enthusiasts rises in the field of Artificial intelligence and Machine Learning as well.

15.6.3   WEAK/LIGHTWEIGHT IOT SECURITY

In most situations, to find a compromise between protection and power usage on the computers, we can only search at lightweight methods and algorithms. Including authentication (Lee et al., 2014; Xuanxia Yao, 2015), cryptography (Xuanxia Yao, 2015; Zhicheng, 2018) access control and key exchange (Raza et al., 2012), such lightweight algorithms can exist.

A common example of such a process for lightweight authentication is for the for low-rate and long-range IoT implementations in which the base station and the IoT modules operate together using a much simpler and lighter authentication system, 802.11ah or "WiFi HaLow."

15.6.4 CYBER DEFENSE BASED IN DECEPTION

Passive security is the bulk of conventional cyber defense strategies such as encryption, authentication and access control, which means that they do not deliberately pursue attackers, but aim to make it more difficult for the attackers to break in. One of the successful cyber defense strategies used to strengthen a single institution's cyber defense capability is deception-based defense (Zhicheng, 2018).

Address jumping, honeypots and network telescopes are common example approaches. Deception and any unexpected ongoing shifts in network configurations may theoretically confuse the attackers and confuse them, or draw them to any pre-deployed honeypot traps (La et al., 2016).

The strategy focused on manipulation will also create a vast number of false credentials on the network of the enterprise, making it impossible for cyber criminals to obtain access to the networks as a collection of valid user identities. If the assailants use these false passwords, security managers can identify and track them. Once they are logged and documented, the traces and archives of the attackers' actions can be more studied to explain how they targeted the target device and the general trends they used. Such expertise is very helpful in improving the security of the organization's networks. The honeypot traps are typically physical or virtual computers that, while under close supervision and logging by security managers, claim to be the actual systems. They try to trick the attackers into taking action in favor of the defenders or merely spending time and money on false goals (Zhicheng, 2018). However, this is not a one-way road in the context that by using various types of tactics ranging from a dubious to an apparently normal one, the perpetrators will also still wish to prevent being identified.

In keeping the large-scale "edge computing + IoT" systems safer, it is expected that deception-based cyber security tactics will play a more significant role.

15.6.5 ISOLATED IDENTITY OF THE DEVICES UNDER THE INTERNET OF THINGS

Today almost all the device we use for internet connectivity are based on IP (IPv4 or Ipv6). In terms of inter-connectivity, this provides immense convenience by encouraging devices to engage with each other and to get content and services. Such rich inter-connectivity also makes it much simpler for administrators via remote setup or management.

The disadvantage, though, is that it effectively exposes all devices, including some vital industrial control systems, such as smart grids, to a risky role that if any basic passwords were taken and claimed to be legitimate users, possible hackers might still get access to them. Few traffic can be filtered by conventional firewalls, but they often use a perimeter-based protection approach and firewall filtering is typically based on arbitrary IP. It does not avoid attacks on the inside either (Zhicheng, 2018).

A common example of this is that the host identifiers used in the Host Identification Protocol (HIP) can be used as a different identity and naming scheme other than IP addresses for IoT applications (Zhicheng, 2018). The link to IoT devices would need host identities instead of IP addresses to be used for such an overlay identity and naming scheme. They need to create a binding for exchanging crypto protocols before two individuals can communicate to each other. This will preclude the outside world from being able to hack sensitive IoT devices and networks directly without stringent security protocols being implemented. Similarly, an independent and protected identification and naming scheme other than IP will be very helpful in the future "edge computing + IoT" environment, particularly for those resource-poor IoT devices in sensitive systems that need additional security. In this respect, we assume that there will be quite a few research opportunities (Zhicheng, 2018).

15.7 CONCLUSION

For several potential IoT implementations, the paradigm changes to the Internet of Things (IoT) and the advent of edge computing are providing major benefits. It also introduces essential challenges to cyber security. We identified and addressed five cyber security issues and five potential prospects for cyber security relevant to this vision throughout this report. The development of some convergence between the "edge computing + IoT" platform and the evolving blockchain and AI technologies could theoretically create many of these opportunities that are beneficial.

Overall, we conclude that there are still challenges left that are to be achieve in the future and the opportunities are still rising which we can take advantage of in a useful and sustainable way.

REFERENCES

Akhunzada, E. A. (2015). Securing software defined networks: taxonomy, requirements, and open issues. *IEEE Communications Magazine*, 53(4), 36–44.

Amoroso, E. 2006. Cyber security. New Jersey: Silicon Press.

Baldwin, D. A. 1997. The concept of security. Review of International Studies, 23(1), 5–26.

Bulakh A., Tuohy E., & Pernik P. 2016. Estonia's developing level playing field for critical energy infrastructure protectors-a model for broader scale platforms?, *Energy Security: Operational Highlights*, 10: 4–10. Available on the Internet: www.icds.ee/fileadmin/media/icds.ee/failid/no_10_20160410.pdf

Buzan, B., Wæver, O., & De Wilde, J. 1998. Security: *a new framework for analysis*. Boulder, CO: Lynne Rienner Publishers.

Canongia, C., & Mandarino, R. 2014. Cybersecurity: the new challenge of the information society. In Crisis *management: concepts, methodologies, tools and applic*ations: 60–80. Hershey, PA: IGI Global. http://dx.doi.org/10.4018/978-1-4666-4707-7.ch003

CID, D. (2016, June). Large CCTV Botnet leveraged in DDoS attacks. Retrieved from https://blog.sucuri.net/: https://blog.sucuri.net/https://blog.sucuri.net/: https://blog.sucuri.net/2016/06/large-cctv-botnet-leveraged-ddos-attacks.html

Council, E. (2008). The European Critical Infrastructures Directive. Council Directive 2008/114/EC. Available on the Internet: http://eurlex.europa.eu/legal-content/DA/ALL/?uri=CELEX:32008L0114.

Craigen, D. D.-T. (2014). Defining cybersecurity. *Technology Innovation Management Review*.

DHS. 2014. A glossary of common cybersecurity terminology. National Initiative for Cybersecurity Careers and Studies: Department of Homeland Security. October 1, 2014: http://niccs.us-cert.gov/glossary#letter_c

Dong Wei, Y. L. et al. (2010). An integrated security system of protecting Smart Grid against cyber attacks. *Innovative Smart Grid Technologies (ISGT)*. doi: 10.1109/ISGT.2010.5434767.

GAO-05-434. (2005). Critical Infrastructure Protection: Department of Homeland Security Faces Challenges in Fulfilling Cybersecurity Responsibilities. Retrieved from www.gao.gov/: http://www.gao.gov/new.items/d05434.pdf

Gibson, W. (2016). *Neuromancer*. UK: Orion. (Reissue of 1984 novel.)

Govindarasu, M. (2017). Cybersecurity of the Power Grid: A Growing Challenge. Retrieved from www.usnews.com/: www.usnews.com/https://www.usnews.com/: www.usnews.com/news/national-news/articles/2017-02-24/cybersecurity-of-the-power-grid-a-growing-challenge

ITU. 2009. Overview of cybersecurity. Recommendation ITU-T X.1205. Geneva: International Telecommunication Union (ITU). www.itu.int/rec/T-REC-X.1205-200804-I/en

Johnson, T. A. (2015). *Cybersecurity: Protecting critical infrastructures from cyber attack and cyber warfare*. CRC Press.

Julian Jang-Jaccard, S. N. (2014). A survey of emerging threats in cybersecurity. *Journal of Computer and System Sciences*, 80(5), 973–993.

Karpaviciute, Ieva. (2017). Securitization and Lithuania's national security change. *Lithuanian Foreign Policy Review*, 36. 10.1515/lfpr-2017-0005.

Keith Stouffer, J. F. (2013). *Guide to Industrial Control Systems (ICS) security*. Supervisory Control and Data Acquisition (SCADA) Systems, Distributed Control Systems (DCS), and Other Control System Configurations such as Programmable Logic Controllers (PLC).

Kemmerer, R. A. (2003). Cybersecurity. Proceedings of the 25th IEEE International Conference on Software Engineering: 705–715. http://dx.doi.org/10.1109/ICSE.2003.1201257

Kizza, J. M. (2002). Computer network security and cyber ethics. Jefferson, NC: McFarland (reference from an article 2011).

La, Q. D. et al. (2016). Deceptive attack and defense game in honeypot-enabled networks for the Internet of Things. *IEEE Internet of Things Journal*, 3(6), 1025–1035.

Lal, S. et al. (2017). NFV: security threats and best practices. *IEEE Communications Magazine*, 55(8), 211–217.

Lee, J. et al. (2014). A lightweight authentication protocol for Internet of Things. International Symposium on Next-Generation Electronics (ISNE), Kwei-Shan, 1–2.

Lewis, J. A. (2006). Cybersecurity and critical infrastructure protection. Washington, DC: Center for Strategic and International Studies. http://csis.org/publication/cybersecurity-and-critical-infrastructure-pr...

Limba, T., Plėta, T., Agafonov, K., & Damkus, M. (2017). Cyber security management model for critical infrastructure. *Entrepreneurship and Sustainability Issues* 4(4), 559–573.

McElhannon, J. P. (2018). Future edge cloud and edge computing for internet of things applications. *IEEE Internet of Things Journal*, 5(1), 439–449.

Moslehpour, K. J. (2016). Cyber security management: a review. *Business and Management Dynamics*, 16–39.

Oxford University, p. (2014, October 1). Oxford Online Dictionary. Retrieved from Oxford: Oxford University Press: www.oxforddictionaries.com/definition/english/Cybersecurity

Pan, J. et al. (2016). HomeCloud: An edge cloud framework and testbed for new application delivery. 23rd International Conference on Telecommunications (ICT), Thessaloniki, 1–6.

Raza, S. et al. (2012). Lightweight IKEv2: a key management solution for both the compressed IPsec and the IEEE 802.15.4 security. In Proceedings of the IETF workshop on smart object security, Vol. 23.

Singer, P. W. (2014). *Cybersecurity: What everyone needs to know.* Oxford University Press.

Tikk, E. (2010). Global cybersecurity – thinking about the niche for NATO. SAIS Review of International Affairs 30(2), 105–119. www.muse.jhu.edu/article/403442.

WaterISAC. (2016, October). 10 Basic Cybersecurity Measures: Best Practices to Reduce Exploitable. Retrieved from www.waterisac.org/: https://www.waterisac.org/sites/default/files/public/10_Basic_Cybersecurity_Measures-WaterISAC_Oct2016%5B2%5D.pdf

Wiener, N., Hill, D., & Mitter, S. (2019). *Cybernetics or control and communication in the animal and the machine*, Reissue of the 1961 Second Edition. United States: MIT Press .(Reissue of 1948 Cybernetics.)

Xuanxia Yao, Z. C. (2015). A lightweight attribute-based encryption scheme for the Internet of Things. *Future Generation Computer Systems*, 49, 104–112.

Zhicheng, P. J. (2018). *Cybersecurity challenges and opportunities in the new "edge computing + IoT" world*. New York: Association for Computing Machinery.

Index

For Product Safety Concerns and Information please contact our EU
representative GPSR@taylorandfrancis.com
Taylor & Francis Verlag GmbH, Kaufingerstraße 24, 80331 München, Germany